CARBON-CENTERED FREE RADICALS AND RADICAL CATIONS

Wiley Series of Reactive Intermediates in Chemistry and Biology

Steven E. Rokita, Series Editor

Quinone Methides
Edited by Steven E. Rokita

Radical and Radical Ion Reactivity in Nucleic Acid Chemistry
Edited by Marc Greenberg

Carbon-Centered Free Radicals and Radical Cations
Edited by Malcolm D. E. Forbes

CARBON-CENTERED FREE RADICALS AND RADICAL CATIONS

Structure, Reactivity, and Dynamics

Edited by

MALCOLM D. E. FORBES

WILEY

A John Wiley & Sons, Inc., Publication

Library of Congress Cataloging-in-Publication Data:

Library of Congress Cataloging-in-Publication Data
Carbon-centered free radicals and radical cations / edited by Malcolm D. E. Forbes.
        p. cm.
    Includes index.
    ISBN 978–0–470–39009–2 (cloth)
    1.   Free radicals (Chemistry) 2.   Carbon, Activated. 3.   Reactivity
(Chemistry) 4.   Cations.   I. Forbes, Malcolm D. E., 1960–
    QD471.C337 2010
    547'.1224–dc22

                                        2009031417

Printed in the United States of America
10 9 8 7 6 5 4 3 2 1

CONTENTS

ABOUT THE VOLUME EDITOR

Born in Belfast, Northern Ireland, and raised in western Massachusetts, Malcolm Forbes completed his university training at the University of Illinois at Chicago, receiving a double major B.S. degree in Chemistry and Mathematics there in 1983. He undertook doctoral studies at the University of Chicago, where he worked with the late Gerhard Closs on the study of unstable spin-polarized biradicals using time-resolved electron paramagnetic resonance spectroscopy. In 1988, his accomplishments in this area were recognized with the Bernard Smaller Prize for Research in Magnetic Resonance. After receiving his doctoral degree, Malcolm was awarded a National Science Foundation Postdoctoral Research Fellowship. From 1988 to 1990 he worked at the California Institute of Technology with Nathan Lewis on interfacial charge transfer kinetics at silicon/liquid junctions.

In July 1990, Malcolm joined the Department of Chemistry at the University of North Carolina at Chapel Hill and was promoted to the position of Professor of Chemistry in 1999. He has received a number of awards: a National Science Foundation Young Investigator Award (1993–1998), a Japan Society for the Promotion of Science Foreign Fellowship Award (1998–1999), the 2000 Sir Harold Thomson Award from Elsevier, and most recently a 2007–2008 J. W. Fulbright Fellowship from the U. S. State Department. Malcolm was co-Chair of the 2008 Gordon Research Conference on Electron Donor–Acceptor Interactions.

Malcolm's research interests span a wide area of physical organic chemistry. His primary focus is studying free radical structure, dynamics and reactivity using a variety of magnetic resonance techniques. Current projects include the fundamentals of "spin chemistry," proton-coupled electron transfer reactions, and the photodegradation and chain dynamics of polymers.

Malcolm lives in Chapel Hill with his wife Natalia and sons Matt, Cameron, and Elliot. Together they enjoy swimming, traveling, and home improvement projects.

PREFACE TO SERIES

Most stable compounds and functional groups have benefited from numerous monographs and series devoted to their unique chemistry, and most biological materials and processes have received similar attention. Chemical and biological mechanisms have also been the subject of individual reviews and compilations. When reactive intermediates are given center stage, presentations often focus on the details and approaches of one discipline despite their common prominence in the primary literature of physical, theoretical, organic, inorganic, and biological disciplines. The *Wiley Series on Reactive Intermediates in Chemistry and Biology* is designed to supply a complementary perspective from current publications by focusing each volume on a specific reactive intermediate and endowing it with the broadest possible context and outlook. Individual volumes may serve to supplement an advanced course, sustain a special topics course, and provide a ready resource for the research community. Readers should feel equally reassured by reviews in their speciality, inspired by helpful updates in allied areas and intrigued by topics not yet familiar.

This series revels in the diversity of its perspectives and expertise. Where some books draw strength from their focused details, this series draws strength from the breadth of its presentations. The goal is to illustrate the widest possible range of literature that covers the subject of each volume. When appropriate, topics may span theoretical approaches for predicting reactivity, physical methods of analysis, strategies for generating intermediates, utility for chemical synthesis, applications in biochemistry and medicine, impact on the environmental, occurrence in biology, and more. Experimental systems used to explore these topics may be equally broad and range from simple models to complex arrays and mixtures such as those found in the final frontiers of cells, organisms, earth, and space.

Advances in chemistry and biology gain from a mutual synergy. As new methods are developed for one field, they are often rapidly adapted for application in the other. Biological transformations and pathways often inspire analogous development of new procedures in chemical synthesis, and likewise, chemical characterization and identification of transient intermediates often provide the foundation for understanding the biosynthesis and reactivity of many new biological materials. While individual chapters may draw from a single expertise, the range of contributions contained within each volume should collectively offer readers with a multidisciplinary analysis and exposure to the full range of activities in the field. As this series grows, individualized compilations may also be created through electronic access to highlight a particular approach or application across many volumes that together cover a variety of different reactive intermediates.

Interest in starting this series came easily, but the creation of each volume of this series required vision, hard work, enthusiasm, and persistence. I thank all of the contributors and editors who graciously accepted and will accept the challenge.

STEVEN E. ROKITA

University of Maryland

ABOUT THE SERIES EDITOR

STEVEN E. ROKITA, PhD, is Professor in the Department of Chemistry and Biochemistry at the University of Maryland. His research interests lie in sequence and conformation specific reactions of nucleic acids, enzyme-mediated activation of substrates and coenzymes, halogenation and dehalogenation reactions in biology, and aromatic substitution and quinone methide generation in bioorganic chemistry.

INTRODUCTION

Carbon radicals and radical cations hold central places in modern organic reactivity, from alkene addition reactions in the synthesis of novel polymers to fundamental studies of electronic distribution of spin and charge in the study of donor–acceptor interactions. The importance of free radicals in biological reactions was recognized initially in fields such as photosynthesis, but they are now of interest in areas of research as diverse as tissue damage and the aging process. The field of biological free radicals has grown to the extent that an entire journal is now devoted to the topic: *Free Radicals in Biology and Medicine*. The ubiquity of radical intermediates in chemistry and biology has commanded attention from chemists, biologists, and physicists, across a variety of subdisciplines, who are seeking to understand the structure, reactivity, and dynamics of radicals in magnetic and chemical environments that are often complex.

To this end, high levels of theory have been developed in conjunction with a sophisticated array of experimental techniques that now make it possible to measure the properties of organic reactive intermediates with extraordinary precision. This volume, on carbon-centered free radicals and radical cations, highlights several of the most advanced computational and experimental methods currently available for such investigations. The chapters within are written by a well-rounded group of experts, who have made a strong effort to explain difficult concepts clearly and concisely. The authors were selected with the intention of providing a broad range of material, from small molecule synthesis to polymer degradation, and from computational chemistry to highly detailed experimental work in the solid, liquid, and gaseous states.

Chapter 1 presents a short history of the field of free radical chemistry. Building on a few earlier summaries in monographs that are now a bit dated, this chapter covers more modern developments in radical reactions, mechanisms, and physical methods since the 1960s. Particular attention is paid to the chemically induced spin polarization phenomena that have a strong presence in this volume. Chapters 2–4 can be considered to have the common theme of mechanistic chemistry, with an emphasis on synthetic utility. Chemists are sometimes surprised to find useful radical reactions in synthesis, and these three chapters summarize many new ideas for the construction of interesting organic structures. In Chapter 2, Wille describes recent experimental and computational results from her laboratory on cascade-type radical additions to alkynes, with mechanistic examples and synthetic applications. Complementary to her work on building carbon skeletons is Poniatowski and Floreancig's description of radical cation fragmentation reactions in Chapter 3, with applications to asymmetric total synthesis. In Chapter 4, Sevov and Wiest discuss chemo-, regio-, and periselectivity trends and solvent effects in radical cation Diels–Alder reactions.

Chapters 5–7 are focused on molecular structure and are therefore mostly from the computational perspective. However, these authors were invited because of their skills in connecting computation to experiment, and they have provided significant insight in many important reactions. In Chapter 5, Coote, Lin, and Zipse provide a summary of stereoelectronic effects governing the stability of carbon-centered radicals, with a detailed discussion of applications to H-atom transfer and olefin addition reactions. Barone, Biczysko, and Cimino present case studies of vibrational and environmental effects on radical stabilities in Chapter 6, with several important biological examples. In Chapter 7, Gescheidt connects the electrochemistry and magnetic resonance of pagodane-type radical cations to their molecular structures. His experimental measurements are strongly supported by computational results.

Chapters 8–11 represent an effort to present the forefront of spectroscopic investigations of radical structure and kinetics. These particular chapters also provide excellent demonstrations of several "spin chemistry" techniques such as CIDNP and magnetic field effects. In this regard, Chapter 8 by Woodward contains an excellent introduction to the physics of geminate radical pair spin state evolution and magnetic field effects, presenting theoretical details clearly and giving numerous experimental examples. Goez, in Chapter 9, also provides background on the radical pair mechanism as applied to the CIDNP experiment. His examples include reactions of radicals, radical ions, and biradicals. This chapter provides a very useful overview of the theory and contains several worthy demonstrations of the mechanistic power of CIDNP spectroscopy. The contributors of Chapter 10, Rawls, Kuprov, Elliot, and Steiner, have combined their experimental and theoretical talents to analyze the magnetic properties of linked donor–acceptor systems that are model systems for artificial photosynthesis, with a particular emphasis on spin relaxation effects. No volume of this type would be complete without a description of modern gas-phase radical reactions. The crossed molecular beam experiments described by Kaiser in Chapter 11 delineate the chemistry of phenyl radicals and other smaller carbon-containing fragments, as related to atmospheric science.

This volume closes with three chapters on different aspects of free radical chemistry in macromolecules. Several photoinitiation reactions that are widely used in polymer synthesis are discussed by Khudyakov and Turro in Chapter 12. This chapter also gives an informative description of how CIDEP can be used to simultaneously study structure and mechanism in photochemical reactions. The reactions of geminal radical pairs created in bulk polymers are presented by Chesta and Weiss in Chapter 13. Of the many possible chemical reactions for such pairs, they are organized here by polymer and reaction type, and the authors provide solid rationalizations for the observed product yields in terms of cage versus escape processes. Chapter 14 contains a summary of the editor's own work on acrylic polymer degradation in solution. Forbes and Lebedeva show TREPR spectra and simulations for many main-chain acrylic polymer radicals that cannot be observed by steady-state EPR methods. A discussion of conformational dynamics and solvent effects is also included.

On a personal note, I would like to thank the series editor Steven Rokita for the invitation to generate this volume. This was a challenging project and he was always at the ready with good advice during the writing process. Becky Ramos and Anita Lekhwani at Wiley were instrumental in getting this volume off the ground; hands off enough to let me shape the volume the way I wanted, and hands on enough to avoid catastrophe. I am indebted to my editorial assistant (and coauthor) Natalia Lebedeva, without whom I would still be choosing authors and daydreaming about potential topics. I also acknowledge the U.S. Fulbright Scholar Program for a fellowship that gave me substantial time away from everyday duties this year in order to complete this volume.

Finally, no project of this magnitude is ever created without authors who share their commitment and are willing to produce great science within a reasonable time frame. I thank them for their patience with me as the initial deadline slipped past, and for working hard over the holiday break of 2008–2009 to get their manuscripts to me for the final push to production. It is not quite enough to let their efforts shine on the pages within, therefore I close this introduction by stating that the authors' perseverance, diligence, and attention to detail are duly recognized by a most grateful taskmaster.

MALCOLM D. E. FORBES

Chapel Hill, North Carolina
December 2009

CONTRIBUTORS

Vincenzo Barone, Scuola Normale Superiore di Pisa, Pisa, Italy

Malgorzata Biczysko, Dipartimento di Chimica, Università Federico II, Napoli, Italy

Carlos A. Chesta, Departamento de Química, Universidad Nacional de Río Cuarto, Río Cuarto, Argentina

Paola Cimino, Dipartimento di Scienze Farmaceutiche, Università di Salerno, Fisciano, Italy

Michelle L. Coote, ARC Centre of Excellence for Free-Radical Chemistry and Biotechnology, Research School of Chemistry, Australian National University, Canberra Australian Capital Territory, Australia

C. Michael Elliott, Department of Chemistry, Colorado State University, Fort Collins, CO, USA

Paul E. Floreancig, Department of Chemistry, University of Pittsburgh, Pittsburgh, PA, USA

Malcolm D. E. Forbes, Department of Chemistry, University of North Carolina, Chapel Hill, NC, USA

Georg Gescheidt, Institute of Physical and Theoretical Chemistry, Graz University of Technology, Graz, Austria

Martin Goez, Institut für Chemie, Martin–Luther-Universität Halle–Wittenberg, Halle/Saale, Germany

Ralf I. Kaiser, Department of Chemistry, University of Hawaii at Manoa, Honolulu, HI, USA

Igor V. Khudyakov, Bomar Specialties, Torrington, CT, USA

Ilya S. Kuprov, Chemistry Department, University of Durham, Durham, UK

Natalia V. Lebedeva, Department of Chemistry, University of North Carolina, Chapel Hill, NC, USA

Ching Yeh Lin, ARC Centre of Excellence for Free-Radical Chemistry and Biotechnology, Research School of Chemistry, Australian National University, Canberra Australian Capital Territory, Australia

Alexander J. Poniatowski, Department of Chemistry, University of Pittsburgh, Pittsburgh, PA, USA

Matthew T. Rawls, National Renewable Energy Laboratory, Golden, CO, USA

Christo S. Sevov, Department of Chemistry and Biochemistry, University of Notre Dame, Notre Dame, IN, USA

Ulrich E. Steiner, Fachbereich Chemie, Universität Konstanz, Konstanz, Germany

Nicholas J. Turro, Department of Chemistry, Columbia University, New York, NY, USA

Richard G. Weiss, Department of Chemistry, Georgetown University, Washington, DC, USA

Olaf Wiest, Department of Chemistry and Biochemistry, University of Notre Dame, Notre Dame, IN, USA

Uta Wille, ARC Centre of Excellence for Free Radical Chemistry and Biotechnology, School of Chemistry and BIO21 Molecular Science and Biotechnology Institute, The University of Melbourne, Parkville, Victoria, Australia

Jonathan. R. Woodward, Chemical Resources Laboratory, Tokyo Institute of Technology, Midori-ku, Yokohama, Japan

Hendrik Zipse, Department of Chemistry and Biochemistry, LMU Munich, Munich, Germany

1

A BRIEF HISTORY OF CARBON RADICALS

MALCOLM D. E. FORBES

Department of Chemistry, University of North Carolina, Chapel Hill, NC, USA

It may seem difficult to believe, but research on carbon-centered free radicals is about to close out its second century. In 1815, Gay-Lussac reported the formation of cyanogen (the dimer of ·CN) by heating mercuric cyanide.[1] Numerous experiments involving pyrolysis of organometallic compounds followed, most notably from Bunsen,[2] Frankland,[3] and Wurtz[4] in the 1840s. All of these reactions suggested the existence of what we now know to be carbon-centered free radicals, but physical methods of detection were still many decades away, and the field became somewhat stagnant in the latter half of the nineteenth century. From high-temperature gas-phase dissociation reactions, it was well accepted that inorganic compounds such as elemental iodine could exist in equilibrium with their atomic "radical" forms. In 1868, Fritzsche's observation of color changes due to formation of charge transfer complexes between picric acid and benzene, naphthalene, or anthracene represented the first evidence for the existence of carbon radical cations in aromatic systems.[5] However, the attempted isolation of neutral compounds with trivalent carbon was an idea that had definitely fallen out of favor by the 1880s. This lack of interest was amplified by the flourish of new ideas surrounding tetravalent carbon (pushed experimentally by the vapor density method for determining molecular weights[6]) and by the geometrical insight provided by van't Hoff's proposal of tetrahedral carbon in 1874.[7]

Because of this status quo, the field was turned completely upside down with Gomberg's report in 1900 of the preparation of the triphenylmethyl radical.[8] Apart from the influential support of Nef,[9] Gomberg's result was met with much skepticism.

Carbon-Centered Free Radicals and Radical Cations, Edited by Malcolm D. E. Forbes
Copyright © 2010 John Wiley & Sons, Inc.

But as is typical in the scientific endeavor, healthy criticism can provoke new experiments to prove or disprove a novel result, and the carbon radical skeptics were slowly won over. The year 1900 marked the beginning of what we might call the "wet chemistry" era of research on carbon-centered free radicals, although it should be noted that there were also many gas-phase experiments that were useful in establishing radical reactivity patterns. For example, Goldstein's experiments with cathode ray tubes provided the earliest physical method of detection of carbon-based radical cations in the gas phase.[10] On the theoretical side, strong support for the existence of carbon-centered free radicals came from G. N. Lewis in 1916, whose ideas about valence shells and the octet rule had just begun to emerge.[11] Lewis was also the first to recognize that molecules with unpaired electrons would exhibit paramagnetism.[12] It is rather astounding to realize that both of these hypotheses predate the advent of the quantum theory; in regard to molecular structure, Lewis had an unmatched level of insight for his time.

The 1920s saw a flurry of activities in both thermal and photochemical investigations of gas-phase organic reactions, and chemists such as H. S. Taylor began to hypothesize carbon-centered radicals as reactive intermediates in certain mechanisms.[13] In 1929, Paneth and Hofeditz reported their ingenious "mirror" experiments involving thermolysis of vapor-phase $Pb(CH_3)_4$ and other organometallic compounds. Their results clearly demonstrated that alkyl radicals were reasonable postulates as reactive intermediates in these reactions.[14] In solution-phase organic chemistry, free radicals were beginning to be proposed as intermediates whenever "forbidden" chemistry was observed. This included reactions such as autooxidation of carbonyl compounds and the sulfite ion studied by Bäckström in 1927,[15] and in other reactions by Haber and Willstätter a few years later.[16] There was even an early suggestion by Staudinger in 1920 that free radicals were involved in the polymerization reactions of olefins.[17] Carbon-centered radical cations were the subject of many gas-phase ion investigations in the early part of the twentieth century, led by the instrumentation developments of Thomson and Aston.[18,19] Their work during these years built directly on Goldstein's cathode ray results and lay the foundation for the emerging field of organic mass spectrometry.

The year 1937 was an auspicious one for free radical chemistry, with the publication of an extensive review on solution-phase free radical mechanisms by Hey and Waters.[20] At about the same time, Kharasch proposed the now well-accepted mechanism for anti-Markovnikov addition of HBr to alkenes in the presence of peroxides,[21] a reaction he had initially reported with Mayo 4 years earlier.[22] Also in 1937 came Flory's definitive paper on the kinetics of vinyl polymerization reactions, confirming the nature of these reactions as free radical propagations.[23] This work on polymers eventually led to one of the largest spurts of industrial growth of the twentieth century. The year 1937 was noted by Walling in his excellent monograph[24] as the beginning of the general acceptance of carbon-centered free radicals as viable reactive intermediates in solution-phase organic reactions at ordinary temperatures.

In terms of physical methods, by 1937 there had been only a few advances beyond the mirror technique of Paneth or the invoking of "forbidden reactivity" in solution to establish that a mechanism involved free radicals (or not). As noted above, mass spectrometry was one field where radicals and radical cations from organic structures were beginning to be postulated and actively studied.[25] Optical absorption, usually carried out in frozen glasses with unstable radicals, was also a common early technique,[26] and this was an important component of Gomberg's description of the triphenylmethyl radical. With Lewis' recognition of the link between radicals and paramagnetism, the magnetic susceptibility experiment came to be used in the study of stable free radicals.[27] A bit later, scavenging studies were carried out to establish radical mechanisms[28], and the early days of flash photolysis allowed coarse structural and kinetic studies of radicals to be performed for the first time.[29] However, prior to World War II (WWII), there were no high-resolution methods available that could definitively establish the structural (and magnetic) properties of carbon-centered radicals. The War would change this situation quickly.

The threat of airborne bombing raids on major cities during WWII led to intense efforts for the early detection of aircraft, and it was quickly recognized that radio and/or microwave frequency electromagnetic radiation could be used for this purpose.[30] This research in radio physics and engineering led to the availability of high-powered RF sources and sensitive detectors, the potential of which was immediately exploited by chemical physicists for the detection of magnetic resonances due to spin angular momentum in atoms and molecules. Such resonances had been predicted from the quantum theory two decades earlier, but had eluded detection.[31] Purcell and Bloch in the United States, and Zavoisky in Russia (then the USSR), refined these RF experimental techniques to demonstrate "proof of principle" magnetic resonance spectroscopies (Purcell and Bloch discovered and reported NMR independently in 1946,[32,33] while Zavoisky reported the first EPR spectrum of a paramagnetic species in 1945[34]). While the EPR experiment is more directly relevant, both experiments played key roles in understanding mechanistic organic chemistry involving carbon-centered free radicals. The impact of these techniques on the field cannot be overestimated, and both spectroscopic methods are widely used in the study of radical reactions to the present day.

Just a few years after the discoveries of electron and nuclear magnetic resonance phenomena, commercial EPR and NMR spectrometers appeared, and the early 1950s can be considered the dawn of the "spectroscopic era" of research on free radicals. In the United States, the research groups of Weissman in St. Louis and Hutchison in Chicago were soon studying the structures and molecular dynamics of radicals and triplet states. Weissman in particular was developing workable models for simulating solution EPR spectra.[35] In 1958, Hutchison and Mangum reported the first EPR spectrum of an organic triplet state,[36] ending years of speculation and argument about the nature of phosphorescence (much earlier, G. N. Lewis had correctly predicted that the phosphorescent state of an organic molecule was the excited triplet).

Activity in magnetic resonance of free radicals has not let up, and a cursory literature search found almost 80,000 publications related to EPR spectroscopy at the time this book went to press. More than half of these papers are devoted to carbon-centered radicals. In 1963, new photochemical techniques and advances in spectrometer sensitivity led to the first direct observations of free radicals in liquid solution at room temperature.[37] Soon after, it was commonplace to see g-factor (chemical shift) and isotropic electron–nuclear hyperfine coupling constants for novel radicals being published on a regular basis in what we now refer to as "high-impact" journals.

Many sophisticated techniques for the isolation and study of free radicals and carbenes in the gas phase were devised during the spectroscopic era,[38,39] most of them in conjunction with the development of high-intensity CW and pulsed lasers. These experiments were not only highly complementary to magnetic resonance methods, but also had the advantage of driving computational and theoretical work because very simple structures could be studied in the absence of solvent effects with high spectroscopic resolution. An example is the landmark photodetachment experiment of Engelking et al. that led to a precise value for the singlet–triplet energy gap in methylene, the simplest carbene.[40] This energy gap had historically been a problem for computational chemists due to its open-shell structure, but the photodetachment method provided much guidance. The electronic structure of methylene remains one of the healthiest examples ever recorded of experiment/theory convergence in physical organic chemistry. The development of pulsed lasers in the 1960s also improved the time resolution and sensitivity of the flash photolysis experiment, and this allowed the kinetics of many radical reactions in solution to be precisely measured in real time.[41] It is fair to say that prior to the development of time-resolved magnetic resonance techniques in the 1970s, laser flash photolysis was the standard method for determining free radical lifetimes in solution.[42]

Research on carbon-centered radical cations in solution accelerated dramatically with the development of time-resolved optical absorption and emission techniques. The research group of Th. Förster in Germany pioneered photochemical methods of production of radical cations and anions, as well as exciplexes.[43] While the Förster group focused on structure and lifetimes, the later work of D. R. Arnold in Canada,[44] and of H. D. Roth in the United States,[45] reported the reactivity of photochemically generated radical cations from a mechanistic perspective. These studies of radical ion chemistry evolved into the field we now know as electron donor–acceptor interactions, a rich area of science in which carbon-centered radical cations are still actively studied.

Another burst of activity in free radical research occurred in the 1960s and 1970s, after several reports of anomalous intensities in the EPR spectra of photochemically or radiolytically produced radicals, and in the NMR spectra of the *products* from free radical reactions in solution.[46,47] These so-called chemically induced magnetic spin polarization (CIDNP and CIDEP) phenomena provided a wealth of mechanistic, kinetic, dynamic, and structural information and were a cornerstone of carbon-centered free radical research for the better part of three decades.[48,49] The umbrella term for this area of research is "spin chemistry," which is defined as the chemistry of spin-selective processes.

Many new physical methods were developed in response to needs of spin chemists. In particular, the time-resolved EPR (TREPR)[50] and time-resolved NMR (CIDNP)[51] techniques were found to be of unparalleled utility in terms of mechanistic understanding of radical chemistry. Theoretical work to explain CIDNP and CIDEP phenomena was able to link, for the first time, the spin physics of radical pairs to their diffusion, molecular tumbling, confinement (solvent cages versus supramolecular environments[52]), and the effects of externally applied magnetic fields.[53–56]

Several chapters of this book show how magnetic field effects, as well as CIDEP and CIDNP spectral patterns, can be used to solve chemical problems. It should be noted that the study of how applied magnetic fields perturb chemical reactivity is a topic that is highly relevant to biological processes involving radical pairs, for example, photosynthesis.[57–59]

Two other major instrumentation developments had a major influence on the study of carbon free radicals. In the 1950s, George Feher developed electron–nuclear double resonance (ENDOR) spectroscopy,[60] which is still used to great advantage in determination of hyperfine coupling constants in biological systems.[61] The experiment is even run in time-resolved mode in some laboratories.[62] Pulsed EPR has emerged in recent years as a valuable technique,[63] but its utility in the study of organic radicals is somewhat limited by the current status of microwave pulsing technology. Only very narrow spectral widths ($\sim$100 MHz) can be excited with uniform power by such pulses without distortions of the signals. Both electron spin-echo envelope modulation (ESEEM)[64] and FT-EPR[65] are used in the study of biological free radicals, and as the microwave technology improves in the modern era, 2D[66] and even 3D[67] pulsed EPR experiments have become a reality.

It is interesting to look back on this historical perspective and note that in the "wet chemistry" era (pre-WWII), the reactivity of radicals (Bäckström) and synthetic applications (Kharasch) were "hot" experimental topics. Polymers were just beginning to be recognized as fertile areas for research on free radicals (Flory), and gas-phase spectroscopy was leading to some of the most insightful experimental observations of the time (Paneth). This book honors the efforts of these pioneers in that, while the experiments have become more complex, the fundamental relationship between structure and reactivity is still driving intellectual curiosity in free radical research. The level of computational precision regarding structure and reactivity of free radicals has grown incredibly since 1950 and now matches the sophistication of the modern experimental arsenal.[68,69] It is clear that the complexity of the systems that can be studied with these computational methods will continue to increase.

The future of the field is bright: carbon-centered free radicals in chemistry and biology continue to be of broad interest and continue to be studied experimentally with high resolution and high sensitivity. Combined with the latest computational techniques, it is now possible to consider the creation of a "cradle to grave" understanding of a free radical reaction, from the characterization of the excited-state precursor by optical techniques to the structure and dynamics of the radicals themselves by EPR spectroscopy, and finally to the kinetics of formation and structures of the products by NMR spectroscopy and other analytical methods.

REFERENCES

1. Gay-Lussac, H. L. *Ann. Chim.* **1815**, *95*, 172.
2. Bunsen, R. H. *Justus Liebigs Ann. Chim.* **1842**, *42*, 27.
3. Frankland, E. *Ann. Chem. Pharm.* **1849**, *71*, 213.
4. Wurtz, C. A. *Compt. Rend.* **1854**, *40*, 1285.

5. Fritzsche, C. J. *Jahresb.* **1868**, 395.

6. Siwoloboff, A. *Ber. Dtsch. Chem. Ges.* **1886**, *19*, 795.

7. van't Hoff, J. H. *Arch. Neerl. Sci. Exactes Nat.* **1874**, *9*, 445.

8. Gomberg, M. *J. Am. Chem. Soc.* **1900**, *22*, 757.

9. Nef, J. U. *Liebigs Ann.* **1901**, *318*, 137.

10. Goldstein, E. *Ann. Phys.* **1898**, *300*, 38.

11. Lewis, G. N. *J. Am. Chem. Soc.* **1916**, *38*, 762.

12. Lewis, G. N. *Valence and Structure of Atoms and Molecules*; Chemical Catalog Co.: New York, **1923**; p 148.

13. Taylor, H. S. *Trans. Faraday Soc.* **1925**, *21*, 560.

14. Paneth, F.; Hofeditz, W. *Ber. Dtsch. Chem. Ges.* **1929**, *62*, 1335.

15. Bäckström, H. L. J. *J. Am. Chem. Soc.* **1927**, *49*, 1460.

16. Haber, F.; Willstätter, R. *Ber. Dtsch. Chem. Ges.* **1931**, *64*, 2844.

17. Staudinger, *Ber. Dtsch. Chem. Ges* **1920**, *53*, 1073.

18. Thomson, J. J. *Philos. Mag.* **1899**, *48*, 547.

19. Aston, F. W. *Philos. Mag.* **1919**, *48*, 707.

20. Hey, D. H.; Waters, W. A. *Chem. Rev.* **1937**, *21*, 169.

21. Kharasch, M. S.; Engelmann, H.; Mayo, F. R. *J. Org. Chem.* **1937**, *2*, 288.

22. Kharasch, M. S.; Mayo, F. R. *J. Am. Chem. Soc.* **1933**, *55*, 2468.

23. Flory, P. J. *J. Am. Chem. Soc.* **1937**, *59*, 241.

24. Walling, C. *Free Radicals in Solution*; Wiley: New York, **1957**.

25. Taylor, D. D. *Phys. Rev.* **1935**, *47*, 666.

26. Walter, R. I. *J. Am. Chem. Soc.* **1966**, *88*, 1930.

27. van Vleck, J. H.; *Electric and Magnetic Susceptibility*; Oxford University Press: New York, **1932**.

28. Noyes, R. M. *J. Am. Chem. Soc.* **1955**, *77*, 2042.

29. Norrish, R. G. W.; Porter, G.; Thrush, B. A. *Proc. R. Soc. Lond. A* **1953**, *216*, 165.

30. Bowen, E. G.; *Radar Days*; Institute of Physics Publishing: Bristol, **1987**.

31. Gorter, C. J.; Broer, L. J. F. *Physica* **1942**, *9*, 591.

32. Purcell, E. M.; Torrey, H. C.; Pound, R. V. *Phys. Rev.* **1946**, *69*, 37.

33. Bloch, F. *Phys. Rev.* **1946**, *70*, 460.

34. Zavoisky, E. K. *J. Phys. (USSR)* **1945**, *9*, 245.

35. Weissman, S. I. *J. Chem. Phys.* **1954**, *22*, 1135.

36. Hutchison, C. A. Jr.; Mangum, B. W. *J. Chem. Phys.* **1958**, *29*, 952.

37. Piette, L. H. A new technique for the study of rapid free radical reactions. In *Sixth International Symposium on Free Radicals*, Cambridge University Press, **1963**.

38. Friderichsen, A. V.; Radziszewski, J. G.; Nimlos, M. R.; Winter, P. R.; Dayton, D. C.; David, D. E.; Ellison, G. B. *J. Am. Chem. Soc.* **2001**, *123*, 1977.

39. Deyerl, H.-J.; Gilbert, T.; Fischer, I.; Chen, P. *J. Chem. Phys.* **1997**, *107*, 3329.

40. Engelking, P. C.; Corderman, R. R.; Wendoloski, J. J.; Ellison, G. B.; O'Neil, S. V.; Lineberger, W. C. *J. Chem. Phys.* **1981**, *74*, 5460.

41. McGarry, P. F.; Cheh, J.; Ruiz-Silva, B.; Hu, S.; Wang, J.; Nakanishi, K.; Turro, N. J. *J. Phys. Chem.* **1996**, *100*, 646.

42. Scaiano, J. C.; Johnston, L. J. *Org. Photochem.* **1989**, *10*, 309.

43. Förster, Th. *Pure Appl. Chem.* **1973**, *34*, 225.

44. Neunteufel, R. A.; Arnold, D. R. *J. Am. Chem. Soc.* **1973**, *95*, 4080.

45. Roth, H. D. *Acc. Chem. Res.* **1987**, *20*, 343.

46. Ward, H. R.; Lawler, R. G. *J. Am. Chem. Soc.* **1967**, *89*, 5518.

47. Bargon, J.; Fischer, H.; Johnsen, U. *Z. Naturforsch. A* **1967**, *22*, 1551.

48. Trifunac, A. D.; Lawler, R. G.; Bartels, D. M.; Thurnauer, M. C. *Prog. React. Kinet.* **1986**, *14*, 43.

49. Bagryanskaya, E. G.; Sagdeev, R. Z. *Prog. React. Kinet.* **1993**, *18*, 63.

50. Forbes, M. D. E. *Photochem. Photobiol.* **1997**, *65*, 73.

51. Closs, G. L.; Miller, R. J. *Rev. Sci. Instrum.* **1981**, *52*, 1876.

52. Turro, N. J.; Buchachenko, A. L.; Tarasov, V. F. *Acc. Chem. Res.* **1995**, *28*, 69–80.

53. Adrian, F. J. *J. Chem. Phys.* **1971**, *54*, 3918.

54. Closs, G. L. *J. Am. Chem. Soc.* **1969**, *91*, 4552.

55. McLauchlan, K. A.; Steiner, U. E. *Mol. Phys.* **1991**, *73*, 241.

56. Steiner, U. E.; Ulrich, T. *Chem. Rev.* **1989**, *89*, 51.

57. Hoff, A. J. In *The Photosynthetic Reaction Center*; Deisenhofer, J.; Norris, J. R., Jr., Eds.; Academic Press: New York, 1993; Vol. 2, pp. 331–382.

58. Norris, J. R., Jr.; Budil, D. E.; Kolaczkowski, S. V.; Tang, J. H.; Bowman, M. K. In *Antennas and Reaction Centers of Photosynthetic Bacteria*; Michel-Beyerle, M. E., Ed.; Springer: Berlin, 1985; pp. 190–197.

59. Bowman, M. K.; Budil, D. E.; Closs, G. L.; Kostka, A. G.; Wraight, C. A.; Norris, J. R. *Proc. Natl. Acad. Sci. USA* **1981**, *78*, 3305

60. Feher, G. *Phys. Rev.* **1956**, *103*, 834

61. Hoganson, C. W.; Babcock, G. T. *Biochemistry* **1992**, *31*, 11874.

62. Jaegermann, P.; Lendzian, F.; Rist, G.; Moebius, K. *Chem. Phys. Lett.* **1987**, *140*, 615.

63. *Modern Pulsed and Continuous-Wave Electron Spin Resonance*; Kevan, L.; Bowman, M. K., Eds.; Wiley: New York, **1990**.

64. Rowan, L. G.; Hahn, E. L.; Mims, W. B. *Phys. Rev.* **1965**, *137*, A61.

65. van Willigen, H.; Levstein, P. R.; Ebersole, M. H. *Chem. Rev.* **1993**, *93*, 173.

66. Goldfarb, D.; Kofman, V.; Shanzer, J. A.; Rahmatouline, R.; van Doorslaer, S.; Schweiger, A. *J. Am. Chem. Soc.* **2000**, *122*, 1249.

67. Blank, A.; Dunnam, C. R.; Borbat, P. P.; Freed, J. H. *Appl. Phys. Lett.* **2004**, *85*, 5430.

68. Fokin, A. A.; Schreiner, P. R. *Chem. Rev.* **2002**, *102*, 1551.

69. Improta, R.; Barone, V. *Chem. Rev.* **2004**, *104*, 1231.

2

INTERMOLECULAR RADICAL ADDITIONS TO ALKYNES: CASCADE-TYPE RADICAL CYCLIZATIONS

UTA WILLE

ARC Center of Excellence for Free Radical Chemistry and Biotechnology, School of Chemistry and BIO21 Molecular Science and Biotechnology Institute, The University of Melbourne, Parkville, Victoria, Australia

2.1 INTRODUCTION

"Cascade" radical reactions, also known as "tandem" or "domino" radical reactions, that proceed through two or more consecutive steps under involvement of both intra- and intermolecular reactions are a synthetically highly attractive methodology because they enable the access of often very complex structural frameworks with very few synthetic steps. The fact that the generally mild conditions for radical reactions are compatible with a large number of functional groups so that time-consuming protection strategies can be minimized, in addition to the recent observation that the principles of stereocontrol that were discovered and developed for ionic chemistry can also be applied to free radical reactions, has resulted in the development of highly stereoselective radical cascade processes.

The great majority of radical cascades involve sequences of intramolecular steps where the overall propagation is a unimolecular process (with the exclusion of the initiation and termination steps), and recent overviews are given in Refs 1–5. However, meanwhile many tandem reactions involving both intra- and intermolecular steps, as well as multicomponent tandem reactions, have been reported in the literature, which are compiled in recent reviews.[3,4]

Cascade reactions that are initiated by intermolecular radical addition to π systems have been mostly studied using C=C, C=N, and C=O double bonds as acceptors for both carbon- or heteroatom-centered radicals.[3,4] Quite remarkably, however, radical cascades that are triggered by intermolecular radical addition to C$\equiv$C triple bonds have received considerably less attention. This is in stark contrast to the overwhelming number of studies that use alkynes as radical acceptors in intramolecular radical cyclizations. Generally, intermolecular radical addition to alkynes is significantly slower than addition to the corresponding alkenes,[6] even though the former are more exothermic. This observation has been explained in several papers by Fischer and Radom, who suggested that diminished polar effects due to smaller electron affinities and larger ionization energies in alkynes, compared to alkenes, render alkyne reactions less selective by increasing the activation barriers for radical additions. These authors also proposed that, in addition, the higher triplet energy in alkynes could also lead to an increase in the activation barrier for radical additions to alkynes.[7,8]

The addition of free radical species to alkynes leads to highly reactive vinyl radicals that can undergo fast cyclization onto π systems,[9] as well as hydrogen abstraction, and the synthetic value of vinyl radicals in cascade reactions has been explored in numerous studies. However, although the intermolecular addition of both C- and heteroatom-centered radicals to alkynes is principally known for almost 75 years (note that the regiochemistry is controlled by steric factors—the radical addition occurs usually at the less substituted site of the C$\equiv$C triple bond),[6] because of the above-mentioned complications, vinyl radicals have often been generated not through radical addition to alkynes, but through homolytic atom abstraction from suitably substituted alkenes, for example, through reaction of vinyl halides with tributyltin hydride, Bu_3SnH, in the presence of a radical initiator.[9]

This chapter provides a brief overview of radical cyclization cascades that are initiated by intermolecular radical addition to alkynes. The specific reactions have been arranged according to the atoms carrying the unpaired electron in the attacking radical, with special emphasis on the reactions involving O- and N-centered radicals. Because of the often highly complex nature of the cascade reactions, the reaction schemes also include the mechanistic steps involved in the respective sequence. It should be noted that it is the intention of the author to present a selection of synthetically and/or mechanistically interesting examples, rather than to provide a comprehensive collection of reactions, and the reader is encouraged to consult the given references for additional details.

2.2 CASCADE REACTIONS INVOLVING RADICALS OF SECOND ROW ELEMENTS

2.2.1 Cascade Reactions Initiated by Addition of C-Centered Radicals to Alkynes

In 1967, Heiba and Dessau reported perhaps one of the earliest examples of a radical cyclization cascade that is initiated by intermolecular addition of C-centered radicals to alkynes. Reaction of carbon tetrachloride with 1-heptynes **1** in the presence of benzoyl peroxide (BPO) as radical initiator resulted, among other products, in the formation of 1,1-dichlorovinylcyclopentane derivatives **2** in moderate yields (Scheme 2.1).[10]

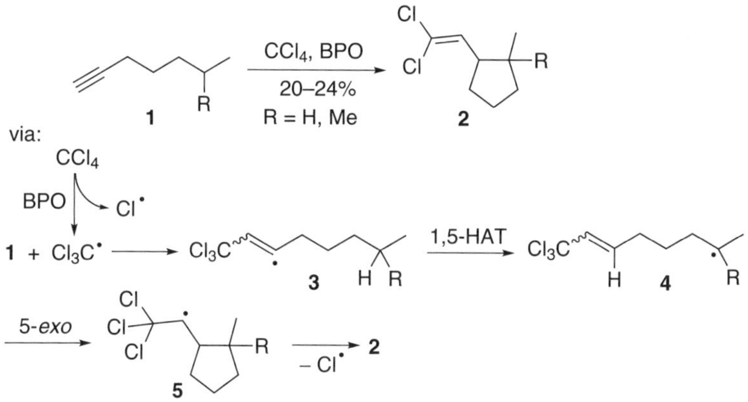

SCHEME 2.1

The reaction is believed to proceed through addition of the initially formed trichloromethyl radical, $Cl_3C^{\bullet}$, to the terminal end of the $C{\equiv}C$ triple bond in **1** to give vinyl radical **3**, which undergoes a 1,5-hydrogen atom transfer (1,5-HAT), followed by 5-*exo* cyclization to the C=C double bond in alkyl radical **4**. The sequence is terminated by homolytic scission of a C–Cl bond in radical intermediate **5**, to form the stable alkene moiety in **2** with release of a chlorine atom, $Cl^{\bullet}$. The terminating β-fragmentation, which is obviously considerably faster than chlorine transfer from CCl_4 by **5** in a radical chain transfer process, is mechanistically quite remarkable, since the released $Cl^{\bullet}$ is a substituent of the cascade-initiating radical $Cl_3C^{\bullet}$. Although $Cl^{\bullet}$ is a highly reactive species that might get involved in radical chain processes, this principle can be regarded as the foundation of "self-terminating radical cyclizations," a recently discovered new concept in radical cyclizations, which will be discussed later.

In another early example, C radicals, which were generated from the reaction of trialkylboranes with dioxygen, O_2, were added to α-epoxy alkynes **6** to give allenes **7** (Scheme 2.2).[11] In this reaction, the initially formed vinyl radicals (or α-oxiranyl radicals) **8** undergo epoxide ring opening to give the allenic alkoxyl radical intermediate **9**, which is trapped by triethylborane to give boronate **10**, the end product of the radical cyclization cascade. The observed allenic alcohol **7** results from hydrolysis of **10**.

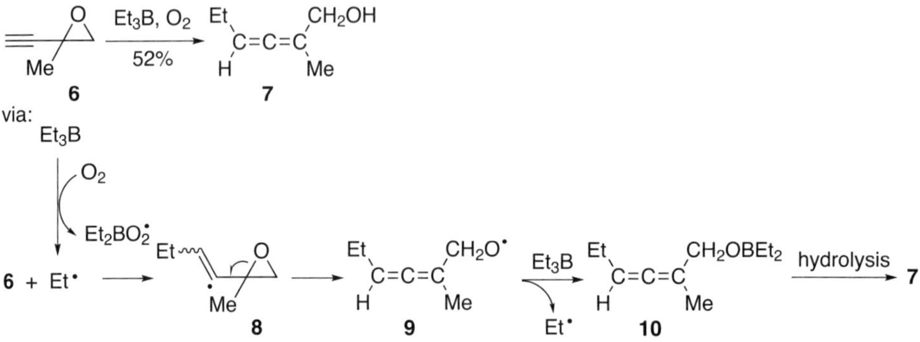

SCHEME 2.2

The intermolecular radical addition of alkyl iodides to terminal alkynes has been used in cyclization cascades performed under radical chain conditions (Scheme 2.3). For example, Oshima and coworkers applied triethylborane as radical initiator to

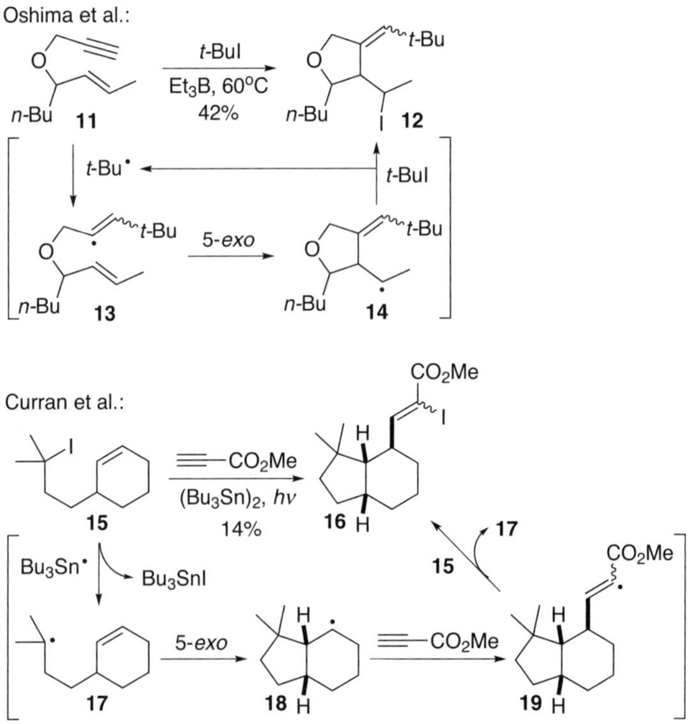

SCHEME 2.3

generate tetrahydrofurans **12** from the respective enyne precursor **11** in a radical cascade consisting of addition of the nucleophilic *t*-butyl radicals, *t*-Bu•, to the less electron-rich π system in **11** (e.g., the C≡C triple bond), followed by 5-*exo* cyclization of vinyl radical **13** and iodine abstraction by the cyclized radical intermediate **14**.[12] In the cyclization described by Curran and Kim, the cyclohexenyl-substituted alkyl radicals **17**, generated from the corresponding halide **15** using tin radicals as mediator, first undergo a *cis* stereoselective 5-*exo* cyclization to yield cyclohexyl-derived radicals **18**, which are subsequently trapped from the sterically less hindered front side by intermolecular addition to the terminal end of methyl propiolate to ultimately produce an *E/Z* mixture of the bicyclic vinyl iodide **16**.[13]

A combination of radical and electron transfer steps mediated by manganese triacetate has been used in the synthesis of 5-acetoxyfuranones **21** through carboxymethyl radical addition to mono- and disubstituted alkynes **20**, followed by oxidative cyclization of the resulting vinyl radicals **22** (Scheme 2.4).[14] The cyclic intermediate **24** is transformed into the furanone **21** through stepwise one-electron oxidation and capture of the resulting allyl cation **26** by acetate.

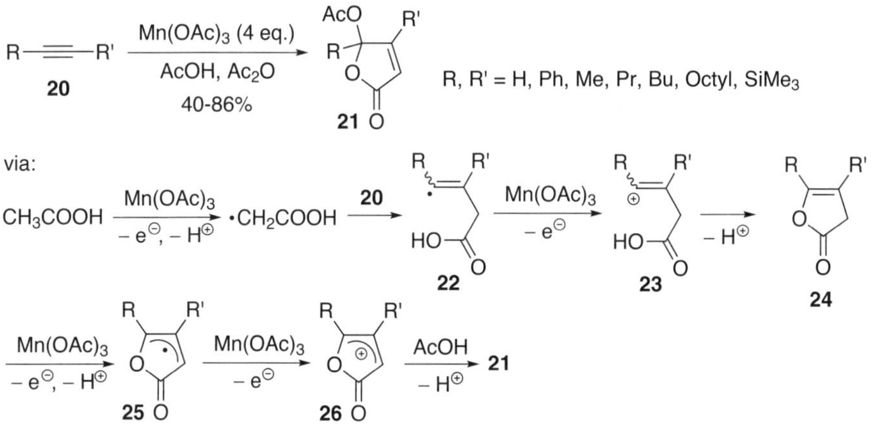

SCHEME 2.4

Zanardi and coworkers reported on a radical cascade that involves intermolecular addition of *ortho* cyanoaryl radicals to terminal alkynes (Scheme 2.5).[15] In this sequence, decomposition of the aryldiazonium salt **27** in pyridine leads to the aryl radical **29**, which undergoes addition to the terminal end of the alkyne to give the vinyl radical **30**. 5-*exo* Radical cyclization onto the cyano group, followed by homolytic aromatic substitution of the intermediate iminyl radical **31**, results ultimately in the formation of the cyclopentaphenanthridine **28**. Although the yield of this sequence is only moderate, the potential synthetic interest lies in the fact that nitrogen heterocycles with elaborate frameworks can be obtained in one single step from readily available starting materials.

Azobisisobutyronitrile (AIBN) is a widely used radical initiator that usually does not get involved as a reagent in radical reactions itself. Montevecchi et al. reported

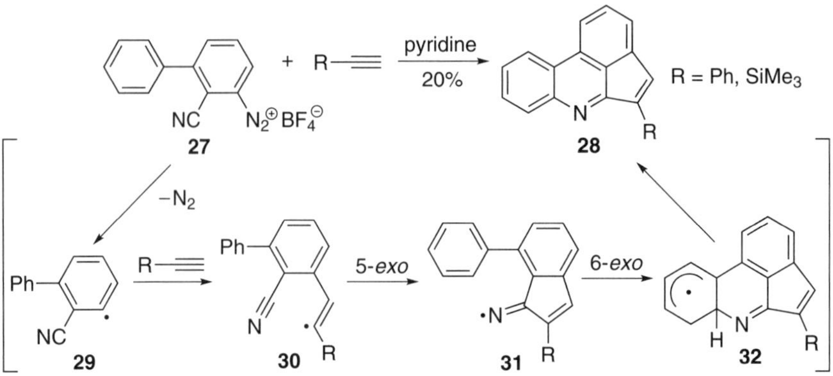

SCHEME 2.5

recently that the transient C-centered 2-cyanoisopropyl radical, $(i\text{Pr})_2\text{C}^\bullet\text{CN}$, obtained from AIBN thermolysis can undergo addition to electron-poor alkynes. However, this reaction leads to a variety of different products in very low yield and appears to be synthetically not useful.[16]

In recent years, *N*-hydroxyphthalimide (NHPI) has been increasingly used as catalyst in radical reactions under tin-free conditions. The example in Scheme 2.6 shows that the phthalimide *N*-oxyl radical (PINO), which is generated by Co(II)-mediated oxidation of NHPI, abstracts a hydrogen atom from isopropanol (**34**) to give the C-centered α-hydroxy radical **37**, which is trapped with good efficiency by the electron-poor acetylene **33**. The resulting vinyl radical **38** abstracts a hydrogen atom from NHPI, leading to the *E/Z* isomers of alkene **35** in a 1 : 1 ratio with simultaneous regeneration of the oxidant. *E*-**35** undergoes subsequent cyclization to lactone **39** that

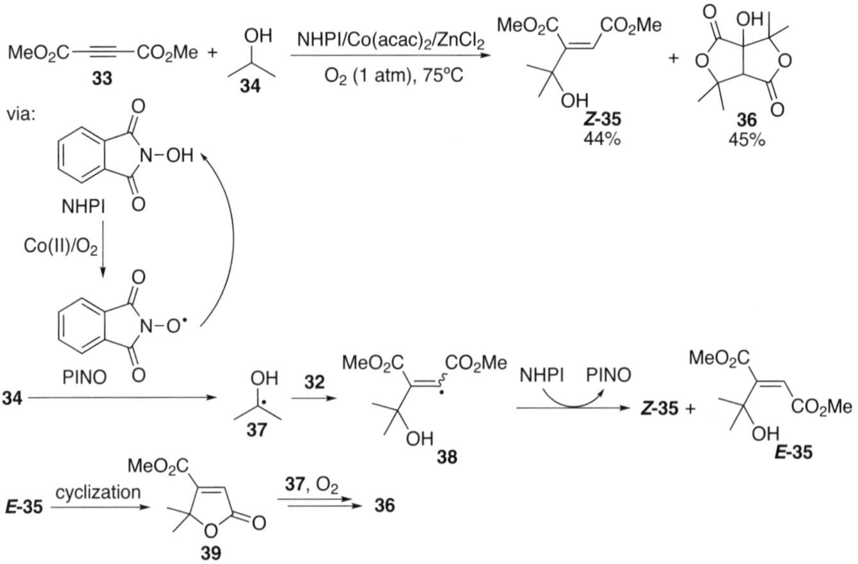

SCHEME 2.6

reacts with a second α-hydroxy radical **37** to give ultimately the bicyclic bislactone **36** through an oxidative cyclization.[17]

A remarkable example for a highly regioselective intermolecular radical addition to an unsymmetrical bis-substituted alkyne is shown in Scheme 2.7. The 1,3-dioxolan-2-yl radical **43**, which can be generated from dioxolane **41** under photosensitized conditions, exclusively undergoes addition to the δ-site of the C≡C triple bond in the ketimine **40**. This triggers an intramolecular cyclization cascade consisting of 6-*exo* cyclization of the vinyl radical **44** to the imine C=N double bond, followed by reduction of radical intermediate **45**, to give the polycyclic compound **42** as sole product in 65% yield.[18,19] However, a similar reaction involving intermolecular addition of both S- and Sn-centered radicals to δ-yne ketimines of type **40** appears to be less selective, since products arising from initial radical addition to both δ and ε sites of the C≡C triple bond were obtained—this might be a result of the reversibility of S and Sn radical addition to alkynes (see below).

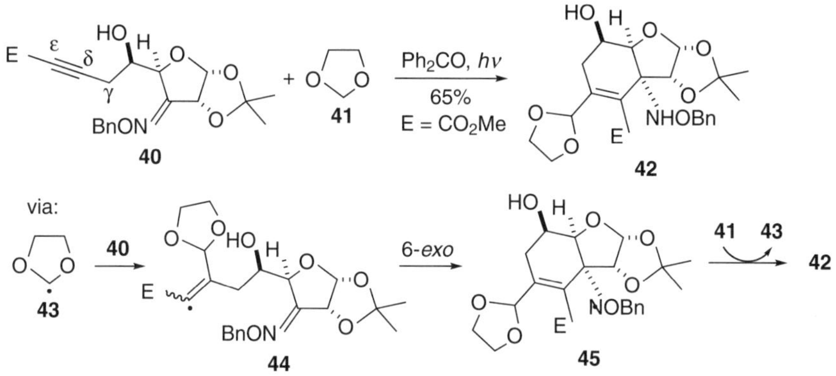

SCHEME 2.7

Cyclopenta-fused pyridines **48** have been synthesized through a cascade initiated by intermolecular addition of C radicals to the C≡N triple bond in vinylisonitriles.[20] The reaction in Scheme 2.8 shows addition of the nucleophilic alkynyl radical **49** to the carbon end of the isonitrile group in **46** to give vinyl radical **50**, which undergoes a

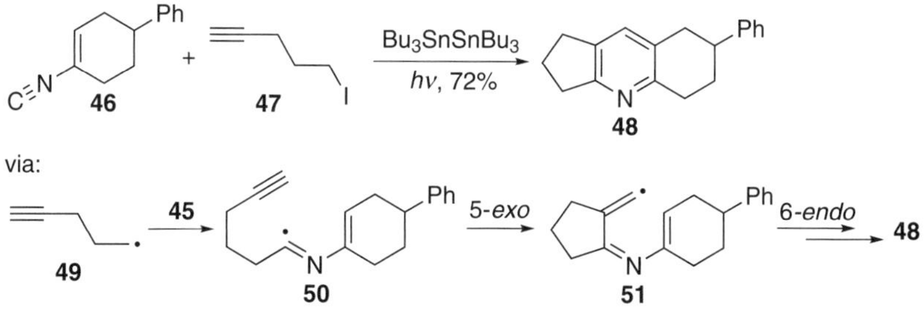

SCHEME 2.8

5-*exo* cyclization to the alkyne moiety, followed by 6-*endo* cyclization of vinyl radical **51**. If, instead of iodoalkynes of type **47**, iodonitriles are used as precursors of C radicals, pyrazines are obtained in an analog cyclization cascade (not shown).

2.2.2 Cascade Reactions Initiated by Addition of O-Centered Radicals to Alkynes (Self-Terminating Radical Oxygenations)

In contrast to intramolecular cyclizations,[4] the intermolecular addition of O-centered radicals to π systems as initiating step in complex radical cyclization cascades has only recently attracted considerable attention. The reason for the low number of papers published on O radical addition to alkenes and alkynes could originate from the perception that O-centered radicals, such as alkoxyl radicals, RO•, or acyloxyl radicals, RC(O)O•, may not react with π systems through addition at rates that are competitive to other pathways, for example, allylic hydrogen abstraction and β-fragmentation in the case of RO• or decarboxylation in the case of RC(O)O• (Scheme 2.9).

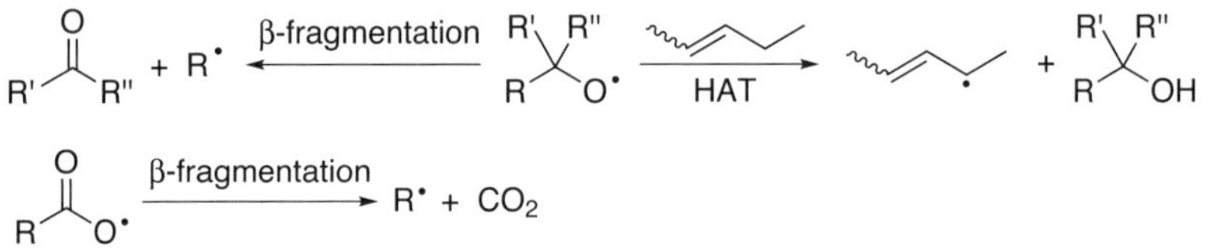

SCHEME 2.9

However, oxidations based on addition of O-centered radicals to unsaturated compounds appear to be a highly desirable synthetic goal, especially when the new C—O bond could be formed under the mild conditions that are typical for radical reactions. If this radical addition would involve C≡C triple bonds, the resulting reactive vinyl radical would be highly suitable for the promotion of intramolecular cascade reactions.

The initial approach to explore addition of O-centered radicals to alkynes was to use radicals that are less prone to fragmentation and/or hydrogen abstraction, compared to RO• or RC(O)O•, for example, O-centered inorganic radicals such as nitrate ($NO_3•$), sulfate ($SO_4•^-$), or hydroxyl radicals (HO•). Using the medium-sized cyclic alkyne **52** as a model system, the reaction with $NO_3•$ generated by anodic oxidation of lithium nitrate led to formation of the *cis* fused bicyclic ketones **53** and **54** in 72% combined yield (Scheme 2.10).[21] BHandHLYP/6-311G** calculations of the potential surface revealed activation barriers, E_a, and reaction energies, ΔE, for each step in the sequence, which are given in Scheme 2.10 (all computed energy data include zero-point vibrational energy correction (zpc)).[22] The vinyl radical **55** formed through addition of $NO_3•$ to the π system in **52** (this reaction has a low E_a of only few kJ/mol and is strongly exothermic)[23] undergoes a 1,5- or 1,6-HAT to the secondary radicals **56a/b**, which subsequently cyclize in a 5-*exo* or 6-*exo* fashion, respectively, to the C=C double bond to give the α-nitrato radicals **57a/b**. The key step in the cyclization cascade is β-fragmentation of the highly labile O—NO_2 bond in the latter radical intermediates, which proceeds practically barrierless, to give the thermodynamically stable ketones **53** and **54** under release of nitrogen dioxide, $NO_2^•$. According to the computations, each single step in this cascade is not only strongly exothermic, but also associated with a low to moderate E_a.

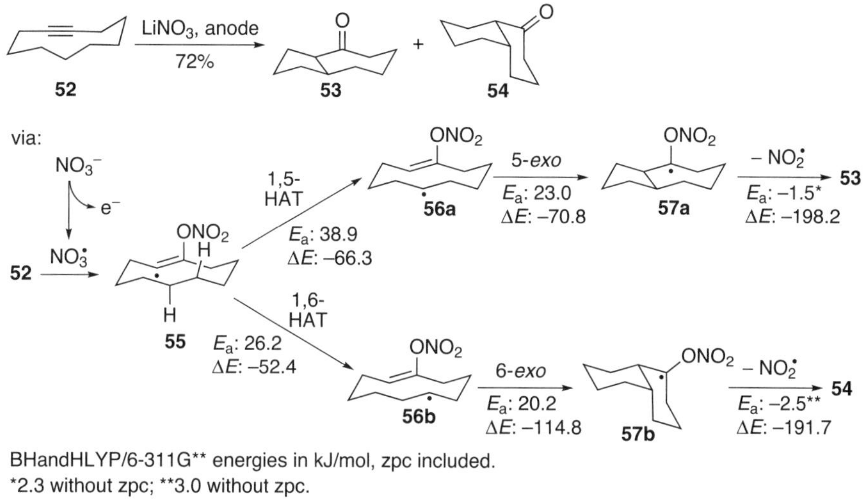

BHandHLYP/6-311G** energies in kJ/mol, zpc included.
*2.3 without zpc; **3.0 without zpc.

SCHEME 2.10

In this sequence, $NO_3^•$ can formally be regarded as a synthon for O atoms in solution. Because the released $NO_2^•$ is a comparatively stable radical that does not initiate a radical chain process under the applied reaction conditions, this sequence is termed a "self-terminating oxidative radical cyclization"—the cascade-initiating radical contains the unreactive leaving group that is released in the final step of the reaction. In contrast to reactions performed under "classical" radical chain conditions and that are nearly always associated with a loss in overall functionality of the molecule, in self-terminating radical cyclizations the net amount of functional groups is retained in the system.

The initially formed vinyl radical can also be trapped through cyclization onto a carbonyl bond. For example, reaction of electro- or photochemically generated $NO_3^•$ with 5-cyclodecynone (**58**) leads to the isomeric epoxy ketones **59** and **60** (Scheme 2.11).[21]

Since the cycloalkyne **58** is not symmetrical, $NO_3^•$ addition to the C≡C triple bond gives the isomeric nucleophilic vinyl radicals **61** and **64**, which undergo 5-*exo* or 6-*exo* cyclization, respectively, to the less electron-rich carbon end of the carbonyl C=O double bond. The resulting allyloxyl radicals **62** and **65** cyclize in a 3-*exo* fashion to yield the α-oxiranyl radicals **63** and **66**, respectively. In accordance with the literature,[24] the calculations show that this latter step is highly reversible, and the only reason why the reverse ring opening does not occur in this cascade is the ease by which the radicals **63/66** undergo β-fragmentation of the O−NO_2 bond to give the respective epoxy ketones **59** and **60** with release of $NO_2^•$. This self-terminating radical cycliza-tion represents one of the very few examples for the synthetic access to epoxides through a 3-*exo* cyclization of allyloxyl radicals. Interestingly, this sequence can also

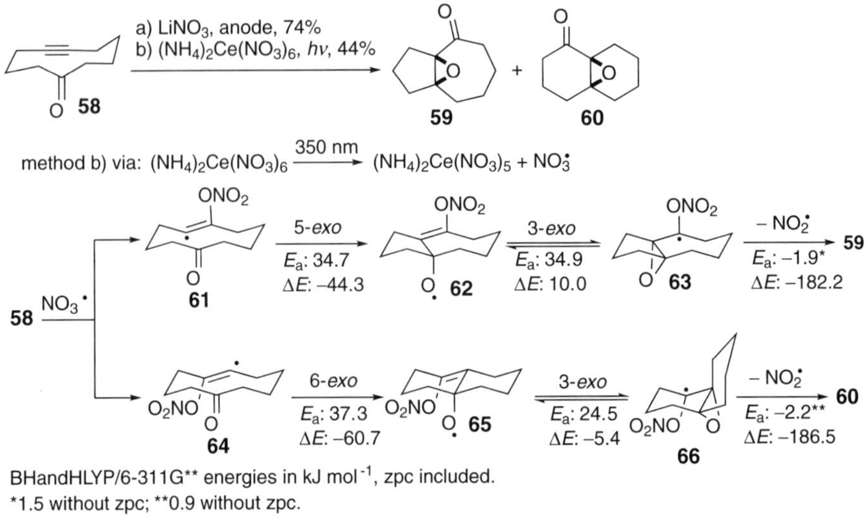

SCHEME 2.11

formally be considered as "retro-Eschenmoser–Ohloff" fragmentation,[25] although the mechanism is significantly different.

Self-terminating radical oxygenations are not restricted to cyclic alkynes. Electro- and photochemically generated $NO_3^•$ can be used for the oxidative cyclization of cycloalkyl-clamped alkynes **67** (Scheme 2.12). This reaction leads to formation of anelated tetrahydrofurans (**68** with X = O)[26–28] or pyrrolidines (**68** with X = NTs, NTf)[29] possessing four asymmetric centers in good to excellent yields through a cascade consisting of radical addition, rate-determining 1,5-HAT **69** → **70**,[26,28] 5-*exo* cyclization **70** → **71**, and terminating β-fragmentation with release of $NO_2^•$. The diastereoselectivity of this sequence ranges from moderate to very high and depends on various factors, such as the nature of the heteroatom X and the substituent R.

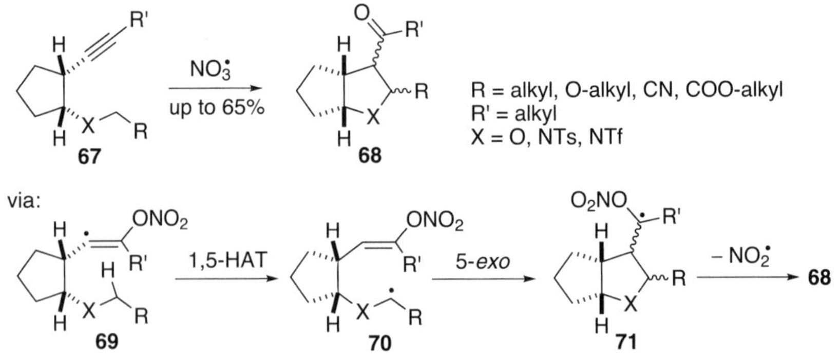

SCHEME 2.12

The role of the cycloalkyl clamp in the starting alkyne **67** is to reduce the conformational degrees of freedom in the intermediate highly reactive vinyl radical

69 and to achieve a favorable alignment between the SOMO on the π system and the σ-orbital of the respective $C-H$ bond for the 1,5-HAT. The experiments revealed that the self-terminating cyclization cascade worked best using a *cis*-1,2-disubstituted cyclopentyl clamp; however, a *trans*-1,2-disubstituted cyclohexyl ring could also be used. With larger ring sizes, both yields and diastereoselectivities were reduced.[26–28] A similar conformational restriction of the intermediate vinyl radical was also obtained using the Thorpe–Ingold effect[30] provided by two geminal ester groups (not shown).[31]

The key step in $NO_3{}^{\bullet}$-induced self-terminating radical oxygenation is the final homolytic cleavage of the weak $O-NO_2$ bond in radicals of types **57, 63, 66**, and **71**, by which a carbonyl group is formed and a stable radical, $NO_2{}^{\bullet}$, is released. Therefore, other O-centered radicals of type $XO^{\bullet}$, where X would form a stable radical upon release in a homolytic β-fragmentation of the $O-X$ bond (Scheme 2.13), should also be suitable for self-terminating radical oxygenations, thus rendering this synthetic concept much more general.

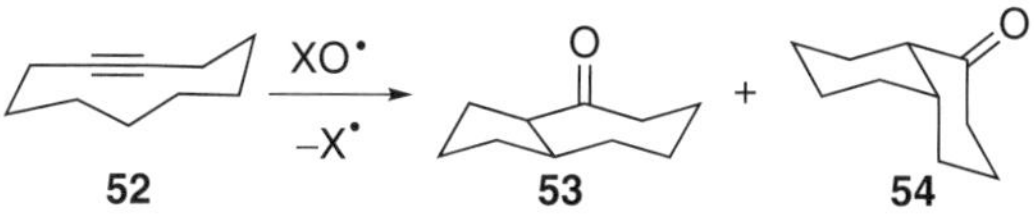

SCHEME 2.13

This is indeed the case. All major classes of inorganic and organic O-centered radicals $XO^{\bullet}$, for example, $SO_4{}^{\bullet-}$ $(X=SO_3{}^{-})$,[32] $HO^{\bullet}$ $(X=H)$,[33] $RC(O)O^{\bullet}$ $[X=RC(O)]$,[34] alkoxycarbonyloxyl $[ROC(O)O^{\bullet}, X=ROC(O)]$,[35] alkoxycarbonyla-cyloxyl $[ROC(O)C(O)O^{\bullet}, X=ROC(O)C(O)]$,[35] carbamoyloxyl $[R_2NC(O)O^{\bullet}, X=R_2NC(O)]$,[35] nitroxyl $[R_2NO^{\bullet}, X=R_2N]$,[35] and $RO^{\bullet}$ $(X=R)$,[36] have been shown to promote oxidation of, for example, cycloalkyne **52** to give the bicyclic ketones **53** and **54**. Although especially the organic O-centered radicals are known to undergo facile β-fragmentation, under optimized conditions they could be very efficiently trapped by the alkyne prior to decomposition. Representative outcomes of these reactions are shown in Table 2.1.

However, although these results show in an impressive way the generality of the concept "self-terminating radical cyclizations," including the discovery of so far unknown reactions of many seemingly well-known radicals, it should be noted that some of the released radicals $X^{\bullet}$ are not considered as stabilized species, for example, the hydrogen atom or the C-centered acyl radicals, and for these, the homolytic $O-X$ fragmentation has little precedent. Indeed, BHandHLYP/6-311G** calculations of the potential surface of the homolytic $O-X$ bond cleavage in a series of representative simplified model reactions (**72** $\rightarrow$ **74**, Scheme 2.14) revealed that, apart from $NO_2{}^{\bullet}$, only the cleavage of the sulfite radical, $SO_3{}^{\bullet-}$, the benzyl radical, $Bn^{\bullet}$, and within limitations, potentially also the methyl radical, $Me^{\bullet}$, requires a reasonably low activation energy, E_a, to give thermodynamically favorable products ($\Delta E < 0$). On the other hand, β-fragmentation of the $O-$acyl or $O-H$ bond is associated with high reaction barriers and significant endothermicity, and should not easily occur. However,

TABLE 2.1   Representative Results for the Reaction of Cyclodecyne (52) with Various O-Centered Radicals XO•

Entry	XO•	Conditions	Yield (GC) 53 + 54	Ref.
1	$SO_4^{•-}$	$S_2O_8^{2-}$, Fe(II)	79%	32
2	HO•	photolysis of	63%	33
3	$4\text{-MeOC}_6H_4C(O)O^•$	photolysis of	89%	34
4	$MeOC(O)O^•$	photolysis of	94%	35
5	$EtOC(O)C(O)O^•$	photolysis of	82%	35
6	$Et_2NC(O)O^•$	photolysis of	69%	35
7		$(NH_4)_2Ce(NO_3)_6$ oxidation of	50%	35
8	BnO•	photolysis of	53%	36

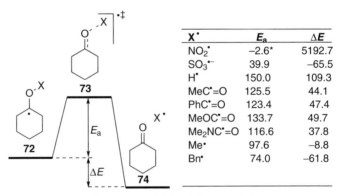

BHandHLYP/6-311G** energies in kJ mol⁻¹, zpc included.
*2.9 without zpc.

X•	E_a	ΔE
$NO_2^•$	−2.6*	5192.7
$SO_3^{•-}$	39.9	−65.5
H•	150.0	109.3
MeC•=O	125.5	44.1
PhC•=O	123.4	47.4
MeOC•=O	133.7	49.7
$Me_2NC•=O$	116.6	37.8
Me•	97.6	−8.8
Bn•	74.0	−61.8

SCHEME 2.14

due to the radical stabilizing effect provided by the lone pair of electrons at oxygen, the α-oxo radicals of type **72** with X = H, C(O)R, C(O)OR, and C(O)NR$_2$ are "trapped" in a local thermodynamic minimum, where potential alternative unimolecular reactions (e.g., ring opening) are also both kinetically and thermodynamically unfavorable.

In these cases, it may be suggested that termination of the radical cyclization sequence could, at least in part, proceed through a different pathway that involves a bimolecular radical chain process, as outlined in Scheme 2.15 using the simplified α-oxy radical **75** as a model system. HO[•] and most of the organic O-centered radicals used during the study of self-terminating radical cyclizations were generated through photolysis of either Barton esters[4] or dithiocarbamates developed by Kim et al. (see Table 2.1).[37] According to calculations at the BHandHLYP/6-311G** level of theory, addition of the α-oxy radical **75** to Barton or Kim esters (using **76** and **79** as representative simplified model compounds) requires only a moderate activation energy of around 40 kJ/mol and is exothermic for both cases. Fragmentation of the adducts **77** and **80** should also both be rapid and exothermic and would lead to the mixed thioacetals **78** and **81**, respectively, which are likely to undergo hydrolytic decomposition to form the carbonyl group in **74**.[22]

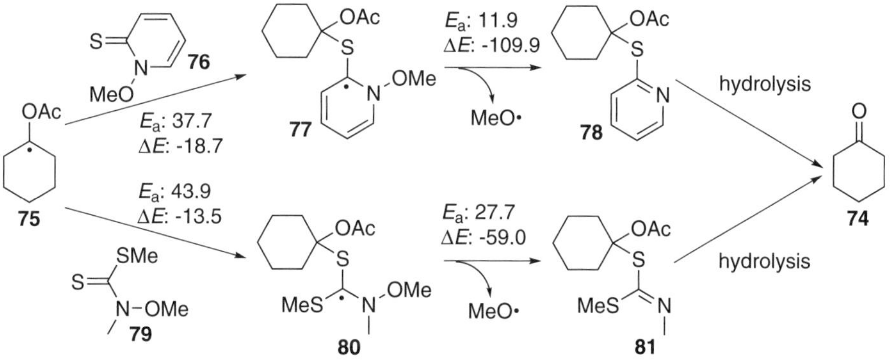

BHandHLYP/6-311G** energies in kJ mol^{-1}, zpc included.

SCHEME 2.15

NO$_3$[•] is a very unique representative of an O-centered radical, since it cannot only transfer one but even two of its oxygen atoms to alkynes to yield α-diketones. The latter reaction occurs when the vinyl radical formed upon addition of NO$_3$[•] to alkynes cannot undergo intramolecular HAT or cyclization. Thus, reaction of the bis-aromatic alkynes **82** with an excess of electrogenerated NO$_3$[•] leads to 1,2-diketones **83** as the major products, which are accompanied by varying amounts of benzophenones **84** (Scheme 2.16).[38] Computational investigation of the mechanism using both *ab initio* and hybrid density functional methods revealed that the diketones **83** are likely formed through a strongly exothermic 5-*endo* cyclization of the intermediate vinyl radical **85**, followed by loss of NO[•]. The fragmentation **86** → **83** is a virtually barrierless process, which can be rationalized by the fact that two stable C=O bonds in **83** are formed, in addition to the stability of the released NO[•]. Thus, this sequence represents another

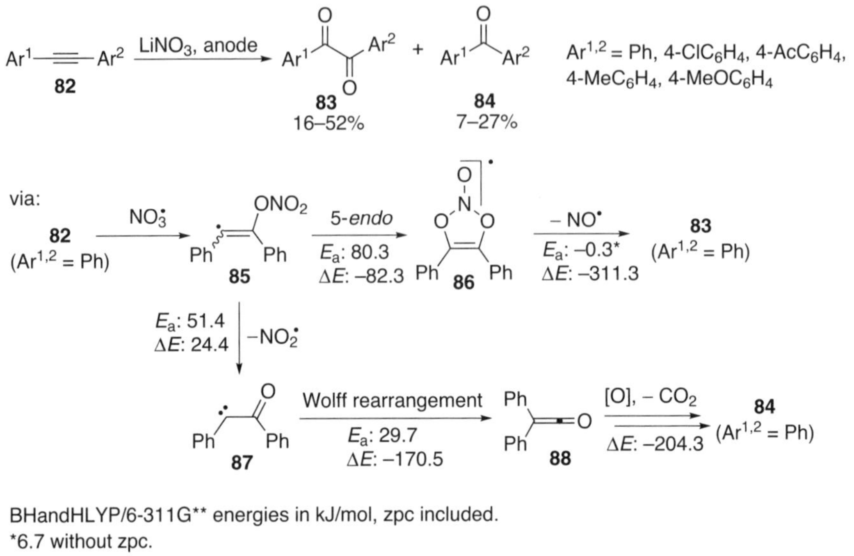

Ar^1 ═══ Ar^2 $\xrightarrow{\text{LiNO}_3,\ \text{anode}}$ Ar^1–C(O)–C(O)–Ar^2 + Ar^1–C(O)–Ar^2 Ar^{1,2} = Ph, 4-ClC$_6$H$_4$, 4-AcC$_6$H$_4$,
82   **83**   **84**   4-MeC$_6$H$_4$, 4-MeOC$_6$H$_4$
                                                   16–52%   7–27%

via:

82 $\xrightarrow{\text{NO}_3^{\bullet}}$ Ph–C(ONO$_2$)=C$^{\bullet}$–Ph $\xrightarrow[E_a:\ 80.3]{\text{5-endo}}$ [isoxazole-type radical] $\xrightarrow[E_a:\ -0.3^*]{-\ \text{NO}^{\bullet}}$ **83**
(Ar^{1,2} = Ph)   **85** ΔE: –82.3   **86** ΔE: –311.3   (Ar^{1,2} = Ph)

E_a: 51.4
ΔE: 24.4 $\Big|$ –NO$_2^{\bullet}$

Ph–C$^{\bullet\bullet}$–C(O)–Ph $\xrightarrow[E_a:\ 29.7]{\text{Wolff rearrangement}}$ Ph$_2$C=C=O $\xrightarrow[\Delta E:\ -204.3]{[O],\ -\ CO_2}$ **84**
       87 ΔE: –170.5   **88**   (Ar^{1,2} = Ph)

BHandHLYP/6-311G** energies in kJ/mol, zpc included.
*6.7 without zpc.

SCHEME 2.16

type of self-terminating radical oxygenations and provides a mild and exciting alternative to the existing nonradical methods for oxidizing alkynes to α-diketones, which usually require toxic reagents and/or significantly harsher reaction conditions that often also lead to complete cleavage of the former alkyne bond.

According to computations, the benzophenones **84**, which contain one carbon atom less than the starting alkyne **82**, could be formed through γ-fragmentation of the relatively labile O−NO$_2$ bond in vinyl radical **85**, followed by Wolff rearrangement of the intermediate α-oxo carbene **87** (and/or oxirene, not shown) to yield ketene **88**, which undergoes decarboxylation upon further oxidation either by NO$_3^{\bullet}$ or directly at the anode.[39] This unprecedented γ-fragmentation of a vinyl radical is a synthetically highly remarkable reaction, as it could potentially provide a very rapid access to α-oxo carbenes simply through addition of suitable radicals to alkynes.

A related reaction was recently discovered when phenylthiyl radicals, PhS$^{\bullet}$, which were generated through autooxidation of thiophenol, were reacted with diphenyla- cetylene (**82**) in the presence of O$_2$ to give unexpectedly diketone **83** (Scheme 2.17).[40] It could be revealed through a combination of experimental and computational studies at the BHandHLYP/cc-pVTZ level of theory that thiylperoxyl radicals **89** are the likely reactive key intermediates in this transformation, which are formed through a reversible reaction of PhS$^{\bullet}$ with O$_2$. This reaction is in competition with the direct addition of PhS$^{\bullet}$ to alkyne **82** (not shown), but becomes the favorable pathway with increasing [O$_2$]. Addition of **89** to **82** gives vinyl radical **90** in a slightly exothermic process, which could be converted into the diketone **83** through various routes. A pathway involving radical recombination with O$_2$ could initially lead to peroxyl radical **91** in a fast and strongly exothermic reaction. Fragmentation of the peroxylic

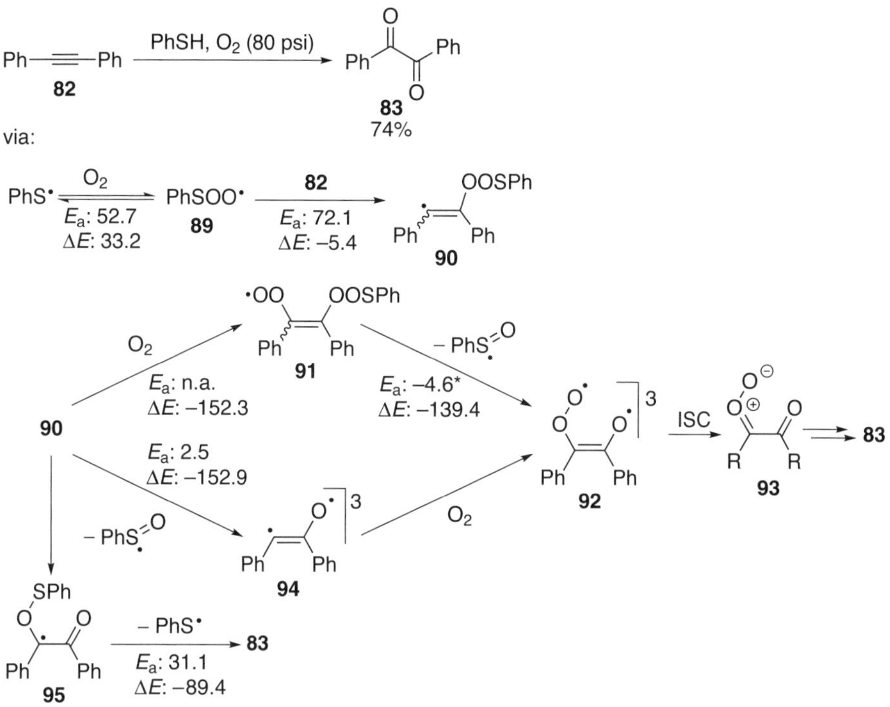

BHandHLYP/cc-pVTZ energies in kJ/mol, zpc included.
*0.8 without zpc.

SCHEME 2.17

O−OSPh bond in **91** leads to biradical **92**, which could isomerize (through intersystem crossing (ISC)) to carbonyl oxide **93**. This fragmentation is kinetically and thermodynamically highly favorable because a comparatively stable sulfinyl radical, PhS$^\bullet$=O, is released. Carbonyl oxides of type **93** are suggested intermediates in the ozonolysis of alkynes, leading to α-diketones.[41,42] Alternatively, **92** could be formed via biradical **94** (which is in resonance with the α-oxo carbene **87**, see Scheme 2.16) that results from exothermic γ-cleavage of the peroxyl O−OSPh bond in vinyl radical **90**, followed by trapping with O$_2$. A pathway that does not require additional O$_2$ could proceed through a concerted or stepwise rearrangement of **90** to radical intermediate **95**, which undergoes β-fragmentation of the O−S bond with release of PhS$^\bullet$ in a subsequent step. However, the experimental observation of a by-product that results from dimerization of PhS$^\bullet$=O would support a pathway proceeding via carbonyl oxide **93**.

Although the detailed mechanism is not yet clear, this reaction is a synthetically highly interesting example for a free radical-mediated "activation" of O$_2$ that enables transformation of alkynes into α-diketones under extremely mild and metal-free conditions.

2.2.3 Cascade Reactions Initiated by Addition of N-Centered Radicals to Alkynes

In intermolecular reactions, neutral aminyl radicals, $R_2N^{\bullet}$, react with π systems preferably by HAT. N-protonation increases the electrophilicity of the radical center, and successful addition of aminium radicals, $R_2NH^{\bullet\,+}$, to π systems, usually alkenes, has been reported in the literature.[43] Compared to aminium radicals, amidyl and imidyl radicals, for example, $RN^{\bullet}C(O)R'$ and $[RC(O)]_2N^{\bullet}$, are less electrophilic. Although they are delocalized π-allyl radicals, they react exclusively at nitrogen.[44] Only very few examples for radical cascades that are initiated by addition of N-centered radicals to alkynes have been reported.

Scheme 2.18 shows that reaction of phthalimidyl radicals, $Im^{\bullet}$, which were generated through photolysis of N-bromophthalimide, with cyclodecyne (52) leads to formation of small amounts of unsaturated bicyclic compounds 96 and 97 among other products (not shown).[45] This sequence is believed to proceed via initial addition of $Im^{\bullet}$ to the $C \equiv C$ triple bond, which initiates a sequence of transannular HAT (98 → 99a/b), followed by transannular radical cyclization (99a/b → 100a/b). In contrast to the self-terminating radical cyclizations shown in the previous section, β-fragmentation of the intermediate α-imidyl radicals 100a/b is not possible, since no stable radical could be released. It is therefore believed that the reaction proceeds through trapping of 100a/b by bromine atoms, which are formed as by-products in the photolysis of N-bromophthalimide, to give the tertiary bromides 101a/b. Elimination of hydrogen bromide leads to the final products 96 and 97.

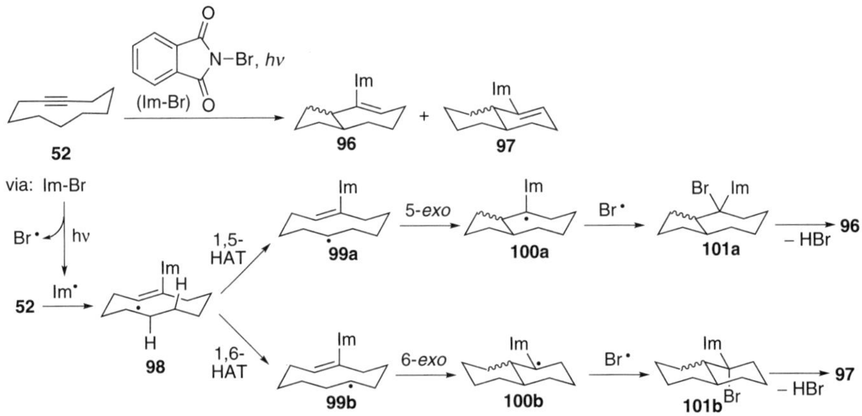

SCHEME 2.18

Self-terminating radical cyclizations have been explored with both aminium and amidyl radicals using the reaction with cyclodecyne (52) as a model system. However, instead of N-containing compounds, the bicyclic ketones 53 and 54 were exclusively obtained. Their formation could be explained by the mechanism that is exemplary shown for reaction of the N-benzyl acetamidyl radical, $AcBnN^{\bullet}$ (Scheme 2.19).[46]

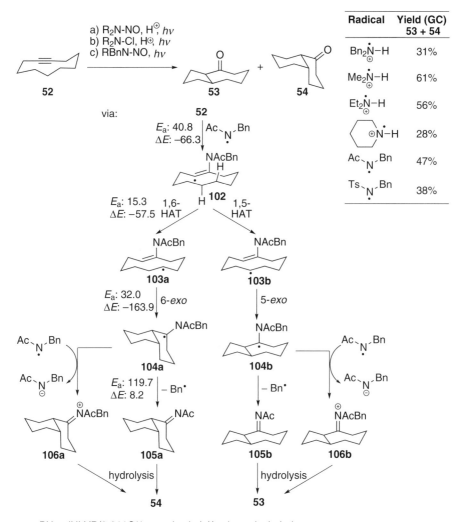

Radical	Yield (GC) 53 + 54
$Bn_2\overset{\cdot}{\underset{\oplus}{N}}$–H	31%
$Me_2\overset{\cdot}{\underset{\oplus}{N}}$–H	61%
$Et_2\overset{\cdot}{\underset{\oplus}{N}}$–H	56%
⊕N–H	28%
$Ac\underset{N}{\diagdown}\diagup Bn$	47%
$Ts\underset{N}{\diagdown}\diagup Bn$	38%

BHandHLYP/6-311G** energies in kJ/mol, zpc included.

SCHEME 2.19

In analogy to the previously shown sequences, N radical addition to the cycloalkyne **52** could initiate the usual radical translocation/cyclization cascade that is terminated by β-fragmentation. The resulting imides **105a/b** could subsequently undergo hydrolysis to form ultimately the observed bicyclic ketones **53/54**. Quite remarkably, however, although both the aminium and the amidyl radicals, which were used in these experiments, were equipped with substituents that should form stable radical leaving groups in the terminating β-fragmentation (e.g., benzyl or tosyl), no apparent correlation between radical stability and yield of cyclized products was found. It even appeared that N radicals with primary alkyl substituents, which should not be released as radicals at all, performed significantly better in this cyclization sequence (see table in Scheme 2.19) than N radicals with seemingly highly suitable substituents. This is a clear indication that the mechanism of radical cyclization cascades involving N radicals must be different.

BHandHLYP/6-311G** calculations of the potential energy surface for the addition of AcBnN˙ to cyclodecyne (**52**) and the subsequent 1,6-HAT/6-*exo* cyclization pathway revealed that the various steps in this sequence are both kinetically and thermodynamically highly feasible, except for the final β-fragmentation (see Scheme 2.19). Systematic computational studies on this β-fragmentation using the simplified model reaction **107** → **109** revealed that, as expected, homolytic cleavage of a N-benzyl bond in an α-amino radical is both kinetically and thermodynamically significantly more favorable than scission of a N-methyl bond (Scheme 2.20). Interestingly, β-fragmentation of a N−C bond at a quaternary, protonated nitrogen, a very likely scenario under the acidic conditions involving aminium radicals, not only lowers E_a for releasing a benzyl radical by about 40 kJ/ mol or a methyl radical by about 25 kJ/mol, but also renders the cleavage thermodynamically more favorable. On the other hand, the required energy for β-fragmentation at an amide or imide nitrogen depends on the nature of the remaining substituent R. Generally, release of an acetyl radical is kinetically and thermodynamically even more unfavorable than release of a methyl radical, and cleavage of a benzyl radical through β-fragmentation of an acetamide N−C bond requires about 20 kJ/mol more energy than β-fragmentation of a sulfonamide N−C bond. Of all the various β-fragmentations studied, the kinetically and thermodynamically most superior leaving group should be the phenylsulfonyl radical (which represents a simplified model for tosylate that was used in the experimental studies).

Since the computational studies support the general perception of good or poor leaving groups in self-terminating radical reactions, but clearly contradict the experimental findings, radical cyclization cascades initiated by N-centered radical addition to alkynes are not terminated by homolytic β-fragmentation.

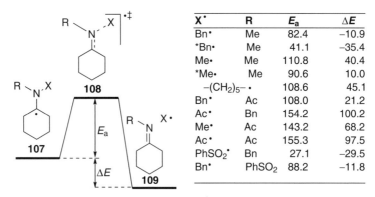

X˙	R	E_a	ΔE
Bn˙	Me	82.4	−10.9
*Bn˙	Me	41.1	−35.4
Me˙	Me	110.8	40.4
*Me˙	Me	90.6	10.0
−(CH₂)₅−˙		108.6	45.1
Bn˙	Ac	108.0	21.2
Ac˙	Bn	154.2	100.2
Me˙	Ac	143.2	68.2
Ac˙	Ac	155.3	97.5
PhSO₂˙	Bn	27.1	−29.5
Bn˙	PhSO₂	88.2	−11.8

BHandHLYP/6-311G** energies in kJ mol⁻¹, zpc included.
*Protonated N.

SCHEME 2.20

On the other hand, α-nitrogen radicals of type **104a/b** could be relatively easily oxidized to iminium ions **106a/b**, which undergo subsequent hydrolysis to the ketones **53/54** (see Scheme 2.19). Indeed, calculation of the reduction potentials, E° (not shown), revealed that in all cases the respective electrophilic N-centered radical, which initiates the radical addition/translocation cascade, is also capable of performing the oxidation **104a/b** → **106a/b**. Thus, in this sequence, the N radicals have a dual role, first as initiator of the radical cyclization cascade and second as terminator of the sequence through a redox process, which is a highly interesting variation of self-terminating radical cyclizations.

2.3 CASCADE REACTIONS INITIATED BY ADDITION OF HIGHER MAIN GROUP (VI)-CENTERED RADICALS TO ALKYNES

Of the higher main group (IV)-centered radicals, so far only cascade reactions initiated by addition of the nucleophilic Sn radicals to alkynes have received considerable attention.[47] It has been shown, however, that hydrogermylation of alkynes via triphenylgermyl radicals, $Ph_3Ge^{\bullet}$,[48] or tris(trimethylsilyl)germyl radicals, $(TMS)_3Ge^{\bullet}$,[49] produces the corresponding alkenylgermanes, but this radical addition has not yet been applied in a radical cyclization cascade.

2.3.1 Cascade Reactions Initiated by Addition of Sn-Centered Radicals to Alkynes

The triethylborane-mediated hydrostannylation has been used to generate the oxolanes **111** from the open-chain precursor **110** in a radical chain process (Scheme 2.21).[50] The reaction proceeds through vinyl radical **112**, resulting from Sn radical addition to the less electron-rich alkyne π system, which undergoes a highly stereoselective 5-*exo* cyclization in accordance with the Beckwith–Houk rules,[51] leading to products possessing (Z)-*trans* geometry. The compounds **111** could be oxidatively destannylated to α-methylene-γ-butyrolactones (not shown), which represent a major class of natural products with important biological activities.

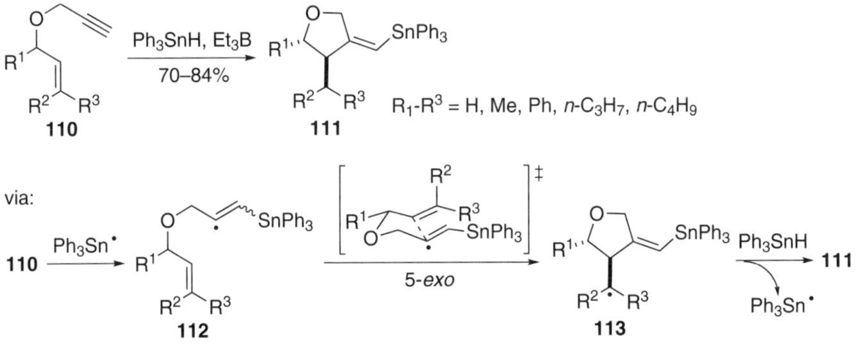

SCHEME 2.21

The facility by which C—Sn bonds can undergo homolytic fragmentation under certain conditions has been used in a number of cyclization cascades involving alkynes, where Sn radicals formally catalyze the rearrangement of the molecular framework without appearing as substituents in the final product. For example, Baldwin and Adlington have shown that in the presence of Sn radicals, 5-cyclodecynone (**58**) undergoes rearrangement to the bicyclic α,β-unsaturated ketone **114** (Scheme 2.22; no yield was reported).[52] The vinyl radical **115** obtained from addition of the Sn radical to the C≡C triple bond undergoes 5-*exo* cyclization to the carbonyl group. However, in contrast to the self-terminating cyclization sequence shown in Scheme 2.11, where the intermediate allyloxyl radicals **62/65** undergo a 3-*exo* cyclization, the allyloxyl radical **116** undergoes α-cleavage to give intermediate **117**, which subsequently cyclizes in a 5-*exo* fashion to produce the allylic radical **118**. The cascade is terminated through β-fragmentation and elimination of the Sn radical.

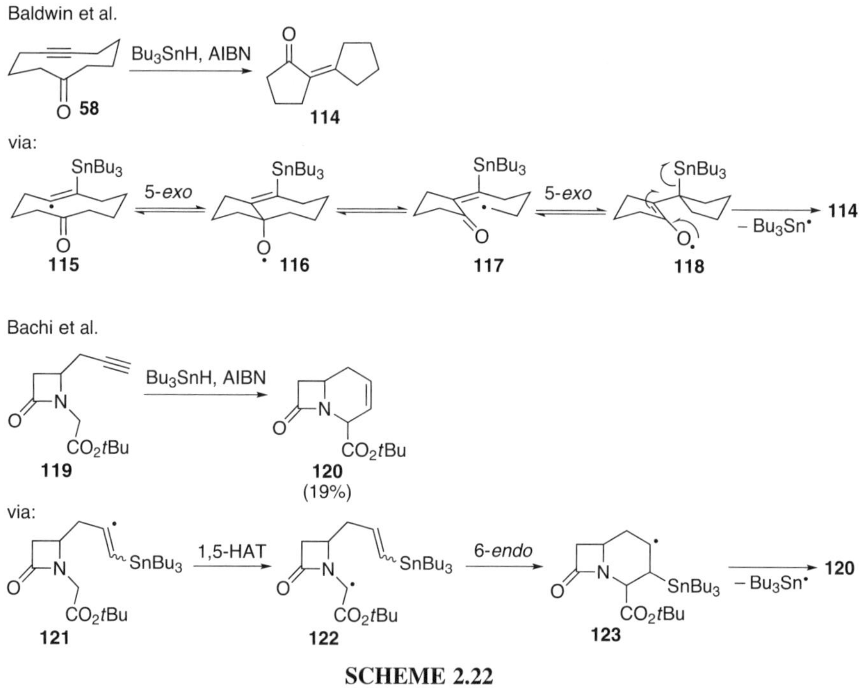

SCHEME 2.22

Bachi and Bosch used Sn radicals as mediator for the cyclization of the β-lactam-substituted alkynes **119** through a sequence consisting of radical addition and 1,5-HAT, followed by a 6-*endo* cyclization of the radical intermediate **122**, to give the bicyclic β-lactam **120** after β-fragmentation and release of Bu₃Sn radicals (Scheme 2.22).[53] The moderately low yield of this cyclization is due to the competing fast reduction of the vinyl radical **121** to the corresponding alkene by the tin hydride (not shown).

A remarkably complex cyclization/twofold ring expansion sequence that is catalyzed by Sn radicals is shown in Scheme 2.23.[54] The vinyl radical **126** formed after radical addition to the terminal end of the alkyne bond in **124** cyclizes to the carbonyl group in a 6-*exo* fashion. The resulting alkoxyl radical **127** undergoes α-cleavage to produce the tertiary radical **128**, which is donor stabilized by the α-heteroatom (it should be noted that capto-stabilizing groups in α-position lead to significant reduction in the overall reaction rate). The first ring expansion occurs in a 5-*exo* cyclization, which is followed by a 3-*exo* cyclization of **129**, leading to the cyclopropyloxyl radical **130**. Opening of the cyclopropyl ring results in the second ring expansion and formation of the bicyclo[6.3.0]undecane framework in **131**. The final product **125** is obtained through β-fragmentation and release of the Sn radical.

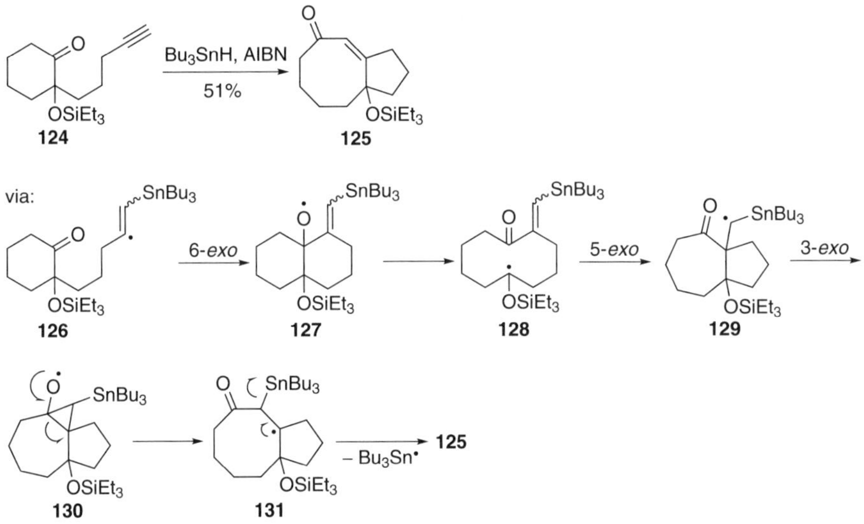

SCHEME 2.23

A double radical cyclization/β-fragmentation of the homolog acyclic ω-yne vinyl sulfides **132a/b** that is catalyzed by Sn radicals and also involves an unusual *endo* radical cyclization is shown in Scheme 2.24.[55] The initially formed vinyl radicals **134a/b** undergo a 5/6-*exo* cyclization to the C=C double bond, which is activated by one electron-withdrawing group (the reaction does not occur if the alkene in **132** carries two electron-withdrawing substituents). β-Fragmentation of **135a/b** leads to the thiyl radical **136a/b**, which in the case of **136a** cyclizes in a highly unusual 5-*endo* fashion (6-*endo* for the larger homolog **136b**) to the remaining C=C double bond to give allylic radical **137a/b**. The reverse ring opening of the latter is prevented by a chain terminating β-fragmentation under release of the Sn radical to give the diene **133a/b**.

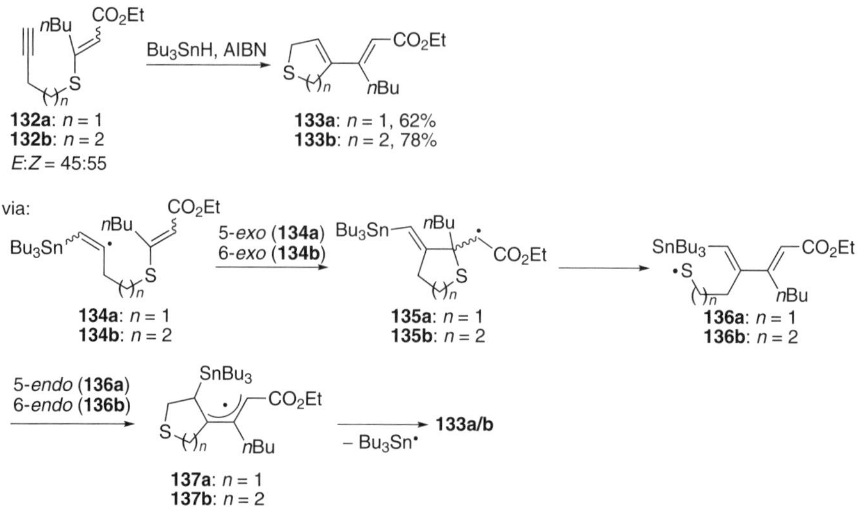

nBu CO$_2$Et

Bu$_3$SnH, AIBN → CO$_2$Et

132a: $n = 1$
132b: $n = 2$
E:Z = 45:55

133a: $n = 1$, 62%
133b: $n = 2$, 78%

via:

Bu$_3$Sn nBu CO$_2$Et

134a: $n = 1$
134b: $n = 2$

5-exo (134a)
6-exo (134b)
→

Bu$_3$Sn nBu CO$_2$Et

135a: $n = 1$
135b: $n = 2$

→

SnBu$_3$ CO$_2$Et nBu

136a: $n = 1$
136b: $n = 2$

5-endo (136a)
6-endo (136b)
→

SnBu$_3$ CO$_2$Et nBu

137a: $n = 1$
137b: $n = 2$

→ 133a/b
− Bu$_3$Sn$^\bullet$

SCHEME 2.24

2.4   CASCADE REACTIONS INITIATED BY ADDITION OF HIGHER MAIN GROUP (VI)-CENTERED RADICALS TO ALKYNES

The vast majority of radical cascades initiated by addition of higher main group (VI)-centered radicals to alkynes focus predominantly on S-centered radicals, mainly thiyl radicals, whereas considerably fewer intermolecular addition reactions involving radicals with the unpaired electron located on selenium or even tellurium are known.

2.4.1   Cascade Reactions Initiated by Addition of S-Centered Radicals to Alkynes

One of the earliest radical cyclization cascades initiated by addition of S-centered radicals to alkynes was reported in 1987 by Broka and Reichert (Scheme 2.25).[56] Thiophenyl radicals, PhS$^\bullet$, which were generated under radical chain conditions, undergo addition to the terminal end of the C≡C triple bond in enyne **138**. The resulting vinyl radical **141** can undergo cyclization in both 6-*endo* (preferred) and 5-*exo* fashion, and reduction of the radical intermediates **142** and **143** leads to the final observed products **139** and **140**, respectively.

   The apparent chemoselectivity for the addition of the electrophilic S-centered radicals to the less electron-rich alkyne moiety in enyne **138** can be rationalized by the fact that addition of S radicals to both alkenes and alkynes proceeds smoothly (the rate constants for addition of S radicals to alkenes are about three orders of magnitude larger than those for the addition to alkynes),[57] but is also reversible. However, the reversibility is less pronounced for the radical addition to alkynes, due to the high reactivity of the vinyl radicals formed (compared to alkyl radicals), which undergo subsequent reactions at faster rates than undergoing fragmentation back to the S

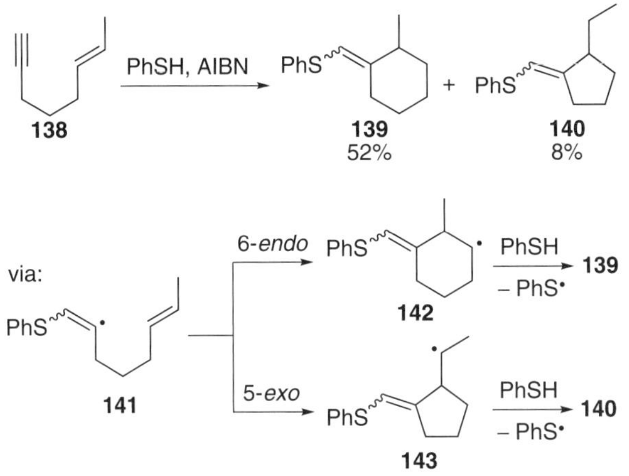

SCHEME 2.25

radical and alkyne. Therefore, S radicals ultimately can promote cyclization of enynes through a mechanism consisting of radical addition to the C≡C triple bond, followed by cyclization to the alkene moiety.[58] A similar behavior is also observed in the reactions of other electrophilic heteroatom-centered radicals with enynes (see below).

A highly complex radical addition/cyclization/rearrangement reaction is shown in Scheme 2.26.[59] Addition of thiyl radicals derived from **144**, under radical chain conditions, to the alkynyl azide **145** leads to vinyl radical **147**, which cyclizes in a 5-*exo* fashion to the aromatic ring to give the spiro radical intermediate **148**. The presence of the *para* cyano group is crucial, as it leads to a significant radical stabilization—indeed, the 5-*exo* cyclization does not occur in systems lacking a radical stabilizing substituent at the aromatic ring; in these cases, the vinyl radical is directly reduced to the corresponding S-substituted alkene (not shown). Reduction of the aryl radical **148**, followed by opening of the S-heterocyclic ring and rearomatization, yields the final product **146** in 65% overall yield.

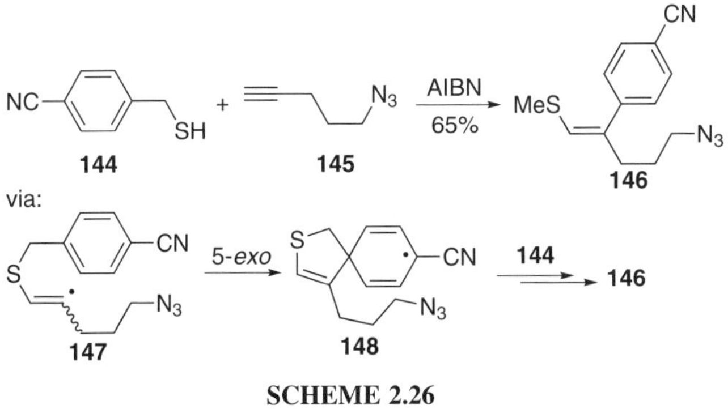

SCHEME 2.26

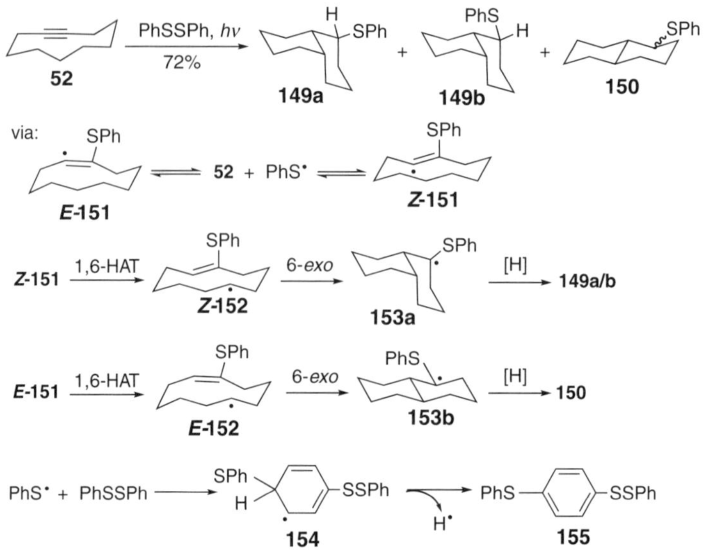

SCHEME 2.27

S-centered radicals were also explored in self-terminating radical cyclizations (Scheme 2.27).[60] Using the 10-membered cyclic alkyne **52** as a model system, reaction with PhS• radicals, which were photogenerated from diphenyl disulfide, leads to the *cis*- and *trans*-fused bicyclic thioethers **149a/b** and **150** possessing a decalin framework, which were identified through a combination of independent synthesis, X-ray analysis and computational studies. Formation of the *cis*-configured diastereomeric thioethers **149a/b** should proceed through the usual pathway involving an initially formed Z-configured vinyl radical Z-**151**, followed by 1,6-HAT and 6-*exo* cyclization to give the α-thio radical **153a**, which undergoes subsequent reduction from both faces of the planar radical center. The *trans*-configured product **150** results from a similar radical addition / translocation cascade involving the isomeric vinyl radical E-**151**. Computational studies revealed that E-**151** is not formed through E/Z isomerization of the vinyl radical Z-**151**, which requires a surprisingly high activation energy due to ring strain in the ten-membered ring, but through reversible addition of PhSl to the alkyne C≡C bond in **52**. According to GC/MS studies, the hydrogen donor in this system is believed to be the radical adduct **154**, which results from addition of PhS• to diphenyl disulfide. Homolytic β-fragmentation of the S–C bond in **153a/b** to yield thioketones, in accordance with the general mechanism of self-terminating radical cyclizations, is not possible in this case because highly unstable phenyl radicals would be released.

Keck and Wagner used a diastereoselective thiyl radical addition/cyclization sequence to generate the key compound **157** for a total synthesis of *ent*-lycoricine **158** (Scheme 2.28).[61] In this sequence, photogenerated PhS• undergo exclusive addition at the β-site of the C≡C triple bond in the starting alkyne **156** to produce the resonance-stabilized vinyl radical **159** (interestingly, with Bu₃Sn•, only addition at the α-carbon was found, which may be a result of the sterical hindrance imposed by the substituents at tin). The subsequent 6-*exo* radical cyclization of **159** proceeds, due to the presence of the *cis* fused dioxolane ring, in a highly diastereoselective fashion through a boat-like transition state, in which the imine C=N double bond and the hydroxy substituent both assume a pseudo-equatorial position. This leads to a *cis* arrangement of the hydroxy and amino substituents in the cyclized intermediate **160**, which is subsequently reduced to **157** in a radical chain process.

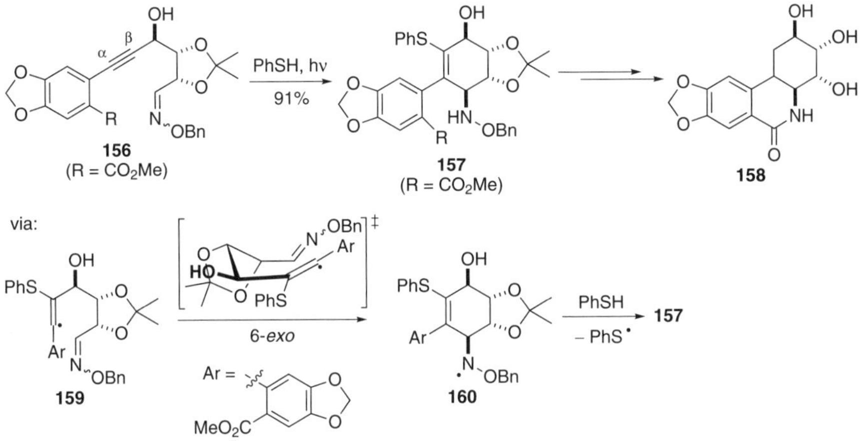

SCHEME 2.28

A diastereoselective formal addition of a *trans*-2-(phenylthio)vinyl moiety to α-hydroxyhydrazones through a radical pathway is shown in Scheme 2.29.[62] To overcome the lack of a viable intermolecular vinyl radical addition to C=N double bonds, not to mention a reaction proceeding with stereocontrol, this procedure employs a temporary silicon tether, which is used to hold the alkyne unit in place so that the vinyl radical addition could proceed intramolecularly. Thus, intermolecular addition of PhS· to the alkyne moiety in the chiral alkyne **161** leads to vinyl radical **163**, which cyclizes in a 5-*exo* fashion, according to the Beckwith–Houk predictions, to give aminyl radical **164** with an *anti*-arrangement between the ether and the amino group. Radical reduction and removal of the silicon tether without prior isolation of the end product of the radical cyclization cascade, **165**, yields the α-amino alcohol **162**. This strategy, which could also be applied to the diastereoselective synthesis of polyhydroxylated amines (not shown), can be considered as synthetic equivalent of an acetaldehyde Mannich reaction with acyclic stereocontrol.

Renaud and coworkers applied a radical addition/translocation/cyclization cascade as a key step in the diastereoselective synthesis of the spirocyclic compound (−)-erythrodiene **168** (Scheme 2.30).[63] Addition of photogenerated PhS· to the terminal

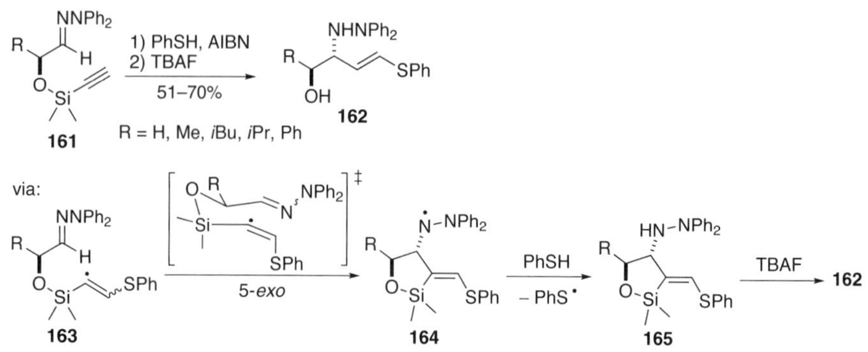

SCHEME 2.29

alkyne in **166** gives vinyl radicals **169**, which undergo a regioselective 1,5-HAT, followed by 5-*exo* cyclization of the intermediate tertiary radical **170** and subsequent reduction of **171** to yield **167** as a mixture of four diastereomers. The high *trans* selectivity of this cyclization can be explained by a transition state where a new axial C−C bond is formed, but it should be noted, however, that the stereochemical outcome was found to be very sensitive to reaction temperature and solvent polarity. A very similar thiophenol-mediated radical cyclization cascade has also been applied to the synthesis of various 1-azabicyclic alkanes from *N*-homopropargylic amines (not shown).[64]

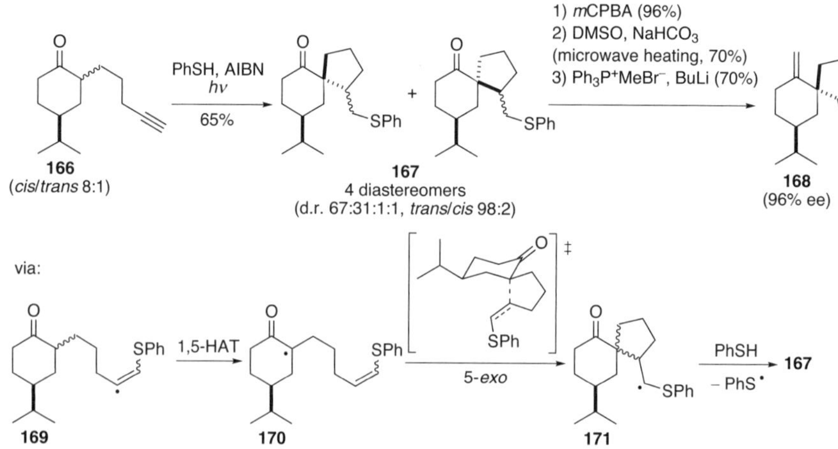

SCHEME 2.30

An interesting four-component radical cascade that leads to formation of β-arylthio-substituted acrylamides **172** and involves thiyl radicals, aromatic acetylenes, and isonitriles, as well as an oxidant, has been reported by Nanni and coworkers (Scheme 2.31).[65] The sequence is initiated by chemoselective addition of photogenerated PhS[.] to the $C \equiv C$ triple bond in phenylacetylene (which is more electron rich than the isonitrile π system), leading to the vinyl radical **173**, which is subsequently trapped by intermolecular addition to the isonitrile $N \equiv C$ triple bond. The resulting imidoyl radical **174** is captured by *m*-dinitrobenzene, which acts as an oxidant to produce the amidyl radical **176** via the intermediate **175**. Subsequent hydrogen transfer, presumably through an intermediate of type **154** (see Scheme 2.27), terminates the cascade.

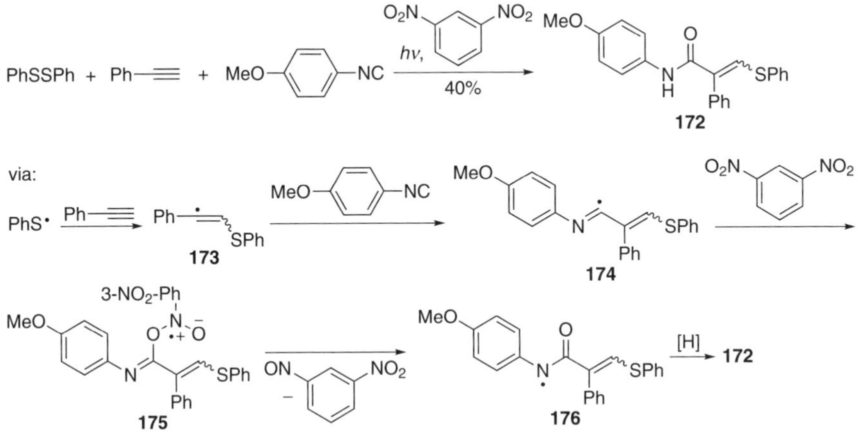

SCHEME 2.31

In the case of S radicals, where sulfur is in a higher oxidation state, sulfonyl radicals $(RSO_2^{.})$ have gained considerable synthetic value, since their addition to π systems, usually C=C double bonds, provides a facile method to introduce the sulfonyl moiety into a molecule. Although in sulfonyl radicals the spin density is delocalized over sulfur and both oxygen atoms, they react with π systems exclusively to form C–S bonds.[57] Simpkins and coworkers used a tandem sequence triggered by the addition of sulfonyl radicals to alkynes of type **177** for the synthesis of N-containing heterocycles **178** (Scheme 2.32).[66] The initially formed vinyl radical **179** undergoes a 5-*exo* cyclization to the remaining C=C double bond, leading to the bicyclic radical intermediate **180**, which is transformed into the final product through homolytic substitution in a chain propagating step. The stereochemistry of the latter step can be understood on the basis that the substitution takes place from the less hindered site of the molecule.

A related radical addition/5-*exo* cyclization cascade involving addition of sulfonyl radicals to alkynes has been used for the diastereoselective synthesis of bicyclic β-lactams **182** from the β-lactamic enyne precursor **181** (Scheme 2.33).[67] In this radical chain sequence, where tosyl bromide is used as source of sulfonyl radicals, the final product is a mixture of epimers (90 : 10) at the newly formed exocyclic chiral center.

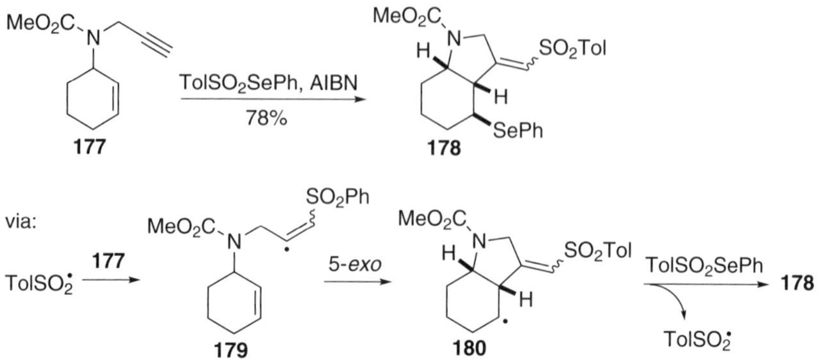

SCHEME 2.32

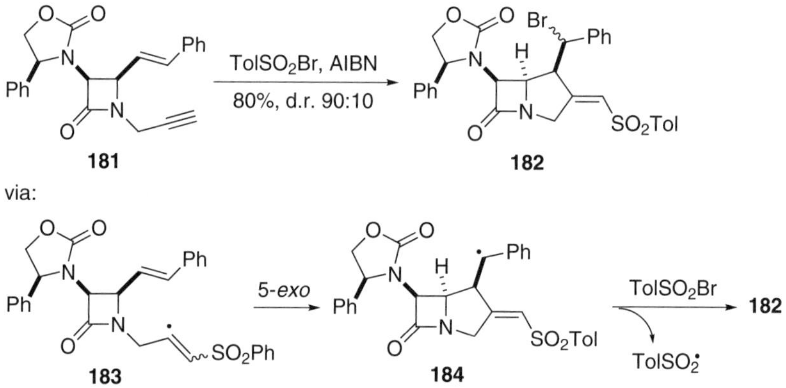

SCHEME 2.33

2.4.2 Cascade Reactions Initiated by Addition of Se-Centered Radicals to Alkynes

In contrast to the large amount of S radical-mediated cascade reactions, only a few examples for radical cascades that are triggered by intermolecular Se radical addition to an alkyne are reported in the literature. Ogawa and coworkers described a highly selective four-component radical coupling reaction of unsaturated compounds that is mediated by phenylselenyl radicals, PhSe• (Scheme 2.34).[68] The reaction of diphenyl diselenide with ethyl propiolate and a large excess of alkenes with an electron-withdrawing group (EWG; e.g., acrylates) and with an electron-donating substituent (e.g., 2-methoxypropene) leads to formation of the highly substituted cyclopentane systems **185a–c** in good to excellent yields. This reaction proceeds through a sequential addition of PhSe• to the alkyne, of the resulting highly electrophilic vinyl radical **186** to the electron-rich olefin, followed by addition of the nucleophilic radical

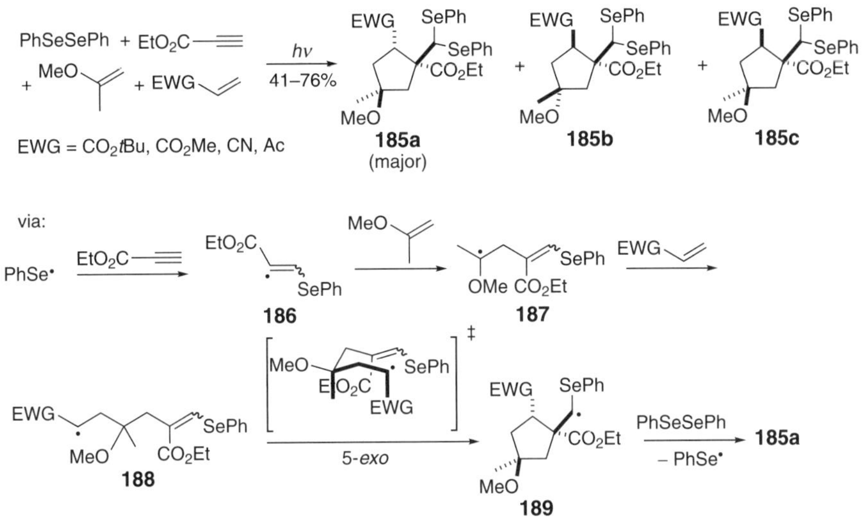

SCHEME 2.34

adduct **187** to the electron-poor olefin. The subsequent 5-*exo* cyclization of **188** proceeds with high preference through a chair-like transition state, according to the Beckwith–Houk predictions, with the electron-withdrawing substituent at the radical center assuming an axial position (to minimize sterical interaction with the C=C double bond in the late transition state, as a result of the radical stabilizing effect of the EWG group). Trapping of the cyclized intermediate **189** by phenyl selenide gives **185a**. The minor diastereomers **185b** and **185c** result from 5-*exo* cyclizations with an axial methoxy group and/or a pseudo-equatorial EWG substituent in the respective transition states (not shown).

2.5   CASCADE REACTIONS INITIATED BY ADDITION OF HIGHER MAIN GROUP (V)-CENTERED RADICALS TO ALKYNES

Of the higher main group (V) elements, only P-centered radicals have been used in intermolecular radical additions to alkynes, whereas radical reactions with the unpaired electron located at the higher metallic elements arsenic, antimony, and bismuth have not been reported.

2.5.1   Cascade Reactions Initiated by Addition of P-Centered Radicals to Alkynes

Phosphorous hydrides, such as phosphites, $(RO)_2P(O)H$, thiophosphites, $(RO)_2P(S)H$, and phosphinates, $R(RO)P(O)H$, have received considerable attention in the recent years as nontoxic alternative to tin hydrides commonly used in radical chain reactions.

However, their intermolecular addition reactions with alkynes are mostly aimed at synthesizing substituted alkenes,[69,70] and only very few cascade reactions that are initiated by P radical addition to $C\equiv C$ triple bonds have been reported. Renaud and coworkers developed a simple one-pot procedure for the cyclization of terminal alkynes mediated by dialkyl phosphites (Scheme 2.35).[71] In this radical chain procedure, dialkyl phosphite radicals, $(RO)_2P^\bullet{=}O$, undergo addition to the $C\equiv C$ triple bond in **190**, which triggers a radical translocation (1,5-HAT)/5-*exo* cyclization cascade. The sequence is terminated by hydrogen transfer from dialkyl phosphite to the intermediate **194** and regeneration of P-centered radicals.

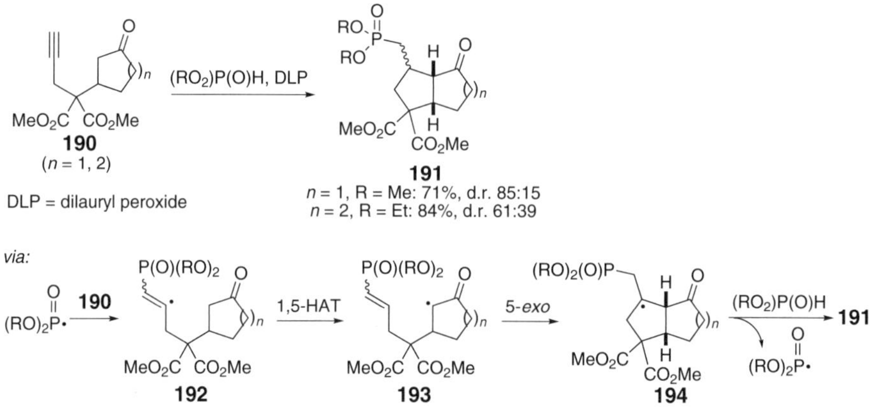

SCHEME 2.35

Diphenylphosphanyl radicals, $Ph_2P^\bullet$, generated from diphenylphosphane in the presence of a radical initiator were used to cyclize the alkynyl-substituted carbohydrate derivative **195** in a radical addition/5-*exo* cyclization sequence to give the bicyclic deoxysugar derivative **196** (Scheme 2.36).[66] $Ph_2P^\bullet$ have also been used as promoters for the cyclization of alkynyl β-lactam **181** for a highly efficient, diastereoselective synthesis of bicyclic β-lactams (see Scheme 2.33).[67]

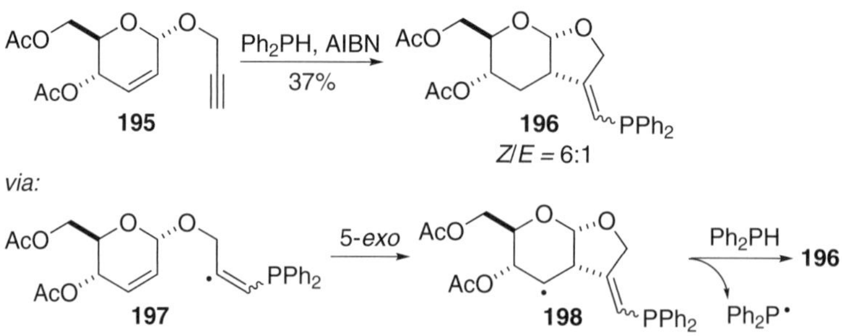

SCHEME 2.36

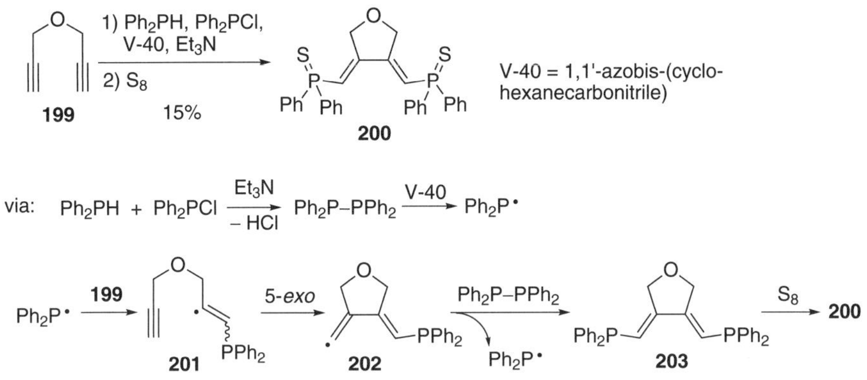

SCHEME 2.37

A radical diphosphanylation of alkynes using tetraorganodiphosphanes as precursors for phosphanyl radicals has been applied to the synthesis of a doubly phosphinated diene **200** (Scheme 2.37).[72] Tetraphenyldiphosphane was generated *in situ* from diphenylphosphane with an excess of chlorodiphenylphosphane in the presence of triethylamine. Addition of the phosphanyl radical to one $C \equiv C$ triple bond in the dialkyne **199** leads to vinyl radical **201**, which undergoes a 5-*exo* cyclization (out of a Z-configured vinyl radical to minimize steric hindrance in the cyclization) to give vinyl radical intermediate **202**. The latter is trapped by a second phosphine moiety in a radical substitution step. The radical cyclization product is ultimately isolated as bis-phosphane sulfide **200** after treatment of the intermediate phosphane **203** with sulfur.

REFERENCES

1. McCarroll, A. J.; Walton, J. C. *Angew. Chem., Int. Ed.* **2001**, *40*, 2224–2248.

2. Parsons, P. J.; Penkett, C. C.; Shell, A. J. *Chem. Rev.* **1996**, *96*, 195–206.

3. Albert, M.; Fensterbank, L.; Lacôte, E.; Malacria, M. *Top. Curr. Chem.* **2006**, *264*, 1–62.

4. *Radicals in Organic Synthesis*; Sibi, M.; Renaud, P., Eds.; Wiley: Weinheim, **2001**; Vols 1 and 2.

5. Yet, L. *Tetrahedron* **1999**, *55*, 9349–9403.

6. Amiel, Y. In *The Chemistry of Functional Groups, Supplement C*; Patai, S.; Rappoport, Z., Eds.; John Wiley & Sons, Ltd., **1983**; pp 917–944.

7. Fischer, H. In *Free Radicals in Biology and Environment*; Minisci, F., Ed.; Kluwer Academic Publishers, **1997**; pp 63–78.

8. Fischer, H.; Radom, L. *Angew. Chem., Int. Ed.* **2001**, *40*, 1340–1371.

9. Beckwith, A. L. J.; O'Shea, D. M. *Tetrahedron Lett.* **1986**, *27*, 4525–4528.

10. Heiba, E. I.; Dessau, R. M. *J. Am. Chem. Soc.* **1967**, *89*, 3772–3777.

11. Suzuki, A.; Miyaura, N.; Itoh, M. *Synthesis* **1973**, 305–306.

12. Ichinose, Y.; Matsunaga, S.; Fugami, K.; Oshima, K.; Utimoto, K. *Tetrahedron Lett.* **1989**, *30*, 3155–3158.

13. Curran, D. P.; Kim, D. *Tetrahedron* **1991**, *47*, 6171–6188.

14. Montevecchi, P. C.; Navacchia, M. L. *Tetrahedron* **2000**, *56*, 9339–9342.

15. Leardini, R.; Nanni, D.; Tundo, A.; Zanardi, G. *Tetrahedron Lett.* **1998**, *39*, 2441–2442.

16. Montevecchi, P. C.; Navacchia, M. L.; Spagnolo, P. *Tetrahedron* **1997**, *53*, 7929–7936.

17. Oka, R.; Nakayama, M.; Sakaguchi, S.; Ishii, Y. *Chem. Lett.* **2006**, *35*, 1104–1105.

18. Fernández, M.; Alonso, R. *Org. Lett.* **2005**, *7*, 11–14.

19. Fernández-Gonzáles, M.; Alonso, R. *J. Org. Chem.* **2006**, *71*, 6767–6775.

20. Lenoir, I.; Smith, M. L. *J. Chem. Soc., Perkin Trans.* **2000**, *1*, 641–643.

21. Wille, U.; Plath, C. *Liebigs Ann./Recueil* **1997**, 111–119.

22. Wille, U. Unpublished.

23. Wille, U.; Dreessen, T. *J. Phys. Chem. A* **2006**, *110*, 2195–2203.

24. Curran, D. P.; Porter, N. A.; Giese, B. *Stereochemistry of Radical Reactions: Concepts, Guidelines and Synthetic Applications*; Wiley-VCH Verlag: Weinheim, 1996.

25. Example: Felix, D.; Schreiber, J.; Ohloff, G.; Eschenmoser, A. *Helv. Chim. Acta* **1971**, *54*, 2896–2912.

26. Wille, U.; Lietzau, L. *Tetrahedron* **1999**, *55*, 10119–10134.

27. Lietzau, L.; Wille, U. *Heterocycles* **2001**, *55*, 377–380.

28. Wille, U.; Lietzau, L. *Tetrahedron* **1999**, *55*, 11465–11474.

29. Stademann, A.; Wille, U. *Aust. J. Chem.* **2005**, *57*, 1055–1066.

30. Beesley, R. M.; Ingold, C. K.; Thorpe, J. F. *J. Chem. Soc.* **1915**, *107*, 1080–1106.

31. Li, C. H. Honours Thesis; The University of Melbourne, 2006.

32. Wille, U. *Org. Lett.* **2002**, *2*, 3485–3488.

33. Wille, U. *Tetrahedron Lett.* **2002**, *43*, 1239–1242.

34. Wille, U. *J. Am. Chem. Soc.* **2002**, *124*, 14–15.

35. Jargstorff, C.; Wille, U. *Eur. J. Org. Chem.* **2003**, 3173–3178.

36. Sigmund, D.; Schiesser, C. H.; Wille, U. *Synthesis* **2005**, 1437–1444.

37. Kim, S.; Lim, C. J.; Song, S. E.; Kang, H. Y. *Synlett* **2001**, 688–690.

38. Wille, U.; Andropof, J. *Aust. J. Chem.* **2007**, *60*, 420–428.

39. See, for example: (a) Koenig, T. W.; Barklow, T. *Tetrahedron* **1969**, *25*, 4875–4886; (b) Dayan, S.; Ben-David, I.; Rozen, S. *J. Org. Chem.* **2000**, *65*, 8816–8818.

40. Tan, K. J.; Wille, U. *Chem. Commun.* **2008**, 6239–6241.

41. Yang, N. C.; Libman, J. *J. Org. Chem.* **1974**, *39*, 1782–1784.

42. Griesbaum, K.; Dong, Y.; McCullough, K. J. *J. Org. Chem.* **1997**, *62*, 6129–4136.

43. Examples: (a) Wolf, M. E. *Chem. Rev.* **1963**, *63*, 55–64; (b) Neale, R. S. *Synthesis* **1971**, 1–15; (c) Minisci, F. *Synthesis* **1973**, 1–24; (d) Chow, Y. L. *Acc. Chem. Res.* **1973**, *6*, 354–360; (e) Minisci, F. *Acc. Chem. Res.* **1975**, *8*, 165–171; (f) Chow, Y. L.; Danen, W. C.; Nelsen, S. F.; Rosenblatt, D. H. *Chem. Rev.* **1978**, *78*, 243–274; (g) Stella, L. *Angew. Chem., Int. Ed. Engl.* **1983**, *22*, 337–350; (h) Chow, Y. L. *Can. J. Chem.* **1965**, *43*, 2711–2714; (i) Chow, Y. L.; Colón, C. *Can. J. Chem.* **1967**, *45*, 2559–2568.

44. Fossey, J.; Lefort, D.; Sorba, J. *Free Radicals in Organic Chemistry*; John Wiley & Sons/ Masson: Chichester, 1995.

45. Wille, U.; Krüger, O.; Kirsch, A.; Lüning, U. *Eur. J. Org. Chem.* **1999**, 3185–3189.
46. Wille, U.; Heuger, G.; Jargstorff, C. *J. Org. Chem.* **2008**, *73*, 1413–1421.
47. Early example: Stork, G., Mook, R. Jr. *Tetrahedron Lett.* **1986**, *27*, 4529–4532.
48. Review: Oshima, K. *Advances on Metal Organic Chemistry*; JAI Press, Ltd., **1991**; Vol. 2, pp 101–141.
49. Bernardoni, S.; Lucarini, M.; Pedulli, G. F.; Vaglimigli, L.; Gevorgyan, V.; Chatgilialoglu, C. *J. Org. Chem.* **1997**, *62*, 8009–8014.
50. Nozaki, K.; Oshima, K.; Utimoto, K. *Bull. Chem. Soc. Jpn.* **1987**, *60*, 3465–3467.
51. (a) Beckwith, A. L. J.; Schiesser, C. H. *Tetrahedron Lett.* **1985**, *26*, 373–376; (b) Spellmeyer, D. C.; Houk, K. N. *J. Org. Chem.* **1987**, *52*, 959–974.
52. Baldwin, J. E.; Adlington, R. M.; Robertson, J. *Tetrahedron* **1991**, *47*, 6795–6812.
53. Bosch, E.; Bachi, M. D. *J. Org. Chem.* **1993**, *58*, 5581–5582.
54. Nishida, A.; Takahashi, H.; Takeda, H.; Takada, N.; Yonemitsu, O. *J. Am. Chem. Soc.* **1990**, *112*, 902.
55. Journet, M.; Rouillard, A.; Cai, D.; Larsen, R. D. *J. Org. Chem.* **1997**, *62*, 8630–8631.
56. Broka, C. A.; Reichert, D. E. C. *Tetrahedron Lett.* **1987**, *28*, 1503–1506.
57. *S-Centered Radicals*; Alfassi, Z. B., Ed.; John Wiley & Sons: Chichester, **1999**.
58. Montevecchi, P. C.; Navacchia, M. L. *Recent Res. Dev. Org. Bioorg. Chem.* **1997**, 1–13.
59. Montevecchi, P. C.; Navacchia, M. L.; Spagnolo, P. *Tetrahedron Lett.* **1997**, *38*, 7913–7916.
60. Tan, K. J. PhD Thesis, University of Melbourne, 2009.
61. Keck, G. E.; Wagner, T. T. *J. Org. Chem.* **1996**, *61*, 8366–8367.
62. Triestad, G. K.; Jiang, T.; Fioroni, G. M. *Tetrahedron: Asymmetry* **2003**, *14*, 2853–2856.
63. Lachia, M.; Dénès, F.; Beaufils, F.; Renaud, P. *Org. Lett.* **2005**, *7*, 4103–4106.
64. Dénès, F.; Beaufils, F.; Renaud, P. *Org. Lett.* **2007**, *9*, 4375–4378.
65. Leardini, R.; Nanni, D.; Zanardi, G. *J. Org. Chem.* **2000**, *65*, 2763–2772.
66. Brumwell, J. E.; Simpkins, N. S.; Terrett, N. K. *Tetrahedron* **1994**, *50*, 13533–13552.
67. Alcaide, B.; Rodríguez-Campos, I. M.; Rodríguez-Lôpez, J.; Rodríguez-Vicente, A. *J. Org. Chem.* **1999**, *64*, 5377–5387.
68. Tsuchii, K.; Doi, M.; Hirao, T.; Ogawa, A. *Angew. Chem., Int. Ed.* **2003**, *42*, 3490–3493.
69. Jessop, C. M.; Parsons, A. F.; Routledge, A.; Irvine, D. J. *Tetrahedron: Asymmetry* **2003**, *14*, 2849–2851.
70. Antczak, M. I.; Montchamp, J.-L. *Synthesis* **2006**, 3080–3084.
71. Beaufils, F.; Dénès, F.; Renaud, P. *Angew. Chem., Int. Ed.* **2005**, *44*, 5273–5275.
72. Sato, A.; Yorimitsu, H.; Oshima, K. *Angew. Chem., Int. Ed.* **2005**, *44*, 1694–2696.

3

RADICAL CATION FRAGMENTATION REACTIONS IN ORGANIC SYNTHESIS

ALEXANDER J. PONIATOWSKI AND PAUL E. FLOREANCIG

Department of Chemistry, University of Pittsburgh, Pittsburgh, PA, USA

3.1 INTRODUCTION

Radical cations, the products of single electron oxidation reactions, are high-energy intermediates that can react through numerous pathways, including cycloaddition, nucleophilic addition, metathesis, and fragmentation processes.[1] The potential for reaction through multiple pathways leads to speculation that radical cations are not viable intermediates for complex molecule synthesis. Careful mechanistic studies, however, have provided an understanding of the structural features that promote selective reactions and applications of this information have provided spectacular transformations. For example, nucleophilic addition reactions to radical cation intermediates have been applied to natural product synthesis and asymmetric bond formation (Scheme 3.1).[2] In this chapter, we will discuss the manner in which fundamental mechanistic studies on radical cation fragmentation reactions have provided the framework for the development of an electron transfer-initiated cyclization reaction[3] that has been applied to the synthesis of a wide array of complex structures.

Carbon-Centered Free Radicals and Radical Cations, Edited by Malcolm D. E. Forbes
Copyright © 2010 John Wiley & Sons, Inc.

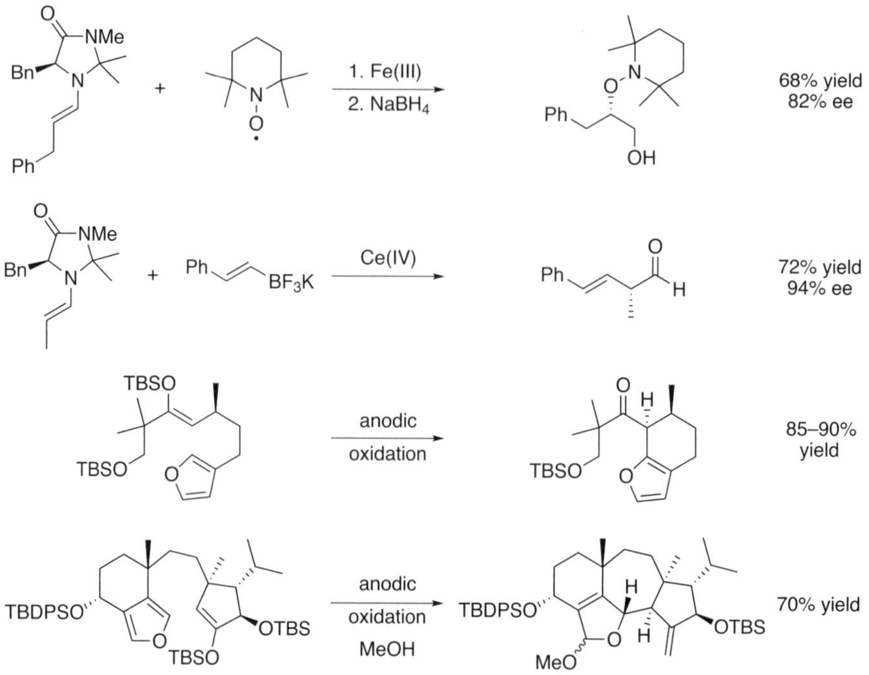

SCHEME 3.1 Addition reactions through radical cation intermediates.

3.1.1 Oxidative Carbon–Carbon Bond Cleavage

Arnold's demonstration[4] that oxocarbenium ion intermediates can be formed through homobenzylic ether radical cation fragmentation reactions shows that mild oxidizing conditions can be used to prepare important reactive intermediates.[5] Scheme 3.2 illustrates a critical observation in the development of an explanatory model that allows for the application of radical cation fragmentation reactions in complex molecule synthesis. In Arnold's seminal work, cleavage of the benzylic carbon–carbon bond in substrate **1** is promoted by 1,4-dicyanobenzene (DCB) with photo-irradiation by a medium-pressure mercury vapor lamp. With methanol as the solvent, the resulting products were diphenylmethane (**2**) and formaldehyde dimethyl acetal (**3**).

Arnold's proposed mechanism[6] for these oxidations is shown in Scheme 3.3. Photoexcitation of 1,4-dicyanobenzene produces a potent single electron oxidant

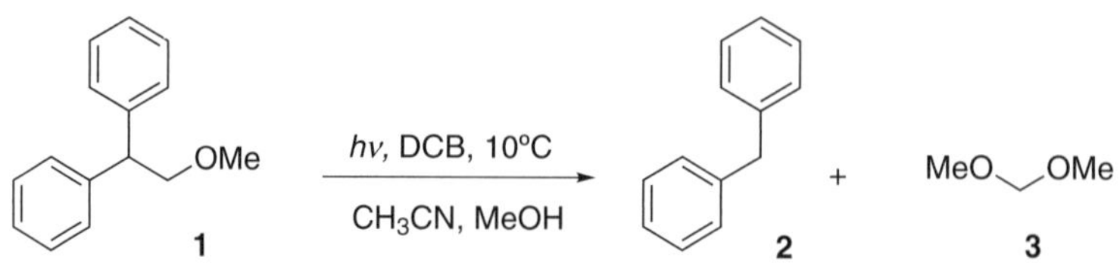

SCHEME 3.2 Arnold's oxidative carbon–carbon bond fragmentation.

(Step 1) that removes an electron from the substrate to form the radical cation (Step 2). The benzylic bonds in the radical cation are substantially weakened relative to those in the neutral substrate, allowing for mesolytic bond cleavage to form the diphenyl-methyl radical and the oxocarbenium ion (Step 3). The oxocarbenium ion is subsequently captured by MeOH to form the acetal product (Step 4). The radical anion of dicyanobenzene is a potent electron donor and can reduce the diphenylmethyl radical to form the diphenylmethyl anion (Step 5), which can be protonated by MeOH to form diphenylmethane (Step 6). Notably, no diphenylmethyl ether product was detected, indicating that the fragmentation reaction proceeds exclusively to form the diphenylmethyl radical and the oxocarbenium ion. This suggests that cleavage provides the more stable cationic fragment.

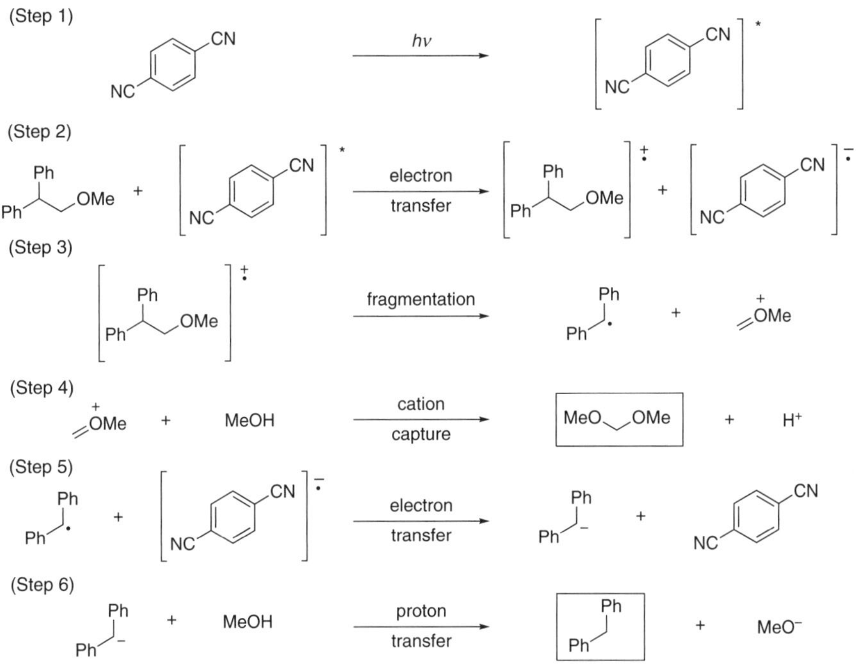

SCHEME 3.3 Mechanism for photosensitized carbon–carbon bond cleavage.

Additional examples of oxidatively initiated cleavage reactions demonstrated that heteroatom stabilization of the cationic fragment is not necessary and that reactivity can be tuned by manipulating the strength of the benzylic carbon–carbon bond (Scheme 3.4). Entry 1 shows that substrate **4** with a benzyl rather than a diphenylmethyl group is completely stable toward oxidative cleavage. Substrate **5**, in which the methoxymethyl group is replaced by a diphenylmethyl group, also does not undergo oxidative cleavage at room temperature (Entry 2) but does cleave at elevated temperature (Entry 3). These experiments show that oxidative cleavage reactivity can be enhanced by weakening the benzylic carbon–carbon bond through the addition of phenyl groups or alkoxy groups and that alkoxy groups are particularly effective at promoting fragmentation.[7]

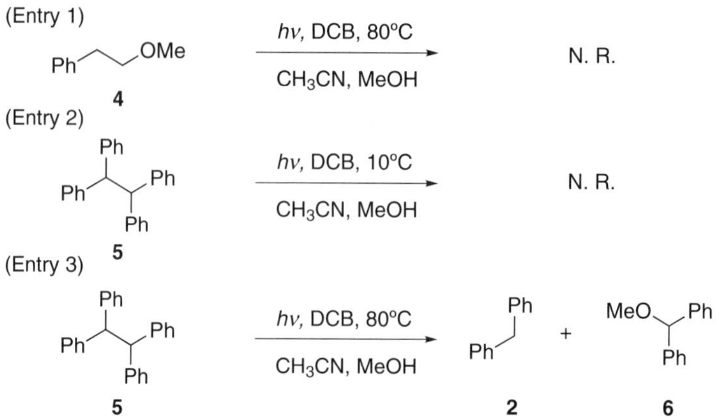

SCHEME 3.4   Tuning reactivity by manipulating benzylic bond strength.

Substitution of the aromatic ring was also shown to be an important factor in determining the facility of bond cleavage (Scheme 3.5).[8] Cyano-substituted ether **7** cleaved under oxidizing conditions, while methoxy-substituted ether **9** did not. These results are contrary to the ease of radical cation formation, but are consistent with fragmentation proceeding more readily from nonstabilized radical cations.

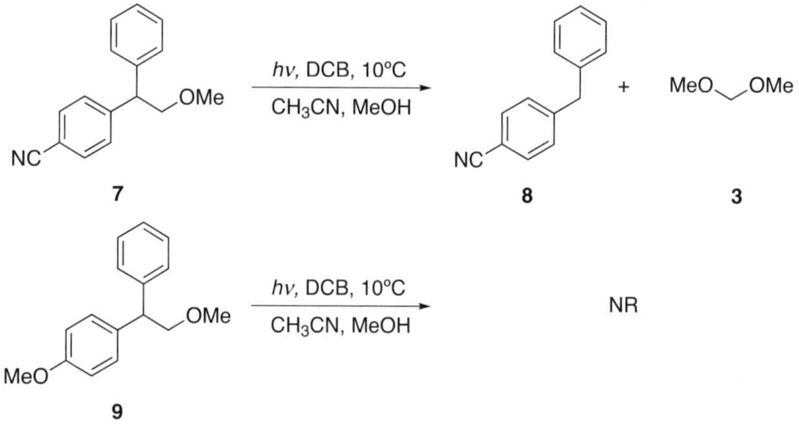

SCHEME 3.5   Reactivity as a function of oxidation potential.

3.1.2   Thermodynamic and Kinetic Considerations

These studies taken together show that, from a thermodynamic perspective, three factors can be utilized to predict the bond dissociation energy of a homobenzylic ether radical cation (BDE(RC)) and thereby its propensity to fragment: (1) the bond dissociation energy of the benzylic carbon–carbon bond in the neutral substrate (BDE(S)), (2) the oxidation potential of the substrate ($E_{pa}(S)$), and (3) the oxidation potential of the radical that corresponds to the eventual electrophilic fragment ($E_{pa}(E)$). These factors combine to yield equation 1 as illustrated in Scheme 3.6 by using a representative Arnold mesolytic cleavage. The same equation explains the enhanced reactivity of the diphenylmethyl-containing substrates relative to their monophenyl counterparts because of their lower bond dissociation energies. The low reactivity of the radical cation of methoxyphenyl-containing substrate is also explained by this equation because of its low oxidation potential. Finally, the selectivity of the cleavage to form the oxocarbenium ion rather than the diphenyl-methyl radical can be explained by the lower oxidation potential of alkoxyalkyl radicals compared to diarylmethyl radicals.[9]

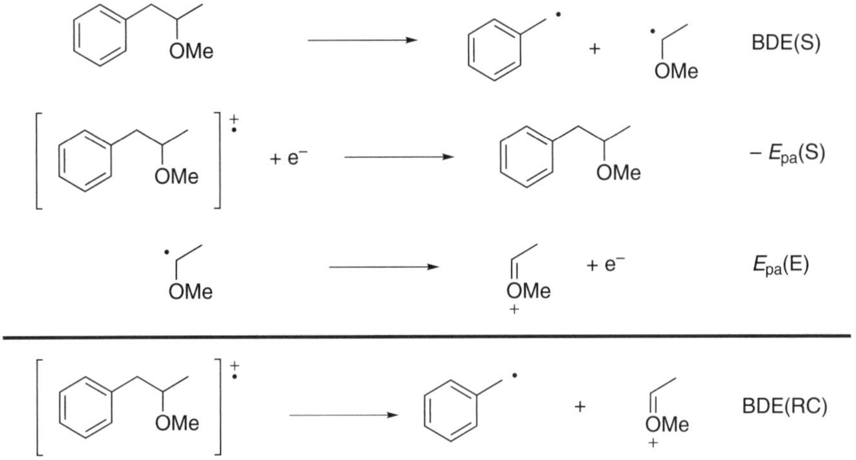

$$BDE(RC) = BDE(S) - E_{pa}(S) + E_{pa}(E) \qquad \text{(Equation 1)}$$

SCHEME 3.6   Thermodynamics of radical cation fragmentation.

A subsequent study[10] from the Arnold group showed an intriguing stereoelectronic effect in oxidative benzylic carbon–hydrogen bond cleavage reactions of substrates **8** and **9** (Scheme 3.7). In this study, electron transfer reactions were conducted in the presence of a nonnucleophilic base. Radical cation formation also weakens benzylic carbon–hydrogen bonds, thereby enhancing their acidity. Deprotonation of benzylic hydrogens yields benzylic radicals that can be reduced by the radical anion of dicyanobenzene to form benzylic anions that will be protonated by solvent. This sequence of oxidation, deprotonation, reduction, and protonation provides a sequence by which epimerization can be effected at the benzylic center. In this study, *trans* isomer **10** showed no propensity to isomerize to *cis* isomer **11** (equation 1 in Scheme 3.7), but **11** readily converted to **10** (equation 2 in Scheme 3.7). The reactions were repeated in deuterated solvents to assure that these observations resulted from kinetic rather than thermodynamic factors. *Trans* isomer **9** showed no incorporation of deuterium (equation 3 in Scheme 3.7) whereas *cis* isomer **11** showed complete deuterium incorporation. The authors attributed this difference in reactivity to conformational effects. The dihedral angle between the phenyl ring and the benzylic carbon–hydrogen bond in **10** was calculated to be 1° while the corresponding dihedral angle in **11** was calculated to be 34°. These studies indicate that benzylic fragmentation is possible only when the cleaving bond can overlap with the π-orbitals of the arene radical cation. The role of orbital overlap in determining fragmentation propensity has also been discussed extensively by other research groups.[11]

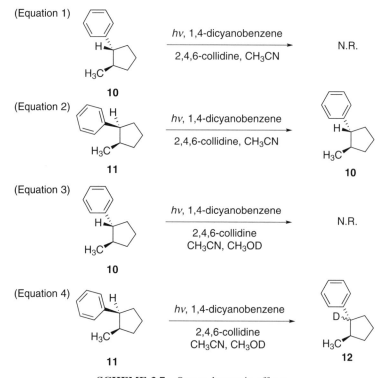

SCHEME 3.7   Stereoelectronic effects.

This work and related studies provided the basis for the stereoelectronic model for homobenzylic ether cleavage in Fig. 3.1. This model includes overlap of the benzylic carbon–carbon bond with the SOMO of the aromatic ring (structure **13**), thereby stabilizing the benzylic radical upon cleavage. Additionally, overlap of a heteroatom lone pair and the benzylic σ* orbital was shown to be necessary for cleavage (structure **14**).

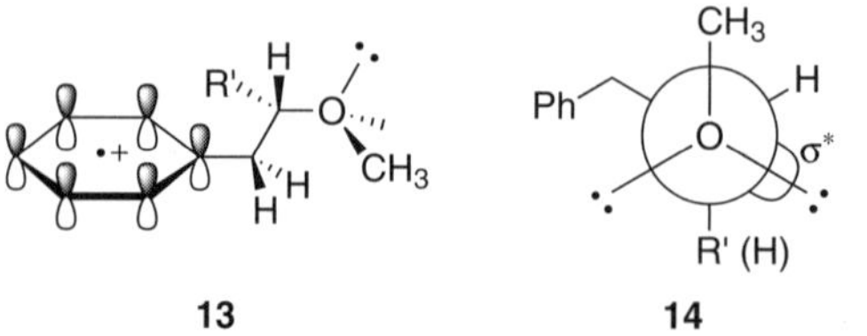

FIGURE 3.1   Preferred conformation for radical cation carbon–carbon bond cleavage.

3.1.3 Reactive Intermediate Lifetime

In addition to thermodynamic and stereoelectronic factors, the lifetime of the radical cation and the efficiency of its formation are important considerations for developing practical reactions based on single electron oxidation. Return electron transfer from the reduced sensitizer to the radical cation is generally a fast and highly thermodynamically favorable process. Slowing this process can dramatically facilitate fragmentation. Dinnocenzo and coworkers demonstrated[12] that promoting electron transfer with the cationic oxidant *N*-methylquinolinium hexafluorophosphate (NMQPF$_6$) in the presence of an aromatic cooxidant such as toluene or biphenyl greatly improved the yield of trialkyl benzylsilane radical cation formation through photooxidation relative to reactions with cyanoarene oxidants (Scheme 3.8). This effect can be attributed to two factors. Cationic oxidants such as the methylquinolinium ion form neutral radicals upon accepting an electron. The neutral reduction product has no electrostatic attraction to the radical cation, in contrast to the strong electrostatic between the radical cation and the radical anion that forms from neutral oxidants such as dicyanobenzene. The lack of an electrostatic attraction allows the reductant and oxidant to diffuse, thereby diminishing the potential for return electron transfer.[13] The cooxidant, being present in large excess, statistically has superior opportunities for encountering the photoexcited oxidant, thereby increasing the radical cation quantum yield.

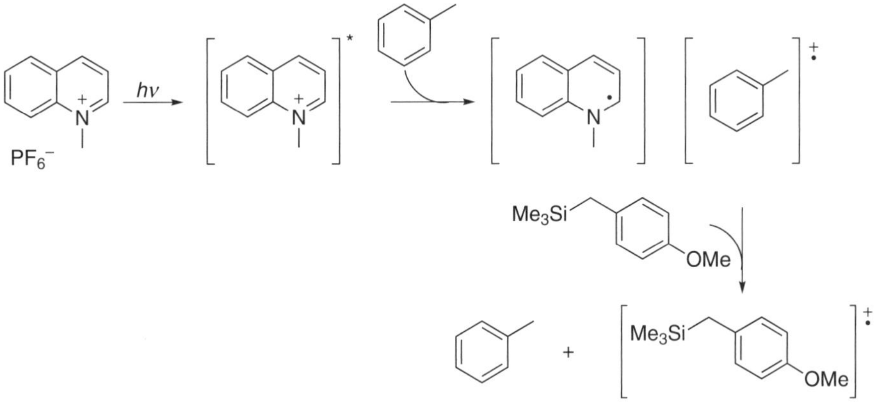

SCHEME 3.8   Cationic oxidant and aromatic cooxidant.

3.2 ELECTRON TRANSFER-INITIATED CYCLIZATION REACTIONS

These impressive mechanistic studies on radical cation cleavage reactions provide the foundations for devising new oxidative methods of increasing molecular complexity. One method for achieving this objective is to append a nucleophile to the homobenzylic ether. This concept was precedented by reports of cyclization reactions that were initiated by oxidative cleavage of stannylalkyl ethers.[14] Intramolecular nucleophilic addition into the oxocarbenium ion that forms upon fragmentation leads to annulation. Kumar and Floreancig showed[15] that applying Arnold's conditions to a homobenzylic ether with a pendent alcohol nucleophile (**15**) provided cyclic acetal **16** in an example of an electron transfer-initiated cyclization (ETIC) reaction (Scheme 3.9). While this reaction was very slow, it provided a proof of concept that justified further research efforts.

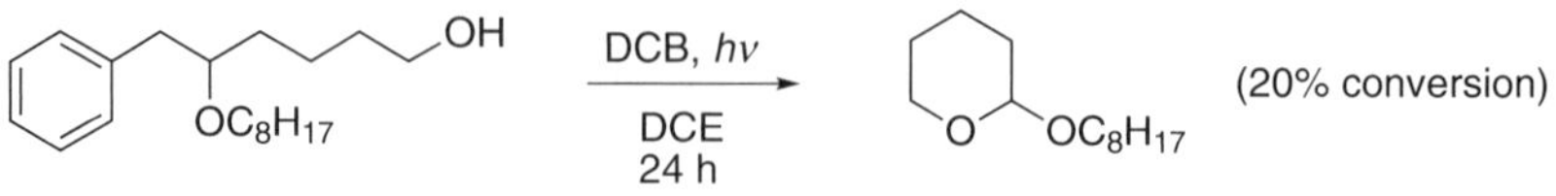

SCHEME 3.9 Initial ETIC reaction.

This low reaction rate was attributed to fragmentation and cyclization not being kinetically competitive with return electron transfer between the homobenzylic radical cation and the dicyanobenzene radical anion. In principle, the kinetics of this reaction could be improved by utilizing a cationic oxidant and an aromatic cooxidant in accord with Dinnocenzo's studies.[12] When **15** was irradiated in the presence of $NMQPF_6$ and the cooxidant *tert*-butylbenzene, shown by Kochi[16] to form a radical cation that stable toward fragmentation, the reaction proceeded to form **16** within 20 min. As illustrated in Scheme 3.10, other examples in the same report demonstrated that five-, seven-, and eight-membered rings could also be formed in good yields.

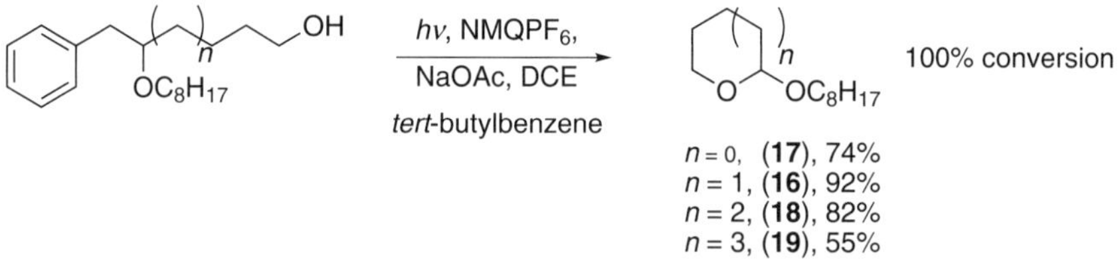

n = 0, (**17**), 74%
n = 1, (**16**), 92%
n = 2, (**18**), 82%
n = 3, (**19**), 55%

SCHEME 3.10 Kinetically viable ETIC reactions.

3.2.1 Rate Enhancement and Mechanistic Studies

The mechanism of these nucleophilic additions was proposed to proceed through a dissociative rather than associative pathway (Scheme 3.11) based on observations that reactions of substrates that contain multiple chiral centers could be stereoselective but were not stereospecific,[15] consistent with the formation of a carbocationic intermediate. Evidence that the dissociation step is rapid and reversible was provided by observations of epimerization at the homobenzylic center in recovered starting materials when reactions were taken to partial conversion.

3.2.2 Development of a Catalytic Aerobic Protocol

The reaction conditions for these cyclization reactions were further improved by devising conditions by which $NMQPF_6$ could be used in catalytic amounts.

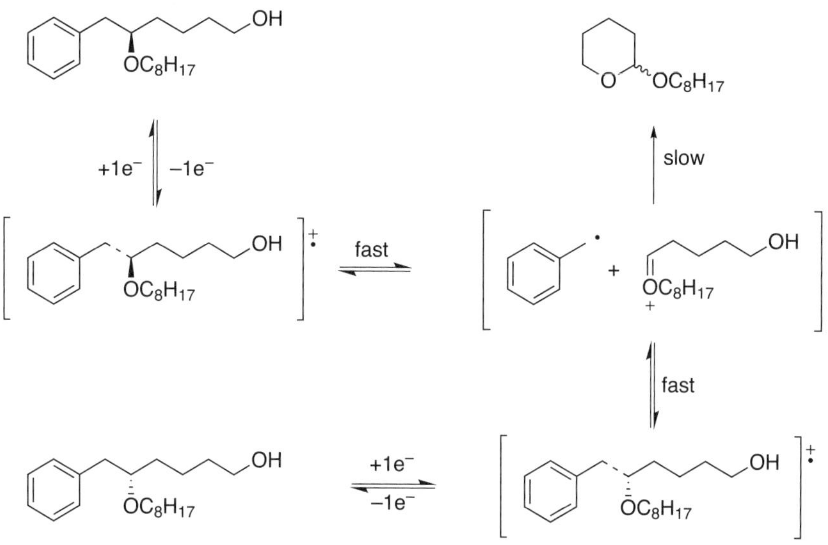

SCHEME 3.11 Mechanism of the cyclization reaction.

Dinnocenzo and coworkers reported[12] that spectroscopic signals from *N*-methyl dihydroquinolyl radicals were suppressed in the presence of O_2, suggesting that methylquinolinuim ions can be regenerated through aerobic oxidation. Conducting the ETIC reaction with continuous aeration[17] allowed the loading of $NMQPF_6$ to be reduced to 2.5 mol percent (Scheme 3.12), in accord with the postulate of quinolinium ion formation through dihydroquinolyl radical oxidation. Aeration also promoted oxidation of benzyl radical leaving group to form benzaldehyde, which facilitated separation by minimizing the number of by-products. Toluene could be used as the cooxidant rather than the less volatile and more expensive *tert*-butylbenzene because toluene oxidation also produces benzaldehyde. Due to superoxide formation, a terminal reducing agent was necessary to guard against unselective overoxidation. While soluble reducing agents inhibited cyclization completely, insoluble sodium

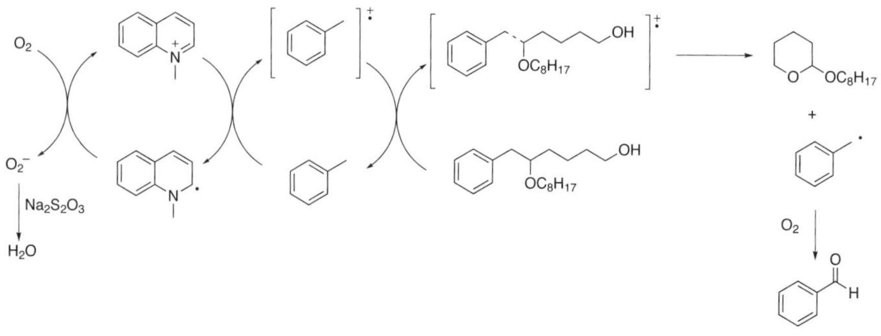

SCHEME 3.12 $NMQPF_6$ catalyzed aerobic photooxidation.

thiosulfate proved effective for scavenging superoxide while not impeding the desired pathway.

3.2.3 Oxidative Cascade Reactions

Oxidatively generated oxocarbenium ions can be trapped by epoxides to form highly electrophilic epoxonium ions under nonacidic conditions. This method, shown in Scheme 3.13,[18] is desirable for designing cascade reactions that employ substrates with multiple acid-sensitive groups because the ordering of electrophile formation is controlled by the kinetics of cyclization rather than bimolecular acid–base interactions. Substrate **20** was subjected to oxidative cleavage conditions to provide polyether **21** through a series of *exo*-cyclizations into epoxonium ion intermediates. A subsequent report[19] showed that fused polyether structures could be formed by structural modifications, as shown in the conversion of **22** to **23**.

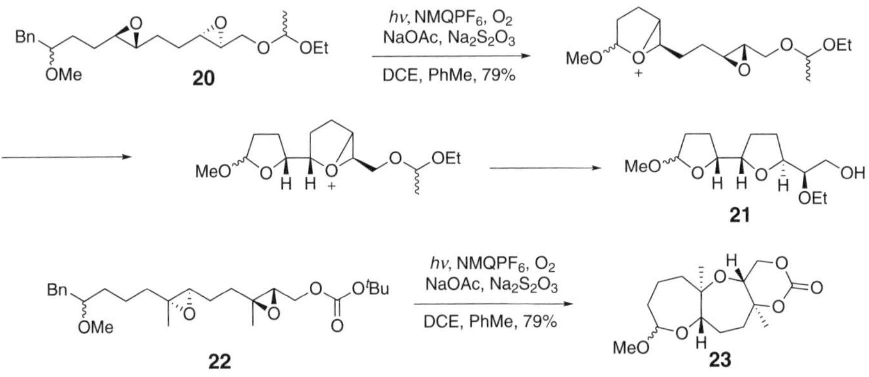

SCHEME 3.13   Epoxonium ion cascade reactions.

3.3   OXIDATIVE ACYLIMINIUM ION FORMATION

Iminium and acyliminium ions are useful nitrogen-containing analogs of oxocarbenium ions that could in principle be formed through oxidative fragmentations of homobenzylic amines and amides. Oxidative fragmentation reactions of silylalkyl amides have been reported by several groups, with notable examples shown in Scheme 3.14. Yoshida and coworkers reported (Entry 1 in Scheme 3.14) that silylalkyl amides could be cleaved under anodic oxidation conditions to form acyliminium ions that react with MeOH to form acylaminals.[20] These studies provided the foundations for the elegant "cation pool" method[21] that was devised by this group. Mariano showed[22] that iminium and acyliminium ions could be formed by cleaving silylalkyl amines and amides, respectively. Cyclization was effected by tethering the electrophile to allylsilane groups (Entry 2 in Scheme 3.14). Moeller reported[23] that the Yoshida conditions could be applied to peptides that contain silylalkyl amide groups (Entry 3 in Scheme 3.14). Oxidative cyclization reactions of these substrates provide a

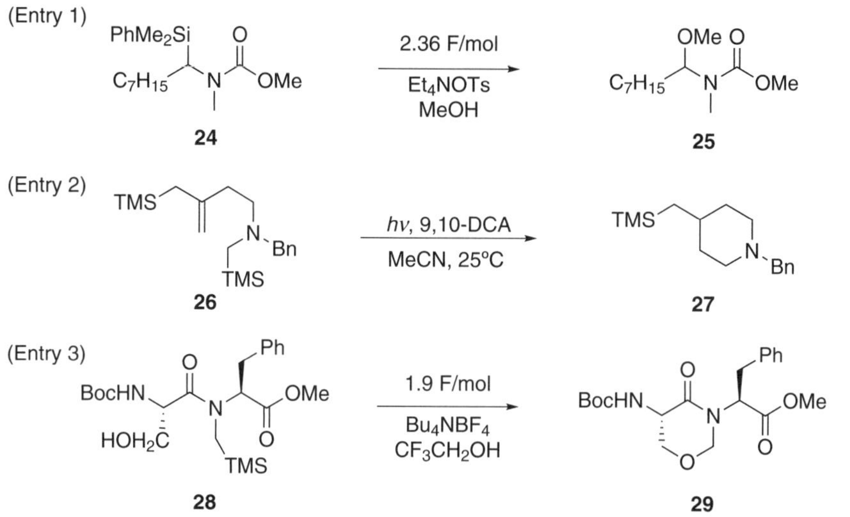

(Entry 1)

(Entry 2)

(Entry 3)

SCHEME 3.14   Previous examples of oxidative iminium and acyliminium ion formation.

powerful method for manipulating peptide structure after completing the synthesis of the linear structure.

Secondary and tertiary amides **30** and **31** were prepared based on the hypothesis that amide groups, such as alkoxy groups, would weaken benzylic carbon–carbon bonds and stabilize the cations that form upon fragmentation of their radical cations. Exposing these substrates to the photochemical aerobic oxidation conditions resulted in the expected cyclizations to form amido tetrahydropyrans **32** and **33** in good yield (Scheme 3.15).[24] Substrates **34** and **35** with methyl substituents at the bis homo-benzylic position cyclized under aerobic ETIC conditions, showing modest diaster-eoselectivity in the formation of secondary amide **36** and high diastereoselectivity in the formation of tertiary amide **37**. The lack of stereospecificity in these reactions suggest that the displacements proceed through discrete acyliminium ion formation occurs rather than S_N2-type displacement. Although electronegative groups at the homobenzylic position could inhibit fragmentation by destabilizing the intermediate acyliminium ions carbamates, **38** cyclized to form **39** in 63% yield and **40** cyclized to form **41** in 89% yield. Notably, the cyclization of **40** showed that THP ethers are nonpolar surrogates for hydroxyl groups in these reactions.

The sensitive amido trioxadecalin framework of the mycalamide family of natural products[25] provides an attractive target for applying the ETIC reaction to complex molecule synthesis. This structure could potentially be prepared by adding a formal-dehyde hemiacetal or its equivalent into an acyliminium ion. Glucose-derived homobenzylic carbamate **42**, in which a tetrahydrofuryl ether was used to prepare a formaldehyde acetal equivalent, was subjected to oxidative cleavage under the standard aerobic conditions to form amido trioxadecalin **43** in good yield (Scheme 3.16).[26] Green and Floreancig recently completed a total synthesis of theopederin D[27] through the cyclization of **44** to **45**. The ability to effect acyliminium

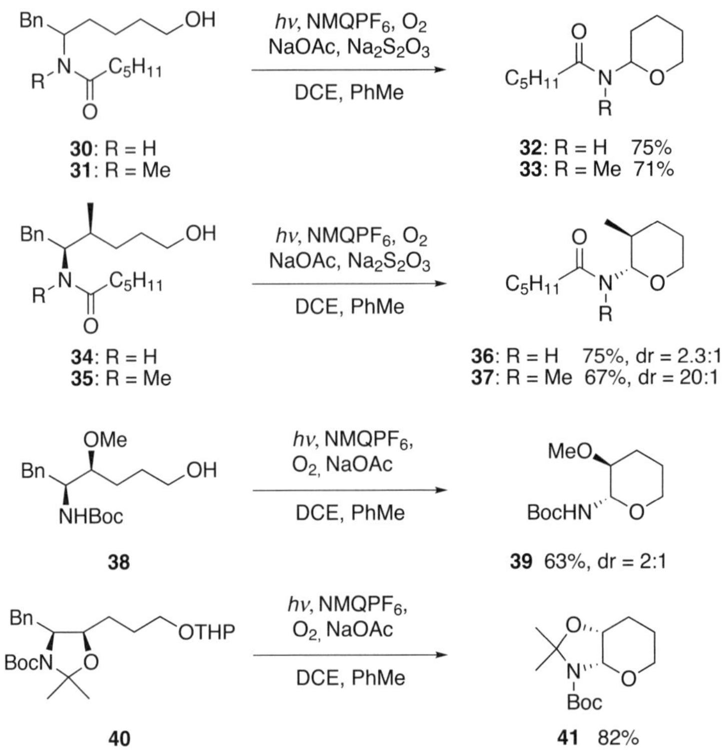
Bn
OH
N
R
C5H11
O
30: R = H
31: R = Me
hv, NMQPF6, O2
NaOAc, Na2S2O3
DCE, PhMe
32: R = H 75%
33: R = Me 71%
34: R = H
35: R = Me
36: R = H 75%, dr = 2.3:1
37: R = Me 67%, dr = 20:1
OMe
NHBoc
38
hv, NMQPF6,
O2, NaOAc
MeO
BocHN
39 63%, dr = 2:1
OTHP
BocN
40
Boc
41 82%

SCHEME 3.15 Oxidative amido tetrahydropyran formation.

ion formation in the presence of acid-sensitive groups such as tetrahydrofuranyl ethers highlights the orthogonality of oxidative cleavage with respect to acid-mediated electrophile formations and highlights the utility of the protocol in complex molecule synthesis.

3.4 CARBON–CARBON BOND FORMATION

3.4.1 Chemoselectivity and Reactivity

Oxidative cleavage of homobenzylic ether substrates in the presence of electron-rich π-nucleophiles presents a significant chemoselectivity challenge to forming carbon–carbon bonds through this process. This problem could be circumvented by lowering the oxidation potential of the arene through appending an electron-donating group. Incorporating a *para*-methoxy group to a substrate that was known to undergo oxidative cyclization (**46**, Scheme 3.17), however, completely suppressed the reaction. From a thermodynamic perspective, this result is consistent with equation 1 in Scheme 3.6, which shows that the bond dissociation energy of the radical cation is raised when the oxidation potential of the substrate is lowered. Reactivity was restored by introducing additional radical stabilizing (bond-weakening) groups at

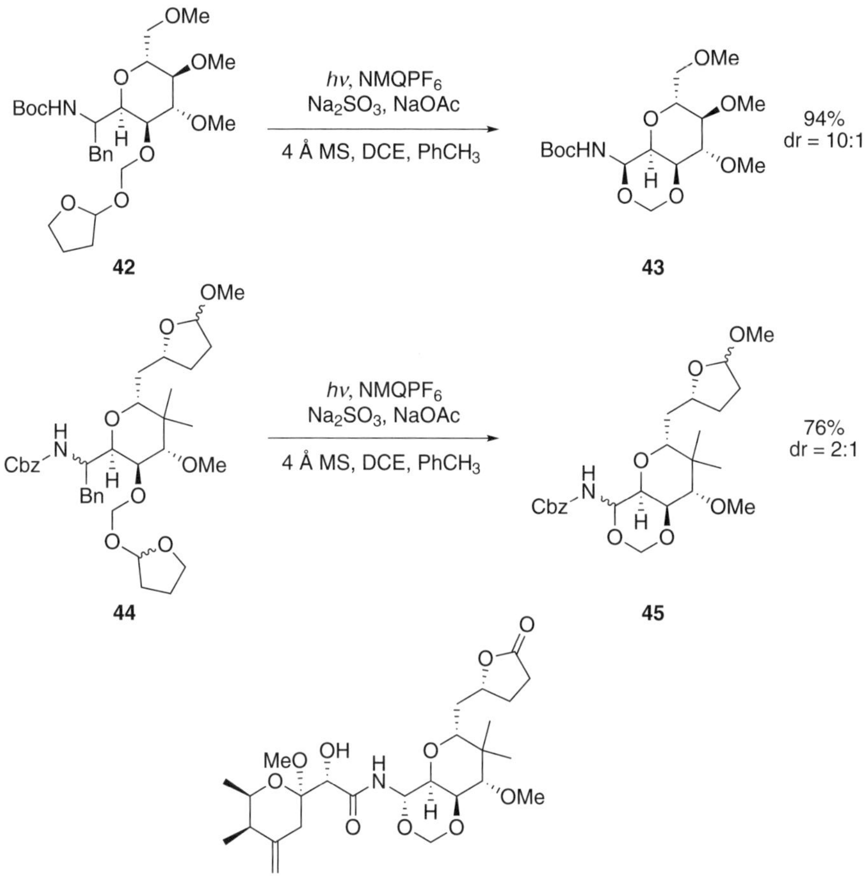

Theopederin D

SCHEME 3.16 Oxidative amido trioxadecalin construction in the synthesis of theopederin D.

the benzylic position, with the phenyl group in **47** being a representative bond-weakening element, in accord with the prediction from equation 1 in Scheme 3.6 that lowering the benzylic bond dissociation energy of the substrate lowers the bond dissociation energy of the corresponding radical cation. The lowered oxidation potential of substrates that contain *p*-methoxyphenyl groups allowed for the use of ground-state oxidants, such as the commonly used reagent ceric ammonium nitrate (CAN), to initiate oxidative cleavage. Utilizing ground-state oxidants facilitates the reactions from a technical perspective and removes the requirement for a photochemical reactor.

3.4.2 Reaction Scope

Cyclizations proceeded smoothly to form carbon–carbon bonds when a variety of π-nucleophiles were appended to electron-rich homobenzylic ethers (Scheme 3.18).[28]

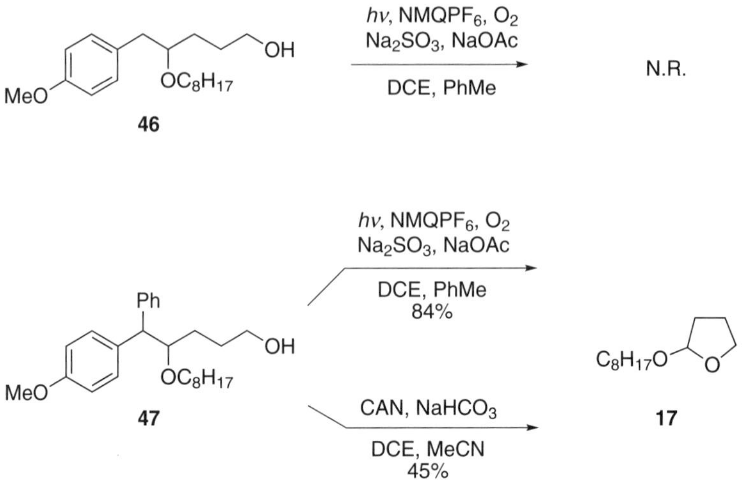

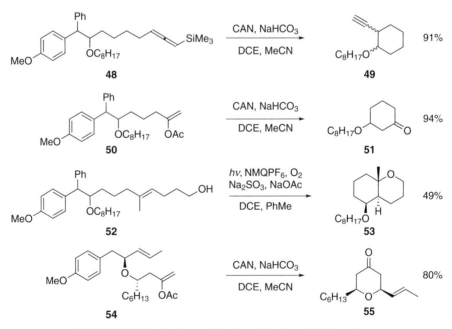

SCHEME 3.17 Cyclizations of substrates with electron-rich arenes.

Silyl allenes such as **48** and enol acetates such as **50** were particularly effective substrates for these reactions. The oxidative cascade reaction of trisubstituted alkene substrate **52** demonstrated a powerful application of ETIC chemistry to form bicycle **53** through consecutive carbon–carbon and carbon–oxygen bond construction. Highly diastereoselective reactions were observed when the nucleophile was

SCHEME 3.18 Carbon–carbon bond forming ETIC reactions.

tethered to the oxygen of the homobenzylic ether, resulting in an *endo*-cyclization rather than an *exo*-cyclization, as shown in the conversion of **54** to **55**. This cyclization was also intriguing because it showed that the alkene substituent that lowers BDE(S) can be appended to the homobenzylic position as well as the benzylic position.

Several of the reactivity patterns that were observed in the development of the oxidative carbon–carbon bond forming reactions were evident in the ETIC reaction of compound **56** (Scheme 3.19).[29] Two homobenzylic ethers that potentially could fragment are present in this substrate. Cleavage in the presence of CAN was observed exclusively at the site that contains the geminal methyl groups, however, resulting in the formation of tetrahydropyran **57**. This cyclization, which was used in the synthesis of a subunit of the cytotoxin apicularen A,[30] demonstrated that fragmentations can be effective in structures that contain multiple groups that undergo single electron oxidation, and that selectivity will arise when differences exist in the kinetic barriers of downstream events.

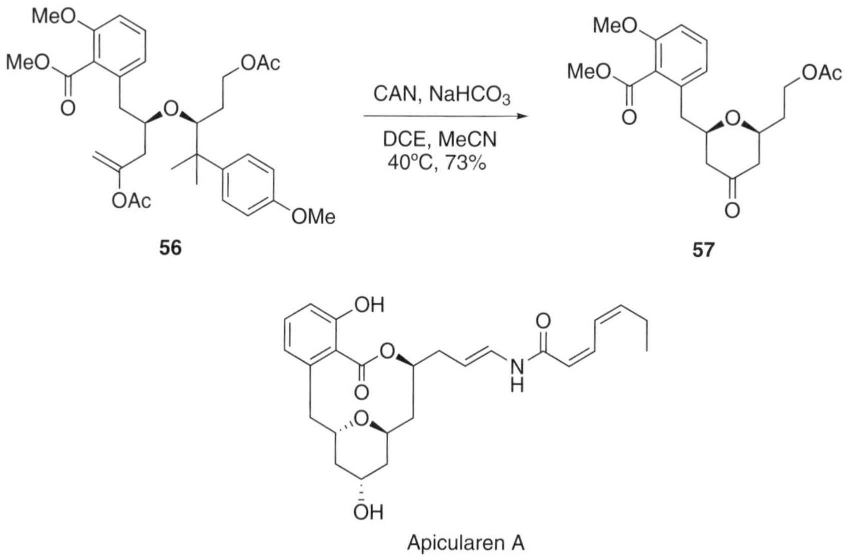

SCHEME 3.19   ETIC reaction en route to apicularen A.

A diastereoselective *endo*-cyclization into an oxidatively generated oxocarbenium ion was a key step in a formal synthesis[31] of leucascandrolide A.[32] Exposing **56** to CAN provided *cis*-tetrahydropyran **57** in high yield and with excellent stereocontrol (Scheme 3.20). This transformation provides further evidence that oxidative electrophile formation is tolerant of several functional groups and can be applied to complex molecule synthesis. The synthetic sequence also utilized a Lewis acid mediated ionization reaction to form an oxocarbenium ion in the presence of the homobenzylic ether (**58**, **59**), illustrating that two carbocation precursors that ionize through chemically orthogonal conditions can be incorporated into the same structure.

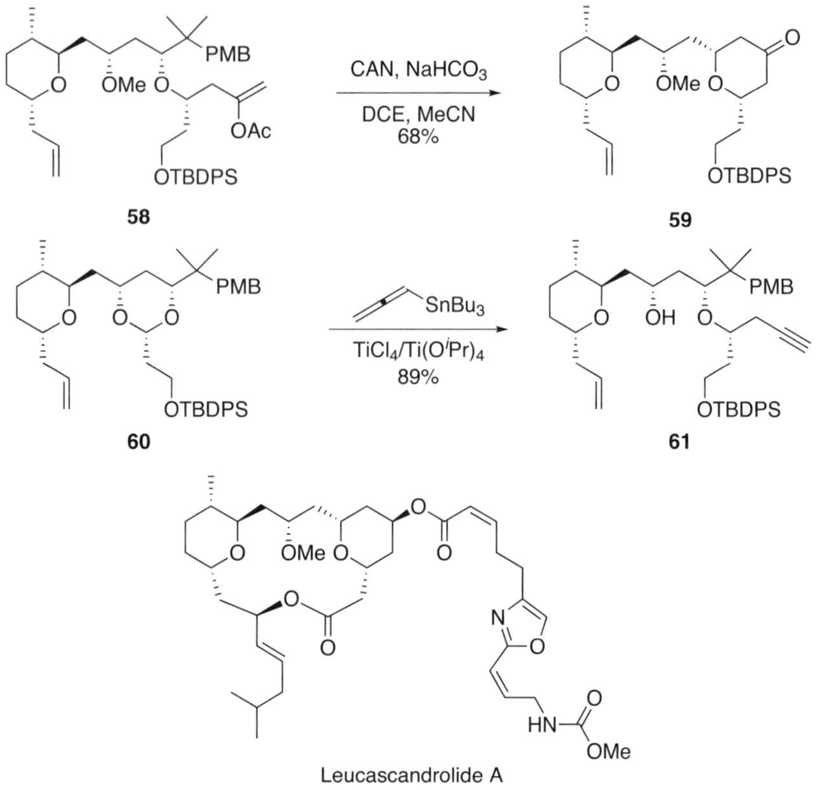

SCHEME 3.20   Ionization reactions in the synthesis of leucascandrolide A.

3.5   SUMMARY AND OUTLOOK

Abundant opportunities for developing new reactions that proceed through radical cation intermediates become available by understanding the structural features that promote the available pathways for these high-energy intermediates. The studies that were detailed in this chapter focused on oxidative cleavage reactions to generate carbocations under nonacidic conditions. These efforts required a basic understanding of the thermodynamic factors that influence bond strengths in radical cations and kinetic factors that impact the lifetime of the intermediates. The basic investment in studying the physical organic chemistry of radical cation fragmentation pathways has resulted in the development of a highly versatile cyclization protocol. The chemo-selective oxidation conditions that are employed to form the radical cation inter-mediates are tolerant of a number of functional groups including but not limited to ethers, silyl ethers, epoxides, acetals, ketones, esters, amides, alkenes, alkynes, and arenes. Thus radical cation fragmentation reactions can be employed to form cations in highly functionalized compounds. Strategically, these reactions create opportunities for forming cations through oxidative cleavage in the presence functional groups that are sensitive to the acidic conditions that are commonly utilized for ionization, and for introducing multiple precursors to electrophiles in molecules that can be released in a perfectly orthogonal manner. With the fundamental reactivity patterns and the utility of the method toward complex molecule synthesis soundly established, continued efforts to devise creative applications of radical cation cleavage processes are highly warranted. Developing new reactions that utilize cationic intermediates in structures where conventional methods would not be viable and making the process "greener" by improving atom economy and utilizing oxygen as the terminal oxidant for ground-state oxidants are exciting and under explored frontiers of research in this area.

REFERENCES

1. Schmittel, M.; Burghart, A. *Angew. Chem., Int. Ed.* **1997**, *36*, 2550.
2. Representative examples: (a) Sibi, M.; Hasegawa, M. *J. Am. Chem. Soc.* **2007**, *129*, 4124. (b) Kim, H.; MacMillan, D. W. C. *J. Am. Chem. Soc.* **2008**, *130*, 398. (c) Mihelcic, J.; Moeller, K. D. *J. Am. Chem. Soc.* **2004**, *126*, 9106. (d) Hughes, C. C.; Miller, A. K.; Trauner, D. *Org. Lett.* **2005**, *7*, 3425.
3. Floreancig, P. E. *Synlett* **2007**, 191.
4. Arnold, D. R.; Maroulis, A. J. *J. Am. Chem. Soc.* **1976**, *98*, 5931.
5. For other examples of benzylic carbon–carbon cleavage through single electron oxidation, see Baciocchi, E.; Bietti, M.; Lanzalunga, O. *Acc. Chem. Res.* **2000**, *33*, 243.
6. (a) Arnold, D. R.; Lamont, L. J. *Can. J. Chem.* **1989**, *67*, 2119. (b) Popielarz, R.; Arnold, D. R. *J. Am. Chem. Soc.* **1990**, *112*, 3068.
7. For comprehensive tables of homolytic bond dissociation energies, see (a) Blanksby, S. J.; Ellison, G. B. *Acc. Chem. Rev.* **2003**, *36*, 255. (b) McMillen, D. F.; Golden, D. M. *Annu. Rev. Phys. Chem.* **1982**, *33*, 493.
8. Arnold, D. R.; Su, X.; Chen, J. *Can. J. Chem.* **1995**, *73*, 307.
9. Wayner, D. D. M.; McPhee, D. J.; Griller, D. *J. Am. Chem. Soc.* **1988**, *110*, 132.
10. Perrot, A. L.; deLijser, H. J. P.; Arnold, D. R. *Can. J. Chem.* **1997**, *75*, 384.
11. (a) Tolbert, L. M.; Khanna, R. K.; Popp, A. E.; Gelbaum, L.; Bottomley, L. A. *J. Am. Chem. Soc.* **1990**, *112*, 2373. (b) Freccero, M.; Pratt, A.; Albini, A.; Long, C. *J. Am. Chem. Soc.* **1998**, *120*, 284.
12. Dockery, K. P.; Dinnocenzo, J. P.; Farid, S.; Goodman, J. L.; Gould, I. R.; Todd, W. P. *J. Am. Chem. Soc.* **1997**, *119*, 1876.
13. Marcus, R. A. *Angew. Chem., Int. Ed.* **1993**, *32*, 1111.
14. (a) Yoshida, J.-i.; Ishichi, Y.; Isoe, S. *J. Am. Chem. Soc.* **1992**, *114*, 7594. (b) Chen, C.; Mariano, P. S. *J. Org. Chem.* **2000**, *65*, 3252.
15. Kumar, V. S.; Floreancig, P. E. *J. Am. Chem. Soc.* **2001**, *123*, 3842.
16. Wightman, R. M.; Kochi, J. K. *J. Am. Chem. Soc.* **1984**, *106*, 3968.
17. Kumar, V. S.; Aubele, D. L.; Floreancig, P. E. *Org. Lett.* **2001**, *3*, 4123.
18. (a) Kumar, V. S.; Aubele, D. L.; Floreancig, P. E. *Org. Lett.* **2002**, *4*, 2489. (b) Kumar, V. S.; Wan, S.; Aubele, D. L.; Floreancig, P. E. *Tetrahedron: Asymmetry* **2005**, *16*, 3570.
19. Wan, S.; Gunaydin, H.; Houk, K. N.; Floreancig, P. E. *J. Am. Chem. Soc.* **2007**, *129*, 7915.

20. Yoshida, J.-i.; Isoe, S. *Tetrahedron Lett.* **1987**, *28*, 6621.

21. Yoshida, J.-i.; Suga, S. *Chem. Eur. J.* **2002**, *8*, 2651.

22. Wu, X. D.; Khim, S. K.; Zhang, X. M.; Cederstrom, E. M.; Mariano, P. S. *J. Org. Chem.* **1998**, *63*, 841.

23. Sun, H.; Moeller, K. D. *Org. Lett.* **2002**, *4*, 1547.

24. (a) Aubele, D. L.; Floreancig, P. E.; *Org. Lett.* **2002**, *4*, 3443. (b) Aubele, D. L.; Rech, J. C.; Floreancig, P. E.; *Adv. Synth. Catal.* **2004**, *346*, 359.

25. Fusetani, N.; Sugawara, T.; Matsunaga, S. *J. Org. Chem.* **1992**, *57*, 3828.

26. Rech, J. C.; Floreancig, P. E. *Org. Lett.* **2003**, *5*, 1495.

27. Green, M. E.; Rech, J. C.; Floreancig, P. E. *Angew. Chem., Int. Ed.* **2008**, *47*, 7317.

28. (a) Seiders, J. R. II; Wang, L.; Floreancig, P. E. *J. Am. Chem. Soc.* **2003**, *125*, 2406. (b) Wang, L.; Seiders, J. R. II; Floreancig, P. E. *J. Am. Chem. Soc.* **2004**, *126*, 12596.

29. Poniatowski, A. J.; Floreancig, P. E. *Synthesis* **2007**, 2291.

30. Kunze, B.; Jansen, R.; Sasse, F.; Höfle, G.; Reichenbach, H. *J. Antibiot.* **1998**, *51*, 1075.

31. Jung, H. H.; Seiders, J. R. II; Floreancig, P. E. *Angew. Chem., Int. Ed.* **2007**, *46*, 8464.

32. D'Ambrosio, M.; Guerriero, A.; Debitus, C.; Pietra, F. *Helv. Chim. Acta* **1996**, *79*, 51.

4

SELECTIVITY IN RADICAL CATION CYCLOADDITIONS

Christo S. Sevov and Olaf Wiest

Department of Chemistry and Biochemistry, University of Notre Dame, Notre Dame, IN, USA

4.1 INTRODUCTION

Ever since its discovery in 1928,[1] the Diels–Alder (DA) reaction has remained one of the most useful reactions in organic synthesis because of its utility in forming two new carbon–carbon bonds. The easy construction of cyclohexenes from acyclic precursors has remained an invaluable method for the synthesis of many important compounds. Attempts to react electron-rich dienes and electron-rich dienophiles reveal a limitation of the DA reaction, namely, the requirement for energetically close frontier orbitals of the reaction partners.[2–4] Low yields or forcing conditions[5,6] often present roadblocks in chemical syntheses. Traditional solutions to this problem have called for the use of Lewis acid catalysts such as transition metals.[7,8] Even with such carefully controlled systems, DA additions fall sometimes short of satisfactory results in terms of atom economy and synthetic applicability. This is, for example, the case for simple hydrocarbons that have large HOMO–LUMO gaps and are often not susceptible to transition metal catalysis. The last few decades have brought about attempts to expand the capabilities of the DA reaction by oxidation of one reacting substrate to the radical cation through single electron transfer (SET),[9–12] providing the desired difference in electron density. This strategy can be classified as a redox umpolung.[13] Product formation and back electron transfer (BET) completes the cycle of electron transfer catalysis (ETC). The application of ETC to pericyclic reactions was first

Carbon-Centered Free Radicals and Radical Cations, Edited by Malcolm D. E. Forbes
Copyright © 2010 John Wiley & Sons, Inc.

suggested by Woodward in 1942,[14] but was not applied until Ledwith's pioneering work 30 years later.[15]

4.2   MECHANISM AND THE ORIGIN OF THE RATE ACCELERATION

Mechanistically, radical cation Diels–Alder reactions are generally considered to be stepwise reactions,[16–18] as demonstrated by experimental[19,20] and computational[21–35] studies. The stepwise mechanism is the result of removing an electron from the HOMO (or, less commonly, the addition of an electron to the antibonding orbital in the case of radical anions), weakening the bond and making the bond easier to break. At the same time, the concerted pathway becomes energetically less favorable because the concerted transition state is no longer subject to aromatic stabilization on account of an odd number of electrons. Furthermore, the symmetry preserving pathway is subject to Jahn–Teller distortions that increase the activation energy even more.[28,36] The overall result is an acceleration of slow or symmetry forbidden cycloadditions,[37–39] electrocyclic reactions,[40–45] sigmatropic shifts,[46–48] cycloreversions[49,50] by up to 13 orders of magnitude.

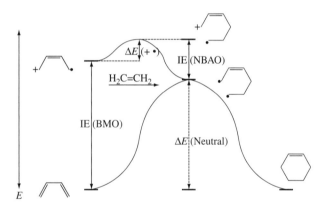

SCHEME 4.1 Thermodynamic representation of radical cation cycloaddition reaction acceleration.

An explanation for the increased reaction rates observed for radical cation Diels–Alder reactions has been proposed by Bauld and is shown in Scheme 4.1.[51] The same principle is applicable for other radical cation pericyclic reactions. The butadiene cation radical is formed by removing an electron from the HOMO of the diene, which is according to Koopman's theorem equal to the ionization energy (IE). This radical cation reacts with neutral ethylene forming the distonic, acyclic radical cation. The energy gained from BET to the acyclic diradical is of equal magnitude to the ionization energy required to remove an electron from the nonbonding 2p atomic orbital of a carbon in a hypothetical 1,6-diradical. Because the electron is removed from a nonbonding orbital, the IE required is much smaller than the energy initially required to remove an electron from the π-bonding molecular orbital of the butadiene.

As a result, the transition state of the radical cation cycloaddition is much more reactant-like and the activation energy will be much smaller than in case of the cycloadditions between two neutral reactants. Finally, the thermodynamic driving force for this radical cation/neutral addition is equal to the large difference in ionization potentials between the bonded and the nonbonded orbitals.

4.3 SELECTIVITY IN RADICAL CATION CYCLOADDITIONS

The Bauld analysis highlights the fact that radical cations are highly reactive species. Surprisingly, they have at the same time been empirically found to be highly selective. In direct comparisons with neutral reactions, the ETC cycloaddition reaction was found to be more selective in both $[4 + 2]^{52,53}$ and $[2 + 2]^{54}$ reactions, which is in contrast to the widely applicable reactivity–selectivity principle. Furthermore, complete facial diastereoselectivity and a high *endo/exo* ratio were observed when chiral dienes were used.[55–57] This remarkable combination of high reactivity and high selectivity has great promise for a variety of processes, but applications remain rare. One possible reason is the lack of predictability of selectivity due to various competing mechanisms as well as the generally low familiarity of many organic chemists with the selectivity determining factors. This review complements the previous, mechanistically oriented reviews[9–11,17,18,58–60] in addressing some of these questions by providing an overview of patterns governing the selectivity that has been observed in ET catalyzed cycloadditions. Using selected literature examples rather than a comprehensive overview, we will discuss the experimental findings and provide an overview of the factors determining them.

We will begin by considering the chemoselectivity of various systems. A cycloaddition reaction between two substrates can in principle lead either to the homodimerization of substrate or to the formation of a cross-adduct. This competition, which we define as chemoselectivity, between the commonly desired cross-adducts and the homodimer side products must be assessed before discussing any higher order selectivity. Once the factors governing the chemoselectivity have been outlined, the regioselectivity, that is, the position of the initial bond formation, will be analyzed. Due to the stepwise nature of the radical cation cycloadditions, the position of the initial bond formation has a significant effect on peri- and stereoselectivity. Periselectivity refers to the position of the ring closure to give either the $[4 + 2]$ Diels–Alder or the $[2 + 2]$ cyclobutane adducts. Due to the breakdown of the Woodward–Hoffmann rules for open shell radical cations, both $[2 + 2]$ and $[4 + 2]$ products have been observed in a number of cases. Finally, a discussion of the stereoselectivity will consider both the *endo/exo* diastereoselectivity as well as the remarkable diastereoselectivity observed with chiral reaction partners.

Several factors that are less important in the case of neutral cycloadditions have to be considered because they can potentially be exploited for greater selectivity. As a result of the charged nature of the radical cations, solvents and the solvation state of the ions as contact ion pairs, solvent separated ions pairs, or free ion pairs involved in the reaction now play a much greater role than for the cycloadditions of a neutral substrate, which for the most part show only a weak solvent dependence. The induction of the reaction by electron transfer also plays a major role in determining both the identity and the concentration of the reactive species and because back electron transfer is a common competing reaction at any point during the reaction. Scheme 4.2 shows some common thermal one-electron oxidants (**2** and **4**) and sensitizers for photoinduced electron transfer (**1**, **3**, **5–7**). In comparison, the formation of the radical cations through electrochemical methods is less common.[61] It is noteworthy that following oxidation of the reacting species, **1–3** will be reduced to neutral while **4–7** will exist in radical anion form leading, to the possibility of an ion pair complex. Together with the substrate concentration and the solvent, this dictates the rate of back electron transfer that is a crucial element in ETC reactions.

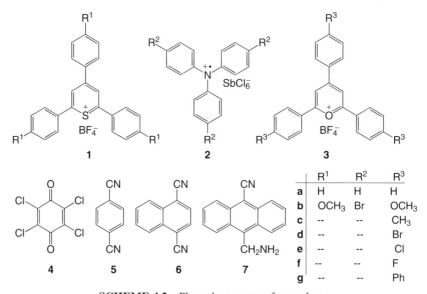

	R^1	R^2	R^3
a	H	H	H
b	OCH$_3$	Br	OCH$_3$
c	--	--	CH$_3$
d	--	--	Br
e	--	--	Cl
f	--	--	F
g	--	--	Ph

SCHEME 4.2   Photoelectron transfer catalysts.

4.4   CHEMOSELECTIVITY

Chemoselectivity is generally defined as the preferential outcome of one instance of a generalized reaction over a set of other possible reactions. In the context of this review, we will narrow this definition to a competition between homodimerizations and cross-additions of starting materials.

4.4.1   Effect of Dienophile Substituents on Chemoselectivity

Most dienes studied in radical cation Diels–Alder reactions are derivatives of 1,3-cyclohexadiene **8**. As shown in Scheme 4.3, the dimerization of **8a** can already lead to four different products **9a–12a** that are commonly observed in many crossed radical cation Diels–Alder reactions. Reacting cyclohexadiene **8a** with styrenes **13a–c** thus results in the unavoidable competition between the homodimerization of **8a** (Scheme 4.3) and formation of the cross-adduct as shown in Scheme 4.4.

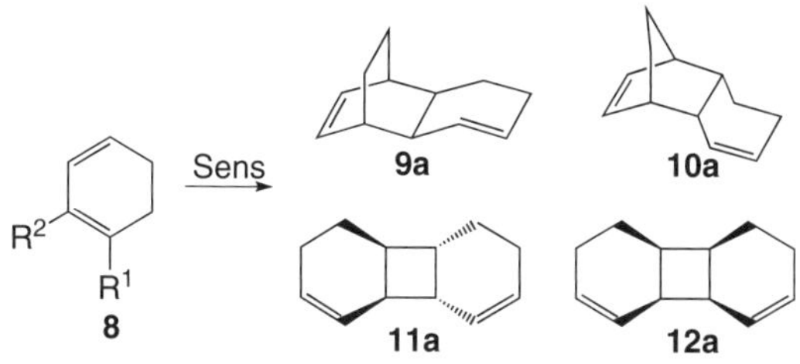

a, $R^1=R^2=H$ c, $R^1=OAc; R^2=H$
b, $R^1=OCH_3; R^2=H$ d, $R^1=H; R^2=OAc$

SCHEME 4.3   Observed products associated with the dimerization of dienes **8a–d**.

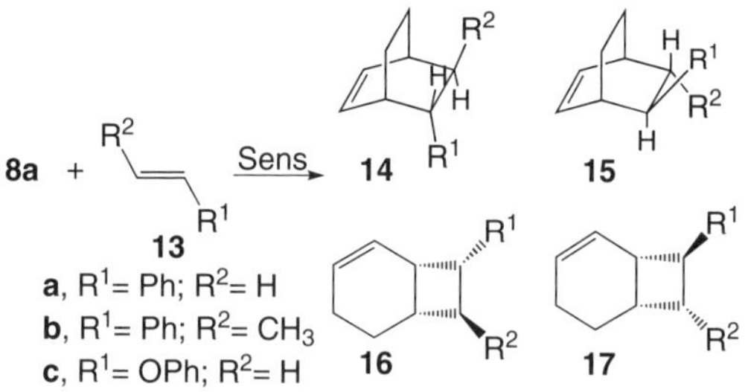

a, $R^1=Ph; R^2=H$
b, $R^1=Ph; R^2=CH_3$
c, $R^1=OPh; R^2=H$

SCHEME 4.4   Cycloadditions of various dienophiles with **8a**.

As shown in Table 4.1, formation of the mixed adduct is favored over homo-dimerization of **8a** with the simple styrene **13a,** but this selectivity is inverted for the case of the more bulky dienophile *trans*-β-methylstyrene **13b**, presumably due to steric effects. Although the overall reaction is highly exothermic on the radical cation surface, the reaction is not insensitive to steric effects. Chemoselectivity in the radical cation cycloaddition is largely a consequence of a substrate's ability to stabilize the radical cation of the oxidized species through the formation of a weakly bound ion-molecule complex.[28] Such complexes have been known for a long time in gas-phase

TABLE 4.1   Chemoselectivity in the Reaction of 8a with 13a-c

Entry	Reagent	Sensitizer	Solvent	Yield (%)	Mixed Product/**8a** Dimer
1[37]	**13a**	**3a**	CH_2Cl_2	49.0	3.46/1
2[37]	**13b**	**3a**	CH_2Cl_2	78.0	0.70/1
4[62]	**13c**	**3a**	CH_3CN	32.3	1.6/1
5[62]	**13c**	**6**	CH_3CN	36.1	3.40/1
6[62]	**13c**	**6**	CH_2Cl_2	27.0	4.40/1
7[62]	**13c**	**6**	THF	3.7	1.06/1
8[62]	**13c**	**6**	Et_2O	13.6	0.94/1

studies of charged species and were studied computationally for radical cation Diels–Alder reactions.[23] Ion-molecule complex formation greatly increases the statistical probability for a successful cycloaddition resulting in either cross-products or homodimers. Steric effects that favor some ion-molecule complexes over others thus become important. As a result, formation of the sterically less constrained ion-molecule complex **8a**$^+$/**8a** is apparently more favorable than forming the less stable **8a**$^+$/**13c** ion-molecule complex.

4.4.2 Effect of Sensitizers and Solvents on Chemoselectivity

In addition to the steric issues in chemoselectivity, which are intrinsic to the reaction of a given set of substrates, solvent and sensitizer effects are of importance for a direct control on selectivity because they can be controlled. The reaction of **8a** and **13c** catalyzed by sensitizer **6** shows an overall increase in cross-product formation with increasing solvent polarity. Entries 4 and 5 in Table 4.1 demonstrate that use of **3a** reduces the formation of cross-products in polar solvents. This difference can be rationalized by the fact that after ET by **6**, a strongly complexed ion pair is formed, whereas in the case of **3a**, the sensitizer is neutral and can dissociate more easily from the radical cation formed.

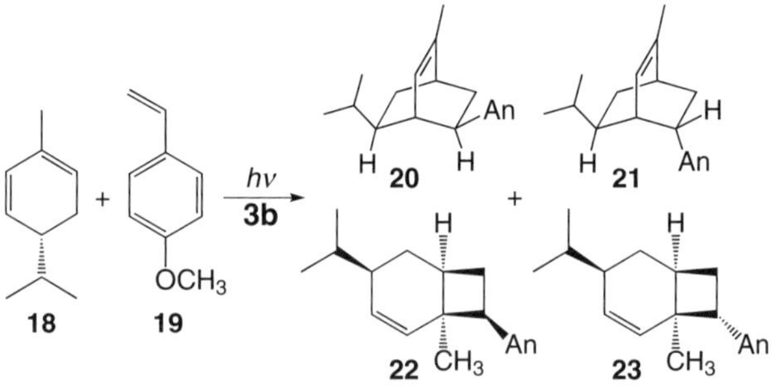

SCHEME 4.5   Cycloadducts of ET catalyzed reaction between **18** and **19**.

These results are in agreement with the ones from a more recent study of the cycloaddition of **18** and **19** catalyzed by **3a**. Ref. 63 Entries 4–7 in Table 4.2 show the loss of cross-product formation with increasing solvent polarity when using **13b** as a sensitizer. Although in direct contrast with results in Table 4.1 catalyzed by **6**, the results of Steckhan et al. using **3a** shows trends in agreement with the results in Table 4.2.[64] Switching catalysts to **7**, which exists as a radical anion upon oxidation of **18**, confirms the inversion of chemoselectivity similar to the one shown in Table 4.1. These differences between sensitizers that are after electron transfer neutral or anionic, respectively, have been rationalized with ion-molecule complex formation.[63,64] Using **3b**$^+$, which is neutral following the oxidation of **18**, is unlikely to complex with **18**$^+$. Weakly polar solvents are unable to stabilize the radical cation, forming an ion-molecule complex with **19** over the more electron-rich diene **18** on account of its steric bias toward **19**. Formation of such ion-molecule complexes is less favored in solvents

TABLE 4.2 Chemoselectivity in the Reaction of 18 and 19[63]

Entry	Concentration (M)	Sensitizer	Solvent	Cross-Product/18 Dimer
1	1.51	3b	CH_2Cl_2	0.33/1
2	0.50	3b	CH_2Cl_2	0.43/1
3	0.15	3b	CH_2Cl_2	2.67/1
4	0.05	3b	CH_2Cl_2	8.12/1
5	0.15	3b	Acetone	2.19/1
6	0.15	3b	CH_3CN	0.99/1
7	0.15	3b	DMSO	0.67/1
8	0.15	7	CH_2Cl_2	4.27/1
9	0.15	7	CH_3CN	8.44/1

of higher polarity due to better solvation of the charged species, leading to the observed decrease in cross-adduct formation. When utilizing neutral sensitizers such as **6** or **7**, the ion-molecule complex is a diene/sensitizer rather than diene/dienophile complex. As the results indicate, solvation in more polar solvents of $7^-/18^+$ frees 18^+ to interact with less sterically hindered **19**, increasing cross-adduct formation.

4.4.3 Effect of Concentrations on Chemoselectivity

Results shown in Scheme 4.5 and Table 4.2 demonstrate a preferred formation of cross-adducts at lower concentrations.[63] Lower concentrations allow radical cations to equilibrate between $18^+/18$ and $18^+/19$. Due to the unfavorable steric bulk discussed earlier between $18^+/18$, a predominant population of $18^+/19$ is present in solution leading to a higher chemoselectivity.

4.4.4 Effect of Electron-Rich Dienophiles on Chemoselectivity

In the cases discussed so far, the charged species has been the diene radical cation due to its generally lower oxidation potential. Consequently, chemoselectivity has a greater dependence on the substituents of the diene rather than the dienophile. Results summarized in Scheme 4.6 and Table 4.3 for the electron-rich dienophile, show an approximately 1 : 1 selectivity when using **8a**.[62] Placing substituents at the 1-position as in **8b,c** completely inhibits the formation of cross-adducts, while substitution at the 2-position of the dienes leads to preferential formation of the cross-adduct.

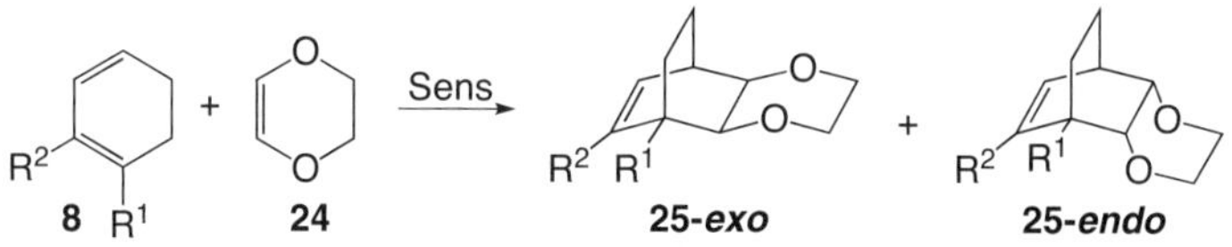

8a, 25a: $R^1=R^2=$ H   **8c, 25c**: $R^1=$ OAc; $R^2=$ H
8b, 25b: $R^1=$ OCH$_3$; $R^2=$H   **8d, 25d**: $R^1=$ H; $R^2=$ OAc

SCHEME 4.6 Diene substituent effects on ETC cycloadditions.

TABLE 4.3   Substituent Effects Upon Reaction of Dienes 8a–d with 24

Entry	Diene	Yield (%)	Cross-Product 25/8a–d Dimer
1	8a	45	0.8/1
2	8b	94	0/1
3	8c	<5	0/1
4	8d	37	4.3/1

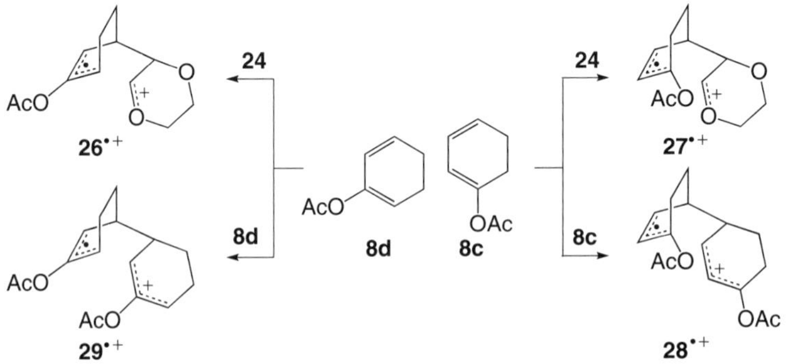

SCHEME 4.7   Comparison of distonic intermediates of cross-adducts and homodimers.

An examination of effects from substituent position is shown in Scheme 4.7.[62] The stability of the singly linked intermediate of the stepwise cycloadditions is key in understanding the selectivity. Comparing the singly linked intermediates of the cross-products 26^+ and 27^+, it is clear that 27^+ will be disfavored due to steric bulk at the bridgehead position and the lack of stabilization of the cation by the acetoxy group at the 2-position of the allyl system. As a result, the intermediate 26^+, where the positive charge can be stabilized as an oxonium ion, is favored. Conversely, intermediate 28^+ will provide greater stabilization of the cation with the acetoxy group at the 1-position rather than the 2-position as in the less stable intermediate 29^+. This 1-position stabilization effect on diene dimerization becomes more pronounced with the use of a methoxy moiety, giving an exclusive 94% yield of homodimers.

4.5 REGIOSELECTIVITY

The presence of the singly linked intermediate resulting from the stepwise nature of the ET catalyzed reaction determines the regioselectivity of the reaction through the radical ion stability following initial bond formation. This has been studied in detail for the reaction of electron-rich indoles with dienes including 8b, 8c, 8d, and 18, via ETC.[20,55,56] Scheme 4.8 shows the singly linked intermediates leading to the experimentally observed regioisomers. This reaction has displayed complete regioselectivity. Dienes with substituents on C-1 led to the exclusive formation of 1-substituted adducts, 33. Dienes with substituents on C-2, on the other hand, resulted in the formation of adducts substituents in 3-position, 34. This observation can be explained by the fact that initial bond formation to

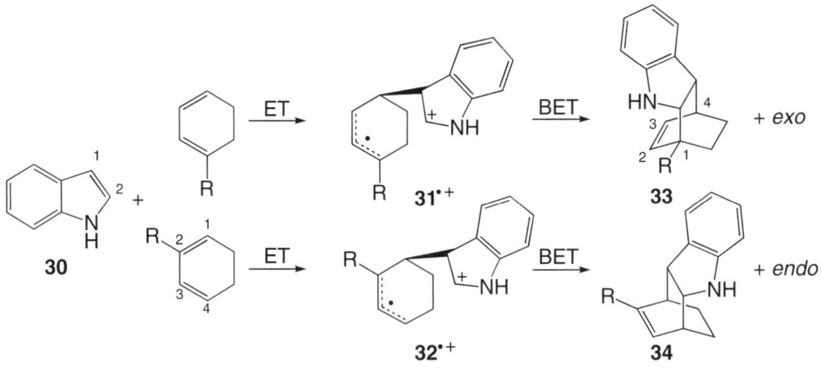

SCHEME 4.8 Indoles reacted with dienes demonstrate the high regioselectivity associated with ETC reactions.

C-1 or C-4 of the diene allow for allylic conjugation and stabilization by substituents at C-1 or C-2 as shown in **31**$^+$ and **32**$^+$, respectively.

Later, a more detailed and quantitative rationalization was given for the selectivity observed in ET catalyzed cycloadditions with indoles and various exocyclic dienes that provided complete regioselectivity. Distonic radical cation intermediates leading to all regioisomers were studied computationally. Potential energy surfaces showed that a reaction path leading to an intermediate with lower energy proceeds with a lower activation barrier, whereas the pathway leading to the high-energy intermediates involved higher energy barriers. Intermediates leading to the observed adducts were more than 10 kcal/mol lower in energy than the other possibilities.[20,65] This deep potential well achieved by the greatest radical cation stabilization is the driving force for the heightened regioselectivity. Similar computational studies were done with methyl-substituted butadienes at different positions undergoing Diels–Alder reactions with ethylene. Regioselectivity was based on the formation of the singly linked intermediate most capable of stabilizing the radical cation species.[66] These trends were also found in substituents effect studies for a number of other pericyclic reactions of radical cations.[67,68]

4.6   PERISELECTIVITY

One of the most interesting aspects of radical cation cycloadditions is that the same substrates can lead to both [4 + 2] and [2 + 2] products under similar conditions. Woodward–Hoffmann controlled, neutral cycloadditions provide only one set of products depending on whether the reaction proceeds thermally or photochemically. Differentiating the forces dictating the periselectivity, that is, the formation of [4 + 2] cyclohexene adducts versus [2 + 2] cyclobutane adducts, is of great practical value and considerable effort has been devoted to such studies. In analogy to the analysis of chemoselectivity and regioselectivity, much of the understanding of periselectivity is derived from the analysis of the singly linked radical cation intermediate. The

TABLE 4.4   **Solvent Effects on Addition of 8a to 13b[69]**

Solvent (ε)	Yield (%)		(14b + 15b)/(16b + 17b)
	14b + 15b	**16b + 17b**	
CH$_3$CN (37.5)	27.5	0.4	68.8/1
CH$_2$Cl$_2$ (9.1)	18.8	3.2	5.9/1
THF (7.4)	0.9	1	0.9/1
Benzene (2.3)	3.3	3.3	1.0/1
1,4-Dioxane (2.2)	4.9	5.3	0.9/1

periselectivity of radical cations has been found to be both solvent and concentration dependent.

4.6.1   Effects of Solvent and Concentration on Periselectivity

Reanalysis of the results shown in Scheme 4.4[63] from the viewpoint of periselectivity demonstrates the influence of external factors such as solvents and sensitizers. Table 4.4 illustrates the trends toward increased [4 + 2] cycloaddition at higher solvent polarity, while [2 + 2] cyclizations are favored at lower solvent polarities.[69]

Kinetic studies of the dimerization of 1,1-diphenylethylene (Scheme 4.9), suggest that different radical ion pairs lead to the observed periselectivity.[70] After oxidation of **52** by **7**, the singly linked intermediate **53**$^+$ is formed. Formation of **56** and **57** results from cage escape of **32**$^+$ leading to a solvent separated ion pair (SSIP). Cyclobutanation to **55** is hypothesized to be the result of an ion-molecule complex formation with sensitizer **7**$^-$ affording a close ion pair (CIP). With a greater probability for BET, formation of a 1,4 biradical and rapid [2 + 2] cyclization is expected. The use of neutral sensitizers produced a 0.65 : 1 of **55** : (**56** + **57**) ratio, while positively charged sensitizers lacking Coulombic attraction to the radical cation following diene oxidation eliminated all formation of [2 + 2] products.

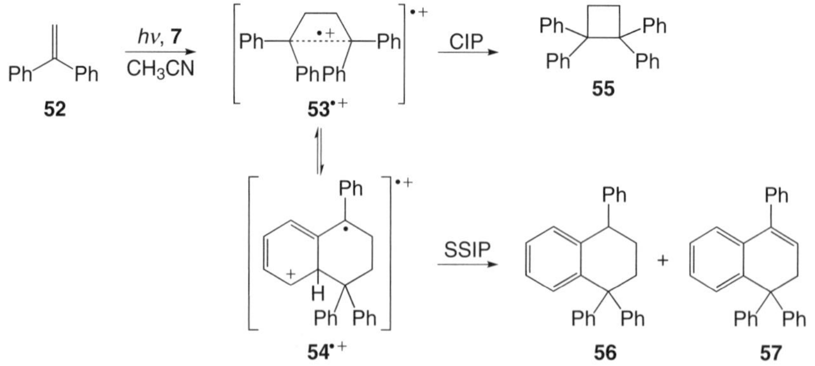

SCHEME 4.9   Different mechanistic pathways associated with periselectivity.

Mechanistic studies of styrene dimerizations have been performed in attempt to elucidate the form of the intermediate, that is, whether it is an acyclic, singly linked intermediate or a "long bond" cyclobutane. Farid suggested an acyclic radical cation intermediate, analogous to classic styrene dimerizations.[71] Later on, a concerted but nonsynchronous [2 + 1] cycloaddition involving a "long bond intermediate" **53**[+] was proposed.[9,51,72,73] Laser flash studies of ET dimerization reactions of vinylanisole and other arylethene derivatives concluded that the rate determining step in the formation of **54**[+] was the cleavage of the long bond intermediate **53**[+].[72] This model was also used to explain the experimentally observed concentration dependence of periselectivity, that is, the formation of [4 + 2] adducts at low substrate concentrations and [2 + 2] products at high substrate concentrations. In the later case, collision of **53**[+] with a substrate would lead to BET and formation of the biradical, which rapidly cyclizes to the cyclobutane. In the case of lower substrate concentrations, collisions and BET are less frequent, allowing the required time for the rearrangement to **54**[+]. However, the distinction between singly linked and "long bond" intermediates might not be clear. Recent computational studies demonstrated that the long bond intermediate is rapidly equilibrating with the acyclic intermediate and that the assignment of the absorption observed in laser flash studies to the "long bond" intermediate might be inconclusive.[74]

Returning to the effect of solvents, two reasons for the frequently observed dominance of [2 + 2] cycloadditions in nonpolar solvents can be proposed. Less polar solvents will be less effective in stabilizing the radical cation species, thus giving rise to formation of CIP with the reduced sensitizer and more likely and more exothermic BET. Additionally, forward electron transfer might not favorable in all cases. The thermodynamically disfavored ET in nonpolar solvents increases the probability of a direct photoexcitation of the substrate, leading to the cyclobutane adduct via the normal Woodward–Hoffman dictated excited-state process.[75–77]

4.6.2 Effect of Diene/Dienophile Redox Potentials on Periselectivity

The dependence of the periselectivity on the relative redox potentials of the reaction partners has been studied for the case of the electron-rich dienophiles **36a–e** and **13b** with the acyclic diene **35**, shown in Scheme 4.10. While the corresponding neutral reaction is not observed, ET efficiently catalyzes this cycloaddition.[51,78] Table 4.5 shows that [2 + 2] selectivity dominates the reaction when incorporating dienophiles with oxidation potentials higher than that of **35** (1.36 eV). In these cases, the reaction will proceed via oxidation of **35**. [4 + 2] addition to acyclic dienes, which

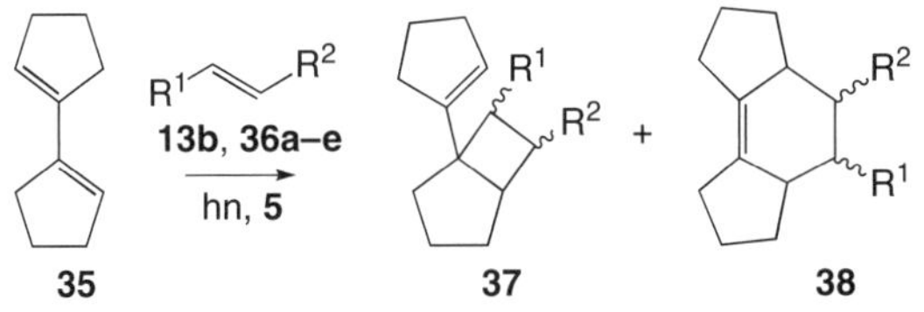

SCHEME 4.10   ET catalyzed additions of dienophiles to acyclic **35**.

TABLE 4.5   Periselectivity in the Reaction of 35 with Various Dienophiles

Dienophile	R^1	R^2	E_{ox} (eV)	[4 + 2] : [2 + 2]
36a[78]	NAcMe	H	1.55	0 : 100
36b[83]	OEt	H	1.60	2 : 98
36d[84]	SPh-p-Br	H	a	10 : 90
13b[83]	OPh	H	1.62	18 : 82
36c[83]	SPh	H	1.42	31 : 69
36e[84]	SPh-p-CH$_2$CH$_3$	H	a	54 : 46
36f[82]	p-An	CH$_3$	1.11	100 : 0

[a]Potentials not given.

predominantly exist in the *trans* conformation, is not observed. Equilibration of the radical cation diene to the s-*cis* conformation is very slow due to the delocalized charge. The high-energy transition state required for bond rotation of a conjugated radical cation species necessary for the [4 + 2] cyclization has been documented.[68,79–81] Therefore, [2 + 2] adducts are the observed products following BET and closure of the biradical species. Conversely, lowering the oxidation potential below that of **35** leads to the [4 + 2] cyclization as shown for the case of **36f**.[82] With an olefin radical cation intermediate, the isomerization to the *trans*-diene is possible, forming the more thermodynamically stable [4 + 2] adduct as the major product.

Although oxidation potentials do not precisely correlate with the observed trend in selectivity, vinyl phenyl sulfide derivatives incorporating p-bromo and p-ethyl moieties were shown to behave as predicted by the oxidation potentials.[84] Absolute oxidation potentials were not given, presumably due to the irreversibility of the redox reactions in cyclic voltammetry, but qualitative estimates show at the increase and decrease in oxidation potentials has a direct effect on the selectivity. p-Bromophenyl vinyl sulfide increases the probability of forming a diene radical cation over its own oxidation, resulting in a higher [2 + 2] selectivity than **36c**. When the oxidation potential is lowered with a p-ethyl moiety, the probability of oxidizing either diene or dienophile becomes equal and an approximately 1 : 1 ratio of [4 + 2] : [2 + 2] adducts is observed. As evidence for the dependence of the periselectivity on the *cis/trans* conformational ratio, the rigid s-*cis* cyclic diene **8a** formed DA adducts exclusively when reacted with **36c** and **13b**.[85] It is noteworthy that **36a** continued to react in a [2 + 2] fashion even with **8a**.[86] This has been tentatively accredited to the highly nucleophilic nature of the vinyl amine .[78]

4.6.3   Substituent and Steric Effects on Periselectivity

A surprising steric sensitivity is frequently observed for these radical cation reactions. As was shown previously during the discussion of chemoselectivity, the variable positioning of bulky substituents has effects on periselectivity as well. DFT calculations on the influence of diene substitutions for the neutral reaction have demonstrated a behavior similar to that of the radical cation reaction.[87] Although

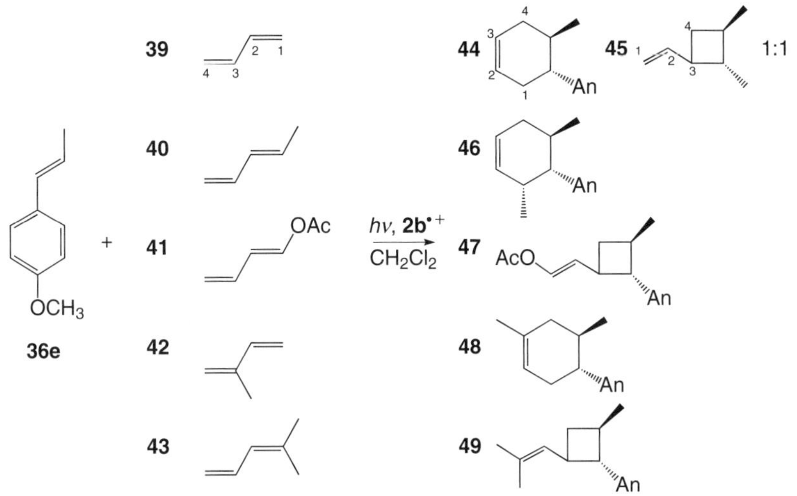

SCHEME 4.11 Acyclic diene substituent effects on periselectivity.

electron-rich (E)-anethole (**36e**) reacts primarily in a [4 + 2] fashion with acyclic **35**, this exclusive periselectivity does not hold true when substituents are introduced at different positions Scheme 4.11.[82] Simple 1,3-butadiene (**39**) was used as a control, providing equal amounts of [4 + 2] and [2 + 2] adducts. Addition of a methyl group to the C-1-position results solely in DA adducts. Scheme 4.12 illustrates the driving force for a [4 + 2] cyclization that is brought about from the stabilizing effect of the electron donating methyl moiety. Radical character localization is reduced at C-3 because of the C-1 methyl that disfavors [2 + 2] addition, as implied by **50**$^+$.

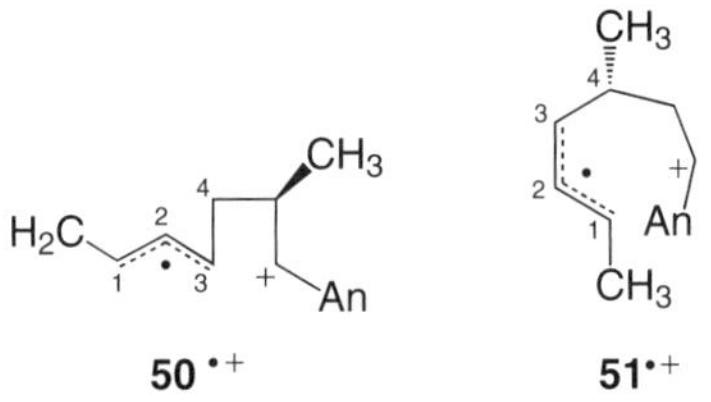

SCHEME 4.12 Acyclic intermediates leading to [2 + 2] (**50**$^+$) and [4 + 2] (**51**$^+$) adducts.

Diene **41** was used to further test this hypothesis. By adding an electron-withdrawing group to the critical C-1-position, one would expect a decrease in [4 + 2] additions on account of destabilization of a positive charge at C-1. A methyl group was added to C-3, depicted as diene **42**, to stabilize intermediate **50**$^+$ and favors [2 + 2] cyclization. The [4 + 2] additions were observed exclusively on account of sterically induced isomerization to the E isomer. Finally, **43** added to **36e** yielded only the [2 + 2] product due to steric bulk at the C-1-position. The (Z) methyl substituent impedes rotation to the s-*cis* diene that would be conducive to [4 + 2] product formation.

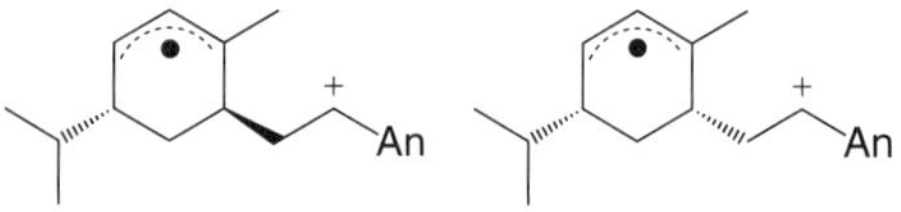

SCHEME 4.13   Singly linked intermediates of ET catalyzed reaction between **18** and **19**.

Another implication of steric effects on the periselectivity was highlighted in a study of the chiral diene phellandrene **18** shown in Scheme 4.5.[63] Addition of **19** can occur on either face of **18**, leading to the to singly linked intermediates shown in Scheme 4.13. Attack of the anisole dienophile on the face, *trans* to the isopropyl group **58**$^{+}$, allows for cyclization to the [4 + 2] products **20** or **21**. In contrast, addition *cis* to the isopropyl group forms **59**$^{+}$, which cannot proceed to the [4 + 2] cyclization on account of steric bulk of the isopropyl group. Consequently, intermediate **59**$^{+}$ can either equilibrate to **58**$^{+}$ by breaking the already formed carbon–carbon bond in the singly linked intermediate or undergo a BET and collapse to cyclobutane products **22** and **23**. Calculations at the B3LYP/6-31G* level of theory suggested that diradical collapse to the [2 + 2] is more likely than the cyclizations of the radical cation to form a cyclobutane intermediate prior to BET. Thus, the facial selectivity (decreased by solvation of the ion-molecule complex in more polar solvents) predetermines the [2 + 2] versus [4 + 2] mode of cyclizations.

4.6.4   Quantifying Periselectivity Through Ion Pair Association

Work by Steckhan attempted to quantify the periselectivity induced by different sensitizers using the Weller equation (Eq. 4.1),[88] as estimation to the degree and direction of charge transfer.[89,90]

$$\Delta G_{ET}(A^{-}D^{+}) = E_{1/2}^{ox}(D) - E_{1/2}^{Red}(A) - \Delta E_{excit} + \Delta E_{Coul} \tag{4.1}$$

where $E_{1/2}^{ox}(D)$ is oxidation potential of donor, $E_{1/2}^{Red}(A)$ is reduction potential of acceptor, ΔE_{excit} is the excitation energy, and ΔE_{Coul} is the Coulombic interaction (incorporates solvent with ε).

The results for cross-cycloadditions between **8a** and **13a** using various pyrylium and thiopyrylium salts are summarized in Table 4.6.[64] The sensitizer **3b**$^{+}$ induces the most periselective reaction and has the lowest oxidation potential although this generalization does not hold true for the remaining sensitizers. Attempting to compute free enthalpies of formation for radical ion pairs demonstrated an unexpected inverse correlation: the more negative the value, the greater the formation of the [4 + 2] product. The most selective sensitizer, salt **3b**$^{+}$, has the highest $\Delta G_{ET}(\textbf{3b/8a}^{+})$ value, indicating that the use of simple relationships such as Equation 4.1 does not fully represent the complexity of the system.

TABLE 4.6   Effect of Sensitizer Variation on Reaction Periselecitivity Between 8a and 13a in CH$_2$Cl$_2$

Sensitizer	Sensitizer E_{red} (eV)	ΔG_{ET} (3$^{\bullet}$/8a$^{\bullet +}$) (eV)	Yield (%) 14 + 15	16 + 17	[4 + 2]: [2 + 2]	14:15
3b	−0.36	−0.29	10.2	—	100:0	24.3:1
3c	−0.28	−0.58	8.3	0.8	10.4:1	14.0:1
1b	−0.19	−0.35	8.2	1.1	7.5:1	15.1:1
3g	−0.18	−0.38	7.2	4.9	1.5:1	8.8:1
3a	−0.13	−0.84	7.2	1.6	4.5:1	9.0:1
3f	−0.10	−0.84	5.8	3.9	1.5:1	19.8:1
1a	−0.05	−0.90	7.5	1.2	6.3:1	11.7:1
3e	−0.05	−0.81	5.1	3.6	1.4:1	16.0:1
3d	−0.03	−0.80	7.4	5.3	1.4:1	11.5:1

4.7   *ENDO/EXO* SELECTIVITY

The relative orientation of the two reaction partners in cycloadditions leads to different *endo* and *exo* diastereomers and is therefore a key event that has been studied extensively in both neutral and ETC cycloadditions. The study of the *endo/exo* selectivity not only is crucial for the stereochemical outcome of the reaction but also has yielded important mechanistic insight into both types of reactions. In the context of this chapter, only a few illustrative examples will be shown.[9,11,59,84,91]

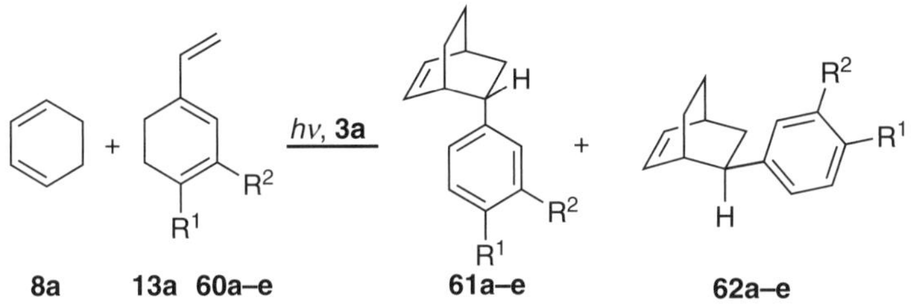

                8a   13a 60a–e   61a–e   62a–e

SCHEME 4.14   *Endo/exo* selectivity when using electron-rich and electron-poor dienophiles.

4.7.1   Effects of Secondary Orbital Interaction and Solvents on *Endo/Exo* Selectivity

In neutral Diels–Alder reactions, the *endo/exo* selectivity is often rationalized through secondary orbital interaction (SOI)[92–94] and differences in the electron density of reacting partners, although this notion is not unchallenged.[95–98] Extending this line of thought to ETC cycloadditions, the reaction shown in Scheme 4.14 allows for an easy control of the electron density difference through appropriate substitution of the

TABLE 4.7   Dienes Reacted with 8a Catalyzed by 3a

Diene	R^1	R^2	E_{pox} (V)	Yield (%)	53 : 54
52a	NO_2	H	1.80	—	—
52b	Cl	H	1.49	7	35 : 1
13a	H	H	1.45	38	10 : 1
52c	CH_3	H	1.35	28	4 : 1
52d	OCH_3	H	0.98	24	2.3 : 1
52e	OCH_3	OCH_3	0.85	6	1.8 : 1

styrene moiety. Table 4.7 shows a good correlation between high oxidation potential of the styrene and high *endo* selectivity. The greater selectivity of **52b** and **13a** has been attributed to their low lying highest occupied molecular orbitals (HOMOs) leading to better SOI with the diene singly occupied molecular orbital (SOMO).[61]

The notion that the electronic structure has a significant effect of the *endo/exo* selectivity through the stabilization of the singly linked intermediate is supported by the observation of a clear solvent dependence of the *endo/exo* selectivity for the reaction of **18** and **19** shown in Scheme 4.15.

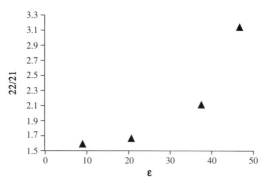

SCHEME 4.15   Solvent dependence of *endo/exo* ratio in the reaction of **18** and **19**.

4.7.2   Effect of Sensitizer on *Endo/Exo* Selectivity

Unlike the correlation between the periselectivity and the physical properties of the sensitizer, the *endo/exo* selectivity does not seem to be dependent on either the measured E_{red} (sensitizer) or the calculated ΔG_{ET} values. An inquiry into the properties of pyrylium salts based on intersystem crossing (ISC) rates was undertaken[99] using the well-studied indole cycloaddition with electron-rich dienes[57] catalyzed by **1a**$^+$ and **3a**$^+$. The efficiency of stereoselection has been hypothesized to be tightly coupled with spin multiplicity.[100] Because of the recognized physiological activity of the carbazole-type products, stereoselective cyclizations of indoles have been of great interest but with limited success under neutral reaction conditions.[101] As an excellent candidate for ET catalysis, the indole dienophiles showed greater *endo* selectivity

(4.6 : 1 versus 1.7 : 1) ET was induced by the latter pyrylium salt $3a^{+.55-57}$ Fluorescence studies established a greater ISC for the thiopyrylium sensitizer $1a^{+}$ ($\Phi_{ISC} = 0.97$ versus 0.47) indicating a larger triplet state population. Photoinduced electron transfer (PET) is thermodynamically feasible to occur with both the triplets as with the singlet excited state of the photocatalyst and quencher. Longer lived triplet states allow reacting partners to diffuse and react, forming *exo* and [2 + 2] adducts. The observation of nearly equal reaction yields (66% and 67%), but lower selectivities for the thiopyrylium salt, implies a higher contribution of the photocatalytic triplet state in these cases.

4.7.3   Ion Pairs and *Endo/Exo* Selectivities

Mattay hypothesized different pathways in forming the *exo* and *endo* products.[62] This hypothesis was the result of observing higher *endo* selectivity at lower concentrations, suggesting that the formation of SSIP would be favored at low substrate concentrations, producing more *endo* adducts. Further evidence was acquired with the addition of $LiClO_4$, EtOH, and 1,2,4-trimethoxy benzene (TMB) to reacting systems. Addition of $LiClO_4$ has shown to reduce formation of the cyclobutane product and enhance *endo/exo* selectivity from 9 : 1 to 15 : 1 in 1,3-cyclohexadiene and styrene ETC reactions. The salt effect observed can be explained by the greater dissociation of the CIP ($3a/8a^{+}$) to ($3b/Li^{+}$). On the other hand, adding radical ion scavengers such as EtOH or TMB more efficiently quenched *endo* adducts, as shown in Table 4.8. The hypothesis that *endo* adducts are formed via SSIP or free radical cation pathways is supported by the fact that TMB and EtOH have the greatest effect on the *endo* adduct. TMB and EtOH can more readily interact with free ion species than those in a CIP. It is noteworthy that $3a^{+}$ does not form any [2 + 2] adducts and has the highest *endo/exo* ratio, presumably due to greater SSIP formation resulting from a lack of Coulomb attraction to the radical cation of the diene.

Work by Turro on dimerizations of **8a** (Scheme 4.3) under variable pressure led to the hypothesis of a comprehensive cause-and-effect scheme for ETC reaction selectivity.[102] Table 4.9 shows an increase in *endo* selectivity at lower concentrations as expected, but a decrease in selectivity with increasing pressure yielding a positive activation volume $\Delta\Delta V^{\neq}$. Reactions in less polar benzene show overall poorer *endo* selectivity in comparison to reactions in CH_3CN. The opposite trend is observed with

TABLE 4.8   Dimerization of 8c

	Yield (%) [Quenched (%)]			
Sensitizer	9c	10c	11c	12c
3a	17.7	3.2	—	—
6	4.6	1.5	1.3	3.8
6/TMB	— [100]	0.5 [67]	0.4 [69]	1.5 [61]
6/EtOH	0.5 [89]	0.6 [60]	0.6 [54]	1.7 [55]

TABLE 4.9   Dimerization of 8a; Ratio 9a/10a in CH₃CN; Sensitized by 6

Pressure (MPa)	1.00 M	0.40 M	0.16 M	0.10 M
0.1	5.7	5.6	9.2	13.0
44	5.6	5.4	9.2	12.6
97	5.4	5.2	8.8	12.0
151	5.3	5.2	8.5	11.7
203	5.1	4.8	8.3	10.9
$\Delta\Delta V^{\neq}$ (cm³/mol)	+ 1.2	+ 1.6	+ 2.4	+ 2.0

higher pressures that increase *endo/exo* ratios. A surprisingly large negative value was found for the activation volume. The higher pressures in nonpolar solvents increase the formation of [4 + 2] adducts (Table 4.10) while the opposite holds true in polar solvents.[103]

Attempts to rationalize periselectivity, stereoselectivity, solvent, and sensitizer effects, ion–ion complexes and ion-molecule complexes have been proposed in a framework as shown in Scheme 4.16 that connects ETC reaction with direct photochemical reactions for the case of the reactants **6** and **8**.[103] A CIP formed following ET in polar solvents can either equilibrate to the SSIP or be captured by another diene forming the [2 + 2] and *exo* products **10a**, **11a**, and **12a** as described in literature.[60,62] Increasing concentration of **8a** would promote capture of the CIP. The experimentally observed pressure effects can then be rationalized with the dissociation of the CIP that requires an increase in volume. Higher pressures favor equilibration to the CIP and decrease *endo* selectivity as SSIP formation is hindered. Nonpolar solvents disfavor an electron transfer event on account of poor radical cation stabilization leading to overall endergonicity as predicted by the Weller equation.[88] An exciplex between donor **8a** and acceptor **6** is formed which can either capture another diene to form a triplex as described by Schuster[104–106] or undergo ISC to the triplet ³**8a***. The triplex path results in formation of the *endo* **9a** while the latter forms the triplet dimers **10a**, **11a**, and **12a**.[107] As demonstrated, increasing concentrations of **8a** enhance triplex formation following capture of the singlet exciplex increasing the *endo/exo* ratio. The increasing pressure disfavors the necessity for an increase in volume associated with dissociation for a triplet sensitized pathway.

TABLE 4.10   Dimerization of 8a; Ratio 9a/10a in benzene; (% 9a + 10a of All Dimers); Sensitized by 6

Pressure (MPa)	1.00 M	0.40 M	0.16 M	0.08 M
0.1	1.62 [56.3]	1.05 [44.1]	0.75 [37.8]	0.59 [36.7]
12	1.74 [58.6]	1.15 [46.0]	0.82 [39.5]	0.64 [37.6]
34	1.86 [61.0]	1.24 [48.0]	0.88 [41.2]	0.70 [39.5]
55	1.93 [62.5]	1.35 [50.1]	0.97 [42.7]	0.74 [41.0]
76	2.02 [64.2]	1.48 [53.5]	0.98 [44.8]	0.83 [42.9]
$\Delta\Delta V^{\neq}$ (cm³/mol)	−11.6	−10.7	−8.7	−11.1

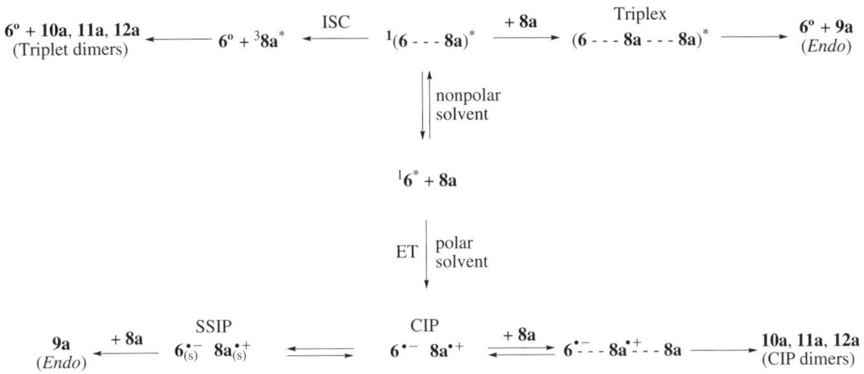

6° + 10a, 11a, 12a ⟵ 6° + ^{3}8a* ⟵^{ISC} ^{1}(6 - - - 8a)* ⟶$^{+8a}$ (6 - - - 8a - - - 8a)* ^{Triplex} ⟶ 6° + 9a
(Triplet dimers)   (Endo)

nonpolar
solvent

^{1}6* + 8a

ET | polar
solvent

9a ⟵$^{+8a}$ 6$_{(s)}^{•-}$ 8a$_{(s)}^{•+}$ ^{SSIP} ⇌ 6$^{•-}$ 8a$^{•+}$ ^{CIP} ⇌$^{+8a}$ 6$^{•-}$- - -8a$^{•+}$- - - 8a ⟶ 10a, 11a, 12a
(Endo)   (CIP dimers)

SCHEME 4.16 Selectivity–reactivity scheme for ETC reactions.

4.8 CONCLUSIONS

Electron transfer catalyzed cycloadditions via radical cations show remarkable selectivity that could be exploited for expanded synthetic methodology. As a complement to the neutral Diels–Alder reaction, ET catalysis fills the void of the electron-rich diene/electron-rich dienophile cyclizations. In attempt to understand the intricate details of the reaction, experimentalists and theorists have uncovered a range of novel factors to control and manipulate these high-energy reactive intermediates. As exemplified by the cases discussed in this contribution, the charged character of the intermediates and the presence of back electron transfer leading to the biradical reaction manifold opens new pathways to control the chemo-, peri-, and stereochemical patterns in these dynamic species.

ACKNOWLEDGMENTS

We gratefully acknowledge the financial support of this research by NSF (CHE-0415344). C. S. S. is a 2008 Pfizer SURF awardee.

REFERENCES

1. Diels, O.; Alder, K. *Justus Liebigs Ann. Chem.* **1928**, *460*, 98–122.
2. Benson, S. C.; Palabrica, C. A.; Snyder, J. K. *J. Org. Chem.* **1987**, *52*, 4610–4614.
3. Davies, D. E.; Gilchrist, T. L. *J. Chem. Soc., Perkin Trans.* **1983**, *1*, 1479–1481.
4. Seitz, G.; Mohr, R. *Chem. Ztg.* **1987**, *111*, 81–82.
5. Wenkert, E.; Moeller, P. D. R.; Piettre, S. R. *J. Am. Chem. Soc.* **1988**, *110*, 7188–7194.
6. Kraus, G. A.; Bougie, D.; Jacobson, R. A.; Su, Y. *J. Org. Chem.* **1989**, *54*, 2425–2428.
7. Fairlamb, I. J. S. *Tetrahedron* **2005**, *61*, 9661–9662.
8. Tsuji, J.; *Transition Metal Reagents and Catalysts: Innovations in Organic Synthesis*; Wiley: Chichester, **2000**.

9. Bauld, N. L. *Tetrahedron* **1989**, *45*, 5307–5363.

10. Fagnoni, M. *Heterocycles* **2003**, *60*, 1921–1958.

11. Schmittel, M.; Burghart, A. *Angew. Chem., Int. Ed.* **1997**, *36*, 2550–2589.

12. Muller, F.; Mattay, J. *Chem. Rev.* **1993**, *93*, 99–117.

13. Seebach, D. *Angew. Chem., Intl. Ed Engl.* **1979**, *18*, 239–258.

14. Woodward, R. B. *J. Am. Chem. Soc.* **1942**, *64*, 3058–3059.

15. Ledwith, A. *Acc. Chem. Res.* **1972**, *5*, 133–139.

16. Saettel, N. J.; Oxgaard, J.; Wiest, O. *Chem. Eur. J.* **2001**, 1429–1439.

17. Wiest, O.; Oxgaard, J.; Saettel, N. J. *Adv. Phys. Org. Chem.* **2003**, *39*, 87–109.

18. Donoghue, P. J.; Wiest, O. *Eur. J. Org. Chem.* **2006**, *12*, 7018–7026.

19. Wiest, O.; Steckhan, E. *Tetrahedron Lett.* **1993**, *34*, 6391–6394.

20. Saettel, N. J.; Wiest, O.; Singleton, D. A.; Meyer, M. P. *J. Am. Chem. Soc.* **2002**, *124*, 11552–11559.

21. Reynolds, D. W.; Lorenz, K. T.; Chiou, H. S.; Bellville, D. J.; Pabon, R. A.; Bauld, N. L. *J. Am. Chem. Soc.* **1987**, *109*, 4960–4968.

22. Bellville, D. J.; Bauld, N. L.; Pabon, R.; Gardner, S. A. *J. Am. Chem. Soc.* **1983**, *105*, 3584–3588.

23. Valley, N. A.; Wiest, O. *J. Org. Chem.* **2007**, *72*, 559–566.

24. O'Neil, L. L.; Wiest, O. *J. Org. Chem.* **2006**, *71*, 8926–8933.

25. Swinarski, D. J.; Wiest, O. *J. Org. Chem.* **2000**, *65*, 6708–6714.

26. Donoghue, P. J.; Wiest, O. *Chem. Eur. J.* **2006**, *12*, 7019–7026.

27. Wiest, O.; Steckhan, E.; Grein, F. *J. Org. Chem.* **1992**, *57*, 4034–4037.

28. Haberl, U.; Steckhan, E., O., W. *J. Am. Chem. Soc.* **1999**, *121*, 6730–6736.

29. Hofmann, M.; Schaefer, H. F. *J. Am. Chem. Soc.* **1999**, *121*, 6719–6729.

30. Hofmann, M.; Schaefer, H. F. *J. Phys. Chem. A* **1999**, *103*, 8895–8905.

31. Hu, H.; Wenthold, P. G. *J. Phys. Chem. A* **2002**, *106*, 10550–10553.

32. Goebbert, D. J.; Liu, X.; Wenthold, P. G. *J. Am. Soc. Mass Spectrom.* **2004**, *15*, 114–120.

33. van der Hart, W. J. *Int. J. Mass. Spectrom.* **2001**, *208*, 119–125.

34. Bouchoux, G.; Salpin, J.-Y.; Yanez, M. *J. Phys. Chem. A* **2004**, *108*, 9853–9862.

35. Bouchoux, G.; Nguyen, M. T.; Salpin, J.-Y. *J. Phys. Chem. A* **2000**, *104*, 5778–5786.

36. Wiest, O. *J. Am. Chem. Soc.* **1997**, *119*, 7513–7519.

37. Mlcoch, J.; Steckhan, E. *Angew. Chem., Int. Ed. Engl.* **1985**, *24*, 412–414.

38. Jia, X. D.; Han, B.; Zhang, W.; Jin, X. L.; Yang, L.; Liu, Z. L. *Synthesis-Stuttgart* **2006**, 2831–2836.

39. Gonzalez-Bejar, M.; Stiriba, S. E.; Domingo, L. R.; Perez-Prieto, J.; Miranda, M. A. *J. Org. Chem.* **2006**, *71*, 6932–6941.

40. Wiest, O. *J. Am. Chem. Soc.* **1997**, *119*, 5713–5719.

41. Kawamura, Y.; Thurnauer, M.; Schuster, G. B. *Tetrahedron* **1986**, *42*, 6195–6200.

42. Pasto, D. J.; Yang, S. H. *J. Org. Chem.* **1989**, *54*, 3544–3549.

43. Sastry, G. N.; Bally, T.; Hrouda, V.; Carsky, P. *J. Am. Chem. Soc.* **1998**, *120*, 9323–9334.

44. Takahashi, Y.; Miyamoto, K.; Sakai, K.; Ikeda, H.; Miyashi, T.; Ito, Y.; Tabohashi, K. *Tetrahedron Lett.* **1996**, *37*, 5547–5550.

45. Barone, V.; Rega, N.; Bally, T.; Sastry, G. N. *J. Phys. Chem. A* **1999**, *103*, 217–219.

46. Oxgaard, J.; Wiest, O. *J. Am. Chem. Soc.* **1999**, *121*, 11531–11537.

47. Dinnocenzo, J. P.; Conlon, D. A. *Tetrahedron Lett.* **1995**, *36*, 7415–7418.

48. Dinnocenzo, J. P.; Conlon, D. A. *J. Am. Chem. Soc.* **1988**, *110*, 2324–2326.

49. Wiest, O. *J. Phys. Chem. A* **1999**, *103*, 7907–7911.

50. Schroeter, K.; Schroder, D.; Schwarz, H.; Reddy, G. D.; Wiest, O.; Carra, C.; Bally, T. *Chem. Eur. J.* **2000**, *6*, 4422–4430.

51. Bauld, N. L. *Chem. Cyclobutanes* **2005**, *1*, 549–587.

52. Rossler, U.; Blechert, S.; Steckhan, E. *Tetrahedron Lett.* **1999**, *40*, 7075–7078.

53. Harirchian, B.; Bauld, N. L. *J. Am. Chem. Soc.* **1989**, *111*, 1826–1828.

54. Galindo, F.; Miranda, M. A. *J. Photochem. Photobiol. A* **1998**, *113*, 155–161.

55. Gieseler, A.; Steckhan, E.; Wiest, O. *Synlett* **1990**, 275–277.

56. Gieseler, A.; Steckhan, E.; Wiest, O.; Knoch, F. *J. Org. Chem.* **1991**, *56*, 1405–1411.

57. Wiest, O.; Steckhan, E. *Angew. Chem., Int. Ed. Engl.* **1993**, *32*, 901–903.

58. Saettel, N. J.; Oxgaard, J.; Wiest, O. *Eur. J. Org. Chem.* **2001**, 1429–1439.

59. Fagnoni, M.; Dondi, D.; Ravelli, D.; Albini, A. *Chem. Rev.* **2007**, *107*, 2725–2756.

60. Miranda, M. A.; Garcia, H. *Chem. Rev.* **1994**, *94*, 1063–1089.

61. Mlcoch, J.; Steckhan, E. *Tetrahedron Lett.* **1987**, *28*, 1081–1084.

62. Mattay, J.; Trampe, G.; Runsink, J. *Chem. Ber. Recl.* **1988**, *121*, 1991–2005.

63. Sevov, C. S.; Wiest, O. *J. Org. Chem.* **2008**, *73*, 7909–7915.

64. Martiny, M.; Steckhan, E.; Esch, T. *Chem. Ber. Recl.* **1993**, *126*, 1671–1682.

65. Haberl, U.; Steckhan, E.; Blechert, S.; Wiest, O. *Chem. Eur. J.* **1999**, *5*, 2859–2865.

66. Valley, N. A.; Wiest, O. *J. Org. Chem.* **2007**, *72*, 559–566.

67. Swinarski, D. J.; Wiest, O. *J. Org. Chem.* **2000**, *65*, 6708–6714.

68. Radosevich, A. T.; Wiest, O. *J. Org. Chem* **2001**, *66*, 5808–5813.

69. Mattay, J.; Vondenhof, M.; Denig, R. *Chem. Ber.* **1989**, *122*, 951–958.

70. Mattes, S. L.; Farid, S. *J. Am. Chem. Soc.* **1986**, *108*, 7356–7361.

71. Flory, P. J. *J. Am. Chem. Soc.* **1937**, *59*, 241–253.

72. Schepp, N. P.; Johnston, L. J. *J. Am. Chem. Soc.* **1994**, *116*, 6895–6903.

73. Schepp, N. P.; Johnston, L. J. *J. Am. Chem. Soc.* **1996**, *118*, 2872–2881.

74. O'Neil, L. L.; Wiest, O. *J. Org. Chem.* **2006**, *71*, 8926–8933.

75. Hoffmann, R.; Woodward, R. B. *J. Am. Chem. Soc.* **1965**, *87*, 2046–2048.

76. Hoffmann, R.; Woodward, R. B. *J. Am. Chem. Soc.* **1965**, *87*, 4388–4389.

77. Woodward, R. B.; Hoffmann, R. *J. Am. Chem. Soc.* **1965**, *87*, 395–397.

78. Bauld, N. L.; Bellville, D. J.; Harirchian, B.; Lorenz, K. T.; Pabon, R. A.; Reynolds, D. W.; Wirth, D. D.; Chiou, H. S.; Marsh, B. K. *Acc. Chem. Res.* **1987**, *20*, 371–378.

79. Oxgaard, J.; Wiest, O. *J. Am. Chem. Soc.* **1999**, *121*, 11531–11537.

80. Oxgaard, J.; Wiest, O. *J. Phys. Chem. A* **2001**, *105*, 8236–8240.

81. Oxgaard, J.; Wiest, O. *J. Phys. Chem A* **2002**, *106*, 3967–3974.

82. Reynolds, D. W.; Bauld, N. L. *Tetrahedron* **1986**, *42*, 6189–6194.

83. Pabon, R. A.; Bellville, D. J.; Bauld, N. L. *J. Am. Chem. Soc.* **1984**, *106*, 2730–2731.

84. Gao, D.; Bauld, N. L. *Electron Transfer Chem.* **2001**, *2*, 133–205.

85. Pabon, R. A.; Bellville, D. J.; Bauld, N. L. *J. Am. Chem. Soc.* **1983**, *105*, 5158–5159.

86. Bauld, N. L.; Harirchian, B.; Reynolds, D. W.; White, J. C. *J. Am. Chem. Soc.* **1988**, *110*, 8111–8117.

87. Robiette, R.; Marchand-Brynaert, J.; Peeters, D. *J. Org. Chem.* **2002**, *67*, 6823–6826.

88. Rehm, D.; Weller, A. *Ber. Bunsen Ges. Phys. Chem.* **1969**, *73*, 834–839.

89. Mattay, J. *Angew. Chem., Int. Ed. Engl.* **1987**, *26*, 825–845.

90. Mattay, J. *Synthesis (Stuttg)* **1989**, 233–252.

91. Schmittel, M.; Wohrle, C. *J. Org. Chem.* **1995**, *60*, 8223–8230.

92. Cohen, T.; Ruffner, R. J.; Shull, D. W.; Daniewski, W. M.; Ottenbrite, R. M.; Alston, P. V. *J. Org. Chem.* **1978**, *43*, 4052–4057.

93. Ohwada, T. *Chem. Rev.* **1999**, *99*, 1337–1375.

94. Salem, L. *J. Am. Chem. Soc.* **1968**, *90*, 543–553.

95. Cordes, M. H. J.; Degala, S.; Berson, J. A. *J. Am. Chem. Soc.* **1994**, *116*, 11161–11162.

96. Imade, M.; Hirao, H.; Omoto, K.; Fujimoto, H. *J. Org. Chem.* **1999**, *64*, 6697–6701.

97. Bakalova, S. M.; Santos, A. G. *J. Org. Chem.* **2004**, *69*, 8475–8481.

98. Bakalova, S. M.; Santos, A. G. *Eur. J. Org. Chem.* **2006**, 1779–1789.

99. Gonzalez-Bejar, M.; Stiriba, S. E.; Miranda, M. A.; Perez-Prietoa, J. *ARKIVOC* **2007**, 344–355.

100. Julliard, M.; Chanon, M. *Chem. Rev.* **1983**, *83*, 425–506.

101. Jones, R. A. *Comprehensive Heterocyclic Chemistry*; Pergamon: New York, **1984**; Vol. 4.

102. Chung, W. S.; Turro, N. J.; Mertes, J.; Mattay, J. *J. Org. Chem.* **1989**, *54*, 4881–4887.

103. Chung, W. S.; Turro, N. J.; Mertes, J.; Mattay, J. *J. Org. Chem.* **1989**, *54*, 4881–4887.

104. Calhoun, G. C.; Schuster, G. B. *J. Am. Chem. Soc.* **1984**, *106*, 6870–6871.

105. Calhoun, G. C.; Schuster, G. B. *J. Am. Chem. Soc.* **1986**, *108*, 8021–8027.

106. Peacock, N. J.; Schuster, G. B. *J. Am. Chem. Soc.* **1983**, *105*, 3632–3638.

107. Valentine, D.; Hammond, G. S.; Turro, N. J. *J. Am. Chem. Soc.* **1964**, *86*, 5202–5208.

5

THE STABILITY OF
CARBON-CENTERED RADICALS

MICHELLE L. COOTE[1], CHING YEH LIN[1], AND HENDRIK ZIPSE[2]

[1]*ARC Centre of Excellence for Free-Radical Chemistry and Biotechnology,
Research School of Chemistry, Australian National University,
Canberra, Australian Capital Territory, Australia*

[2]*Department of Chemistry and Biochemistry, LMU Munich, Munich, Germany*

5.1   INTRODUCTION

Carbon-centered radicals play an important role in organic synthesis, biological chemistry, and polymer chemistry. The radical chemistry observed in these areas can, to a good part, be rationalized by the thermodynamic stability of the open shell species involved. Challenges associated with the experimental determination of homolytic bond dissociation energies (BDEs) have lead to the widespread use of theoretically calculated values. These can be presented either directly as the enthalpy for the C–H bond dissociation reaction described in Equation 5.1, the gas-phase thermodynamic values at the standard state of 298.15K and 1 bar pressure being the most commonly reported values.

$$R_3{'''}\!\!\underset{R_1}{\overset{H}{\diagup}}\!\!C{\diagdown}R_2 \quad \xrightarrow{\Delta H_{rxn}} \quad R_3{'''}\!\!\underset{R_1}{\overset{\bullet}{\diagup}}\!\!C{-}R_2 \; + \; H\bullet \tag{5.1}$$

1   **2**

$$CH_4 \xrightarrow{\Delta H_{rxn}} \bullet CH_3 + H\bullet \qquad (5.2)$$

$$\underset{\mathbf{3}}{} \qquad \underset{\mathbf{4}}{}$$

$$CH_4 + \underset{R_1}{\overset{R_3}{C}}\text{-}R_2 \xrightarrow{\Delta H_{rxn}} \bullet CH_3 + \underset{R_1}{\overset{R_3}{C}}\underset{R_2}{\overset{H}{}} \qquad (5.3)$$

$$\underset{\mathbf{3}}{} \qquad \underset{\mathbf{2}}{} \qquad \underset{\mathbf{4}}{} \qquad \underset{\mathbf{1}}{}$$

Alternatively, the BDE values may be reported relative to the C–H bond dissociation energy in methane (**3**) as the reference.[1] This is quantitatively described in Equation 5.3 as a formal hydrogen transfer process between methane (**3**) and a substituted carbon-centered radical **2**. The reaction enthalpy for this process is often interpreted as the stabilizing influence of substituents R_1, R_2, and R_3 on the radical center and thus referred to as the radical stabilization energy (RSE). When defined as in Equation 5.3, positive values imply a stabilizing influence of the substituents on the radical center. The RSE energies are connected to the BDE values in Equations 5.1and 5.2 as described in Equation 5.4.

$$RSE(\bullet CR_1R_2R_3) = BDE(CH_4) - BDE(HCR_1R_2R_3) \qquad (5.4)$$

It has been pointed out by Zavitsas that RSE values calculated according to Equation 5.3 reflect the influence of substituents R_1, R_2, and R_3 on both the radical and its parent hydrocarbon.[2] This latter effect will be particularly large for all bond dissociation processes, in which the cleaved bond has substantial polar character, and Equation 5.5 has therefore been suggested as an alternative approach for the determination of RSE values.

$$H_3C\text{-}CH_3 + 2\,\underset{R_1}{\overset{R_3}{C}}\text{-}R_2 \xrightarrow{\Delta H_{rxn}} 2\,\bullet CH_3 + \underset{R_1\quad R_3}{\overset{R_3\quad R_1}{R_2}}\overset{}{R_2} \qquad (5.5)$$

$$\underset{\mathbf{5}}{} \qquad \underset{\mathbf{2}}{} \qquad \underset{\mathbf{4}}{} \qquad \underset{\mathbf{6}}{}$$

$$RSE(\bullet CR_1R_2R_3) = 0.5(BDE(CH_3\text{-}CH_3) - BDE(CR_1R_2R_3\text{-}CR_1R_2R_3)) \qquad (5.6)$$

The RSE is calculated here as the difference between the homolytic C–C bond dissociation energy in ethane (**5**) and a symmetric hydrocarbon **6** resulting from dimerization of the substituted radical **2**. By definition the C–C bonds cleaved in this process are unpolarized and, baring some strongly repulsive steric effects in symmetric dimer **6**, the complications in the interpretation of substituent effects are thus avoided. Since two substituted radicals are formed in the process, the reaction enthalpy for the process shown in Equation 5.5 contains the substituent effect on radical stability twice. The actual RSE value is therefore only half of the reaction enthalpy for reaction 5.5 as expressed in Equation 5.6.

Even in the absence of repulsive steric effects, the use of Equation 5.6 is not free from complications as substituents R_1, R_2, and R_3 in the parent dimer **6** may still interact in a through-space fashion (e.g., hydrogen bonding interactions) or in a through-bond fashion (e.g., stereoelectronic effects leading to the *gauche* effect). These latter problems can be minimized by calculating RSE values as the difference between the C–C bond dissociation energy in ethane and in hydrocarbon **7**. This latter closed shell reference compound contains substituents only on one side of the cleaved C–C bond and may thus avoid some of the complications with reference compound **6**. The actual RSE values are then obtained as the difference of the respective C–C BDE values as defined in Equation 5.8.

$$H_3C-CH_3 \; + \; \underset{R_1}{\overset{R_3}{}}\!\!\dot{C}\!-\!R_2 \quad \xrightarrow{\Delta H_{rxn}} \quad \bullet CH_3 \; + \; H_3C\!-\!\!\!\overset{R_1}{\underset{R_3}{\diagdown}}\!\!R_2 \qquad (5.7)$$

$$\quad\; 5 \qquad\qquad 2 \qquad\qquad\qquad 4 \qquad\qquad 7$$

$$RSE(\bullet CR_1R_2R_3) \;=\; (BDE(CH_3-CH_3) \;-\; BDE(CH_3-CR_1R_2R_3)) \qquad (5.8)$$

5.1.1 The Consequences of Different Stability Definitions: How Stable Are Ethyl and Fluoromethyl Radicals?

The differences in RSE values obtained from Equations 5.3, 5.6 and 5.8 will be illustrated here using the ethyl radical ($CH_3CH_2\bullet$, **8**) and the fluoromethyl radical ($FCH_2\bullet$, **9**) as examples. Both systems are not burdened by steric effects and the RSE values can thus be interpreted as the consequences of electronic substituent effects. Also, experimentally measured C–H and C–C bond dissociation energies are available for both systems, allowing for a side-by-side evaluation of experimental and theoretical results (Table 5.1).

The C–H BDE in methane is 18.8 kJ/mol larger than in ethane, implying an equal amount of radical stabilization energy for the ethyl radical according to Equation 5.3.[3] A significantly smaller RSE value of 7.1 kJ/mol is found when comparing the C–C BDEs in ethane and *n*-butane and dividing the difference by half as defined in Equation 5.6. Practically the same value of RSE = 7.1 kJ/mol is found for the ethyl radical when comparing the C–C BDE in ethane and propane as defined in Equation 5.8. Following the arguments proposed by Zavitsas we can interpret the difference between the results obtained from C–H BDEs (as in Eq. 5.3) and those obtained from C–C BDEs (as in Eqs. 5.6 and 5.8) of 11.7 kJ/mol as a reflection of the small, but notable polarity of the $C(sp^3)$–H bond in ethane. Theoretical results obtained from two high-level compound methods (G3(MP2)-RAD[4] and G3B3[5]), designed to reproduce thermochemical data with high accuracy, closely parallel the experimental data and thus lend some credibility to this type of interpretation.

The model proposed by Zavitsas also suggests that RSE values calculated from Equation 5.3 for C–H bond cleavage and Equations 5.6 and 5.8 for C–C bond cleavage will diverge more strongly for systems carrying more electronegative substituents.

TABLE 5.1   RSEs Calculated According to Equations 5.3, 5.6 and 5.8

System	RSE(3)	RSE(6)	RSE(8)
•CH$_3$ (**4**)	0.0	0.0	0.0
•CH$_2$CH$_3$ (**8**)			
Exp.[3]	+ 18.8 ± 1.7	+ 7.1 ± 2.3	+ 7.1 ± 2.9
G3B3	+ 13.8	+ 2.3	+ 2.5
G3(MP2)-RAD	+ 13.5	+ 2.0	+ 2.2
•CH$_2$F (**9**)			
Exp.[3,6]	+ 15.5 ± 4.6	+ 4.6 ± 9.2	− 10.9 ± 9.2
G3B3	+ 13.4	− 8.0	− 13.8
G3(MP2)-RAD	+ 12.8	− 8.2	− 13.9

This is indeed found when RSE values are calculated for the fluoromethyl radical (FCH$_2$•, **9**) that, according to Equation 5.3, is stabilized by RSE(3) = + 15.5 kJ/mol relative to methyl radical. Different results are obtained when using the definitions given in Equations 5.6 and 5.8 with RSE(6) = + 4.6 kJ/mol and RSE (8) = −10.9 kJ/mol). The RSE(3)- and RSE(8)-values obtained for radical **9** from experimental data are closely matched by data from G3B3 and G3(MP2)-RAD calculations, but deviations are somewhat larger for the RSE(6) value for radical **9**, casting some doubt on the accuracy of the experimentally determined C–C BDE in 1,2-difluoroethane of 368.2 ± 8.4 kJ/mol.[6] Aside from these technical problems, it is interesting to note that the defining Equations 5.3 and 5.8 for this particular example only differ in the choice of reference compounds: Equation 5.3 uses the CH$_4$/CH$_3$F pair as a reference for the •CH$_2$F/•CH$_3$ radicals, while Equation 5.8 uses the CH$_3$CH$_3$/CH$_3$CH$_2$F pair. The difference in RSE(3) and RSE(8) for radical **9** of 26.4 kJ/mol is thus solely due to differences in the thermochemical stability of these closed shell reference compounds. We may go ahead at this point to dissect bonding phenomena in the four closed shell reference compounds into individual components to further narrow down the origin of the thermochemical differences, but any such discussion will lead very far away from the original objective of using Equations 5.3, 5.6 and 5.8: the quantification of properties of radicals. Given the fact that hydrogen transfer reactions represent an important part of experimentally observed radical chemistry, we will thus opt for defining radical stability through Equation 5.3 for the remainder of this chapter, keeping in mind the minefield connected to the interpretation of these data as manifestations of substituent effects.

5.2   THEORETICAL METHODS

Theoretical methods used for the calculation of BDE and RSE values in recent studies can be divided into two larger groups. The first corresponds to the class of density functional theory (DFT) methods. The most commonly used functional is the B3LYP hybrid functional, usually in combination with a triple zeta, multiply polarized basis set.[7] More recent studies also include double-hybrid functionals such as B2-PLYP[8] and B2K-PLYP[9] or meta-hybrid functionals such as M05-2X or M06-2X.[10]

The second group contains all theoretical approaches involving the determination of stability values from single or multiple calculations at *ab initio* level. These single point calculations are often, but not always, based on geometries optimized with one of the DFT methods mentioned before. The simplest approach, referred to here as "ROMP2," is based on geometry optimizations at the (U)B3LYP/6-31G(d) level, addition of thermochemical corrections to 298.15K using the rigid rotor/harmonic oscillator model and a scaling factor of 0.9806, and finally a single point energy calculation at the (RO)MP2(FC)/6-311 + G(3df,2p) level.[1,11,12] The more reliable "G3(MP2)-RAD method" developed by Radom et al.[4] is again based on (U)B3LYP/6-31G(d) geometries and thermal corrections to 298.15 K, but also involves single point calculations at the ROMP2(FC)/G3MP2large//UB3LYP/6-31G(d) and the URCCSD (T)/6-31G(d)//UB3LYP/6-31G(d) level of theory. In the spirit of the G3(MP2)B3 compound method developed by Curtiss et al.,[5] higher level corrections (HLCs) are also included with the following definitions: $E(HLC) = -An_b - B(n_a - n_b)$ with $A = 9.413 \times 10^{-3}$ a.u., $B = 3.969 \times 10^{-3}$ a.u., n_a = number of α valence electrons, and n_b = number of β valence electrons. It is worth noting, however, that the empirical HLC term in this and other G3 methods always cancels entirely from RSE calculations; hence, in such cases these composite methods can be viewed as wholly *ab initio*. Finally, a comparison of the performance of several compound energy methods in reproducing thermochemical data of small open shell species has shown that the "G3B3" method developed by Baboul et al.[5] performs somewhat better than G3(MP2)-RAD, but is also significantly more expensive.[4b] More recent developments of accurate compound energy schemes include procedures for extrapolating the Hartree–Fock and/or correlation energy to the infinite basis set limit. Extrapolation schemes can be based on components of variable complexity and the W1 model by Martin et al. represents one of the more economical versions of the approaches using correlation energies calculated at coupled cluster level.[13] The accuracy obtained with this approach for open shell systems is significantly better than at G3 level, but this improved performance comes at a significantly increased computational cost.[14]

5.2.1 Testing the Performance of Different Theoretical Approaches: How Stable Are Allyl and Benzyl Radicals?

A comparison of the stability values for allyl radical (**10**) and benzyl radical (**11**) will be used here to illustrate the performance of these methods (Table 5.2). Earlier compilations of C–H BDE data put the allylic C–H bond in propene (369.0 ± 8.8 kJ/mol) and the benzylic C–H bond in toluene (370.3 ± 6.3 kJ/mol) at almost identical values. Together with the C–H BDE in methane of 438.9 ± 0.4 kJ/mol this equates to RSE(3) values of $+68.6 \pm 6.7$ and $+69.9 \pm 9.2$ kJ/mol for the allyl and benzyl radical, respectively.[15] New measurements of C–H bond strengths in the gas phase[16,17,19] and in solution[17] led to a slightly larger difference with a C–H BDE in propene of 371.5 ± 1.7 kJ/mol and a C–H BDE in toluene of 375.7 ± 2.5 kJ/mol.[3,18]

TABLE 5.2 RSEs at 298.15K Calculated According to Equation 5.3 of Allyl Radical (10) and Benzyl Radical (11) (in kJ/mol) at Various Levels of Theory

Method	•CH₂CHCH₂ (10)	•CH₂Ph (11)
Exp.[15]	$+68.6 \pm 6.7$	$+69.9 \pm 9.2$
Exp.[3,18]	$+67.4 \pm 2.1$	$+63.2 \pm 2.9$
UB3LYP/6-31G(d)	$+83.4$	$+71.9$
UB3LYP/6-31G(d) (scaled, 0K)	$+82.1 \, (-1.3)$	$+69.8 \, (-2.1)$
UB3LYP/6-311 + + G(d,p)[7]	$+79.4$	$+68.2$
UB3LYP/6-311 + + G(3df,2p)[7]	$+80.4$	$+69.4$
UM05-2X/6-311 + + G(3df,2p)[7]	$+76.7$	$+63.1$
UB2-PLYP/TZVPP//UB3LYP/6-31G(d)	$+72.6$	$+56.3$
UB2-PLYP/6-311 + G(3df,2p)//RB3LYP/6-31G(d)[a]	$+71.8$	$+55.2$
RB2-PLYP/6-311 + G(3df,2p)//RB3LYP/6-31G(d)[a]	$+70.5$	$+59.1$
ROMP2	$+78.8$	$+52.5$
G3(MP2)-RAD	$+72.0$	$+61.0$
G3X(MP2)-RAD[a]	$+72.4$	$+61.3$
G3[7]	$+72.4$	$+54.9$
G3B3	$+70.5$	$+55.1$
CBS-Q[7]	$+78.2$	$+62.8$
CBS-QB3[a]	$+76.1$	$+62.0$
ROCBS-QB3[a]	$+75.7$	$+65.3$
W1[a]	$+71.6$	$+61.2$

[a]Data from Ref. 14 and addition of thermal corrections from G3(MP2)-RAD.

In combination with the unchanged C–H BDE in CH_4 this equates to RSE(3) values of $+67.4 \pm 2.1$ and $+63.2 \pm 2.9$ kJ/mol for the allyl and benzyl radicals, respectively. This implies a stability difference of these two systems of just over 4 kJ/mol.

The results obtained at a variety of different theoretical levels are ordered in Table 5.2 approximately by computational effort. The most expensive methods are those at the bottom of the table with W1 being the reference method of choice for systems composed of first row elements.[13] At this latter level, the stability of allyl radical (10) is 4.2 kJ/mol higher than the experimental value, while that for benzyl radical (11) is 2.0 kJ/mol lower. As a consequence the stabilities of these two radicals differ by 10.4 kJ/mol, significantly more than derived from experimental data. This finding also extends to practically all other theoretical methods compiled in Table 5.2, including expensive compound methods from the G3 family as well as the much more economical hybrid DFT methods such as B3LYP. Absolute RSE values are somewhat too large at the B3LYP level, regardless of the basis set used, but can be reduced substantially with either the double-hybrid B2-PLYP or the meta-hybrid UM05-2X functionals. The two compound methods optimized for the treatment of open shell systems G3(MP2)-RAD and G3X(MP2)-RAD yield results that are practically identical to those obtained at W1 level. From a price/performance perspective, the G3(MP2)-RAD level thus represents one of the most attractive procedures for the calculation of radical stability values.

5.2.2   The Application of IMOMO Schemes: How Stable Are Benzyl and Diphenylmethyl Radicals?

The treatment of large molecular systems with G3-type compound methods is still challenging from a technical perspective due to the unfavorable scaling of computational effort with system size. One of the strategies to deal with this problem in a systematic manner is the ONIOM scheme developed by Morokuma and coworkers, in particular in its IMOMO incarnation (integrated MO/MO method) combining two different molecular orbital-based theoretical methods.[28] Energies are in this case calculated at a high level of theory for a small core region, while the overall system is calculated at somewhat lower level. If, for example, we choose to treat the core region at G3(MP2)-RAD level and the overall system at ROMP2 level, the total energies are calculated in the IMOMOM scheme as

$$
\begin{aligned}
E(IMOMOM) &= E(core,G3(MP2)-RAD)) \\
&\quad - E(core, ROMP2) \\
&\quad + E(overall, ROMP2)
\end{aligned}
\tag{5.9}
$$

Combination of these two theoretical methods is particularly straightforward since both are based on the same level of geometry optimization (UB3LYP/6-31G (d)). For the open shell systems discussed here, the core region will most likely be defined as the radical center together with the most important substituents.[29,30] Bonds dissected when cutting a core region out of the overall system are typically saturated with hydrogen atoms in order to arrive at the correct electronic state. The calculation of RSE values with IMOMO schemes involves, in principle, the calculation of three energies for all four species shown in Equation 5.3 and we will use the stability of the diphenylmethyl radical (**12**) as an example here. This system presents a worst-case scenario for many IMOMO schemes as the spin is delocalized over a large π-system and the definition of a meaningful small core region is not immediately obvious. One possibility involves the selection of the radical center and one phenyl substituent as the core region as shown in Equation 5.10. Definition of this core region cuts through the sigma bond connecting the radical center and the second phenyl substituent. After saturation of the cleaved bond with hydrogen atoms, this leaves us with the benzyl radical **11** as the "core" system on the open shell side and with toluene as the "core" system on the closed shell side.

With this choice, the IMOMO(G3(MP2)-RAD,ROMP2) radical stabilization energy of radical **12** is calculated as expressed in Equation 5.11.

```
RSE(IMOMO,Ph₂CH•) =
E(G3(MP2)-RAD,CH₃•) + E(G3(MP2)-RAD,PhCH₃) - E(G3(MP2)-RAD,CH₄) -
E(G3(MP2)-RAD,PhCH₂•) - E(ROMP2,CH₃•) - E(ROMP2,PhCH₃) +
E(ROMP2,CH₄) + E(ROMP2, PhCH₂•) + E(ROMP2,CH₃•) + E(ROMP2,Ph₂CH₂) -
E(ROMP2,CH₄) - E(ROMP2, Ph₂CH•)
```

$$(5.11)$$

Even though this may not be visible immediately from this lengthy list of energies, the first four energy terms in Equation 5.11 equate to the RSE of benzyl radical (**11**) at G3(MP2)-RAD level, the second four energy terms equate to the negative of the RSE of benzyl radical (**11**) at ROMP2 level, and the last four energy terms equate to the RSE of radical **12** at ROMP2 level. Equation 5.12 may thus be rewritten more simply as

```
RSE(IMOMO,Ph₂CH•) = RSE(G3(MP2)-RAD,PhCH₂•)

                    - RSE(ROMP2,PhCH₂•)

                    + RSE(ROMP2,Ph₂CH•)
```

$$(5.12)$$

We can thus conclude that the use of IMOMO schemes for the calculation of RSE values equates to combining absolute RSE values calculated at a high level of theory for a small core system with changes in RSE values through introduction of new substituents at some lower theoretical level. The results obtained with this strategy for radical **12** are summarized in Table 5.3. The RSE of radical **12** can be determined directly at G3(MP2)-RAD level and amounts to $+85.5$ kJ/mol. Using the IMOMO approach, the RSE of benzyl radical **11** at G3(MP2)-RAD level of $+61.0$ kJ/mol is combined with the difference of the RSE values of radicals **12** and **11** at ROMP2 level ($79.8 - 52.5 = 27.3$ kJ/mol) to yield RSE(IMOMO, **12**) $= +88.3$ kJ/mol. Considering the immense savings in computer time and the challenging nature of the diphenylmethyl radical **12** this is a remarkably good result.

TABLE 5.3   RSEs at 298.15 K of Benzyl Radical (11) and Diphenylmethyl Radical (12) (in kJ/mol) at Various Levels of Theory

Method	•CH₂Ph (**11**)	•CHPh₂ (**12**)
ROMP2	$+52.5$	$+79.8$
G3(MP2)-RAD	$+61.0$	$+85.5$
IMOMO	$+61.0$	$+88.3$
Exp.	$+69.9 \pm 9.2$[15]	$+85.8 \pm 2.5$[3]
	$+63.2 \pm 2.9$[3,18]	

5.3   RSE VALUES FOR CARBON-CENTERED RADICALS

Radical stabilization energies calculated according to Equation 5.3 at 298.15 K have been compiled in Table 5.4 for a variety of monosubstituted methyl radicals at either G3(MP2)-RAD or higher level. In those cases in which results are available at more than one theoretical level, good agreement between these results is usually found. Overall, the mean absolute deviation of the G3(MP2)-RAD values from experiment is 8.3 kJ/mol, comparable to that reported in an earlier assessment of the G3 and CBS-Q methods.[7] However, exceptionally large deviations (>20 kJ/mol) occur for the following radicals: $\cdot CH_2SO_2CH_3$, $\cdot CH_2SOCH_3$, $\cdot CH(CH_3)CH_2OH$, $\cdot CH(CH_3)$ $CH=C(CH_3)_2$, $\cdot C(CH_3)_2OC(O)CH_3$, $\cdot C(CH_3)_2CCH$, $\cdot C(O)OH$, and $\cdot C(O)OCH_3$. In those cases, reexamination of the experimental values may be advisable.

Substituent effects on RSE values can be interpreted as a combination of three different molecular orbital interactions: (a) resonance stabilization through interaction of the radical center with π-systems; (b) stabilization through hyperconjugation of the radical center with adjacent C–H bonds; and (c) stabilization through interaction of the radical center with high lying orbitals describing lone pair electrons. A detailed discussion of these effects in more quantitative terms using molecular orbital interaction diagrams has been the subject of earlier compilations and will therefore not be repeated here.[1,22]

An overview of the RSE(3) values compiled in Tables 5.4–5.7 has been given in a graphical manner in Fig. 5.1. It is easily seen from this overview that substituent effects in primary, secondary, and tertiary alkyl radicals are rather similar in most cases, indicating a comparable mechanism of interaction between substituents and the radical center in all these systems. RSE values for sigma radicals derived from alkynes, alkenes, aldehydes, and carboxylic acids through C–H bond cleavage cover a much larger range than those of alkyl radicals and show a rather different dependence on the substitution pattern.

5.4   USE OF RSE VALUES IN PRACTICAL APPLICATIONS

The analysis of substituent effects on RSE values does not only aid our understanding,[1] but also holds a degree of predictive power, allowing one to design and select species with optimal radical stabilities for specific practical applications. Indeed, provided due attention is given to the effects of substituents on the other species involved, RSEs can even provide a qualitative guide to the thermodynamic stability of radicals in other types of chemical reaction, such as addition and beta-scission. In this section, some practical applications of RSE values are illustrated using some selected case studies from the literature.

5.4.1   Susceptibility to Hydrogen Atom Abstraction

An important practical application of RSEs is establishing, at least from a thermodynamic perspective, whether certain species are vulnerable to hydrogen atom

TABLE 5.4   RSEs for Monosubstituted Methyl Radicals at 298.15 K Calculated According to Equation 5.3 Together with BDE(C–H) Energies for Selected Systems (All in kJ/mol)

Radical	RSE(3) (G3(MP2)-RAD)	RSE(3) (Other)	RSE(3) (Exp.)	BDE (C–H) (Exp.)[a]
$\cdot CH_2\text{–}CF_3$	-8.0^{22b}	-8.0 (G3)[c] -6.4 (W1)[14b]	-7.1	446.4 ± 4.5
$\cdot CH_2\text{–}CF_2CF_3$	-6.2	-5.2 (G3(MP2)-RAD)[22b]		
$\cdot CH_2\text{–}CF_2H$	-3.1		$+6.3$	433.0 ± 14.6
$\cdot CH_2\text{–}SO_2CH_3$	-2.4		$+25.1$	414.2
$\cdot CH_2\text{–}H$ (**4**)	0.0	0.0	0.0	439.3 ± 0.4
$\cdot CH_2\text{–}CCl_3$	$+0.4$			
$\cdot CH_2\text{–}CCl_2H$	$+4.6$			
$\cdot CH_2\text{–}CH_2F$	$+5.8$	$+12.4$ (G3)[c]	$+5.8$	433.5 ± 8.4
$\cdot CH_2\text{–}C(CH_3)_3$	$+7.1$	$+6.8$ (G3)[c]	$+20.5$	418.8 ± 8
$\cdot CH_2\text{–}S(O)CH_3$	$+8.4$		$+46.0$	393.3
$\cdot CH_2\text{–}CH_2Cl$	$+10.2$	$+10.8$ (G3)[c]	$+16.2$	423.1 ± 2.4
$\cdot CH_2\text{–}CH_2OH$	$+10.3$	$+11.6$ (CBS-QB3)[20]	$+15.5$	423.8
$\cdot CH_2\text{–}CH_2\text{–}C_6H_5$	$+10.4^{23b}$			
$\cdot CH_2\text{–}CH(CH_3)_2$	$+10.6$	$+10.1$ (G3)[c]	$+20.1$	419.2 ± 4.2
$\cdot CH_2\text{–}CH_2CHCH_2$	$+11.1$		$+28.4$	410.9
$\cdot CH_2\text{–}SiH_3$	$+11.8$			
$\cdot CH_2\text{–}CH_2CH_2CH_3$	$+12.2$	$+10.5$ (G3)[c] $+12.6$ (G3B3)	$+18.0$	421.3
$\cdot CH_2\text{–}Si(CH_3)_3$	$+12.2$	$+12.6$ (G3)[c]	$+21.3$	418 ± 6.3
$\cdot CH_2\text{–}CH_2CH_3$	$+12.2$	$+12.1$ (W1)[14b] $+11.5$ (G3)[c] $+11.9$ (G3(MP2)-RAD)[22b]	$+17.1$	422.2 ± 2.1
$\cdot CH_2\text{–}NO_2$	$+12.4^{22b}$	$+14.4$ (W1)[14b]	$+23.9$	415.4
$\cdot CH_2\text{–}OCF_3$	$+12.7$		$+12.5$	426.8 ± 4.2
$\cdot CH_2\text{–}F$ (**9**)	$+12.8$	$+15.1$ (W1)[14b] $+13.4$ (G3B3)	$+15.5$	423.8 ± 4.2
$\cdot CH_2\text{–}CH_3$ (**8**)	$+13.5$	$+15.1$ (W1)[14b] $+13.8$ (G3B3) $+12.8$ (G3)[c]	$+18.8$	420.5 ± 1.3
$\cdot CH_2\text{–}OPO_3H_2$		$+15.1$ (G3B3)[d]		
$\cdot CH_2\text{–}Br$	$+15.3^{22b}$		$+14.2$	425.1 ± 4.2
$\cdot CH_2\text{–}OCHO$	$+17.2$			
$\cdot CH_2\text{–}OC(O)CH_3$	$+18.4^{22b}$	$+17.9$ (W1)[14b]	$+34.7$	404.6
$\cdot CH_2\text{–}Cl$	$+20.2^{22b}$	$+22.2$ (W1)[14b] $+20.5$ (G3)[c]	$+20.3$	419.0 ± 2.3
$\cdot CH_2\text{–}C(O)N(CH_3)_2$	$+21.0$			
$\cdot CH_2\text{–}C(O)N(CH_2CH_3)_2$	$+22.6$			
$\cdot CH_2\text{–}COOH$	$+22.7^{22b}$	$+25.2$ (W1)[14b]	$+40.6$	398.7 ± 12.1

Table 5.4 (*Continued*)

Radical	RSE(3) (G3(MP2)-RAD)	RSE(3) (Other)	RSE(3) (Exp.)	BDE (C–H) (Exp.)[a]
$\cdot CH_2$-C(O)NHCH$_3$	$+23.1$	$+23.0$ (G3X(MP2)-RAD)[25b]		
$\cdot CH_2$-COOCH$_3$	$+23.2^{22b}$	$+25.0$ (W1)[14b]	$+30.0$	406.3 ± 10.5
$\cdot CH_2$-C(O)NH$_2$	$+23.4$	$+23.4$ (G3X(MP2)-RAD)[25b]		
$\cdot CH_2$-CH(CH$_2$)$_2$	$+23.4$		$+31.8$	407.5 ± 6.7
$\cdot CH_2$-COOCH$_2$CH$_3$	$+23.4$	$+23.4$ (G3(MP2)-RAD(+))[26]	$+37.6$	401.7
			$+39.8$	399.5
$\cdot CH_2$-COOC(CH$_3$)$_3$	$+23.5$			
$\cdot CH_2$-PH$_2$	$+23.5^{22b}$	$+27.1$ (W1)[14b]		
$\cdot CH_2$-P(CH$_3$)$_2$	$+24.9$			
$\cdot CH_2$-CO-C$_6$H$_5$	$+30.4$		$+34.3$	405.0
			$+50.2$	389.1
			$+41.2$	390.4
			$+48.9$	402.8 ± 3.6
$\cdot CH_2$-OCH$_3$	$+31.5^{22b}$	$+35.6$ (W1)[14b] $+32.8$ (G3)[c]	$+37.2$	402.1
$\cdot CH_2$-OCH$_2$CH$_3$	$+31.6$		$+50.2$	389.1
$\cdot CH_2$-OH	$+32.3$	$+36.0$ (W1)[14b] $+33.5$ (G3B3)[d] $+33.2$ (G3)[c]	$+37.4$	401.9 ± 0.6
$\cdot CH_2$-COCH$_3$	$+32.4$	$+33.3$ (G3)[c]	$+38.1$	401.2 ± 2.9
$\cdot CH_2$-CN	$+32.5^{22b}$	$+33.7$ (W1)[14b] $+36.2$ (G3)[c] $+32.5$ (G3(MP2)-RAD(+))[26]	$+37.6$	401.7
$\cdot CH_2$-SC(CH$_3$)$_2$CN	$+35.6$			
$\cdot CH_2$-SH	$+36.2^{22b}$	$+41.5$ (W1)[14b] $+37.7$ (G3)[c]	$+46.4$	392.9 ± 8.4
$\cdot CH_2$-CHO	$+36.7$	$+38.2$ (W1)[14b] $+36.9$ (G3)[c]	$+44.7$	394.6 ± 9.2
$\cdot CH_2$-SCH$_2$COOCH$_3$	$+37.0$			
$\cdot CH_2$-SCH$_2$-C$_6$H$_5$	$+38.5$			
$\cdot CH_2$-BH$_2$	$+40.9^{22b}$	$+41.7$ (W1)[14b]		
$\cdot CH_2$-SCH$_3$	$+41.0$	$+43.0$ (G3)[c]	$+47.3$	392.0 ± 5.9
$\cdot CH_2$-NHCHO	$+42.5$	$+42.5$ (G3X(MP2)-RAD)[25b]		
$\cdot CH_2$-NHC(O)CH$_3$	$+43.0$	$+43.0$ (G3X(MP2)-RAD)[25b]		
$\cdot CH_2$-NH$_2$	$+44.9^{22b}$	$+50.0$ (W1)[14b] $+46.7$ (G3)[c] $+44.9$ (G3X(MP2)-RAD)[25b]	$+46.4$	392.9 ± 8.4

(*Continued*)

Table 5.4 (*Continued*)

Radical	RSE(3) (G3(MP2)-RAD)	RSE(3) (Other)	RSE(3) (Exp.)	BDE (C–H) (Exp.)[a]
•CH_2–$N(CH_3)_2$	+46.1	+48.0 (G3)[c]	+49.5	389.8
•CH_2–$NHCH_3$	+46.6	+48.6 (G3)[c]	+45.2	394.1
•CH_2–CCH	+52.8[22b]	+54.2 (W1)[14b] +53.8 (G3)[c]	+67.3	372.0 ± 4.2
•CH_2–C_6H_4–pNO_2	+61.0		+65.9	383.4
•CH_2–C_6H_5 (**11**)	+61.0	+61.2 (W1)[14b] +54.9 (G3)[c] +55.1 (G3B3)	+69.0	370.3 ± 6.3
•CH_2–C_6H_4–pCN	+62.1		+71.3	368.0
•CH_2–C_6H_4–pOH	+63.0			
•CH_2–C_6H_4–$pOCH_3$	+63.3		+76.8	362.5
•CH_2–$C(CH_3)$=CH_2		+68.4 (G3)[c]		
•CH_2–CH=CH_2 (**10**)	+72.0	+71.6 (W1)[14b] +72.4 (G3)[c] +70.5 (G3B3)	+70.7	368.6 ± 2.9
•CH_2–CH=CH–CH_3 (E)	+73.9	+73.0 (G3)[c]	+82.5	356.8
•CH_2–CH=$C(CH_3)_2$	+77.3			
•CH_2–$C(CH_3)$=$C(CH_3)_2$		+75.9 (G3)[c]	+86.5	352.8
•CH_2–CH=CH–CH=CH_2	+93.71	+95.1 (G3)[c] +91.3 (G3B3)	+92.0	347.3 ± 12.6

[a]All BDE data at 298.15K taken from Ref. 3, if not specified otherwise.
[b]Reference data given at 0 K and corrected to 298.15K by thermal corrections calculated at (U)B3LYP/6-31G(d) level using a scaling factor of 0.9806.
[c]Calculated from BDE(C–H) data at 298.15 K in Ref. 7.
[d]Calculated from BDE(C–H) data at 298.15 K in Ref. 24.

abstraction reactions. By definition, a hydrogen atom abstraction reaction is thermodynamically favorable if the RSE of the product radical is greater than that of the reactant radical. Moreover, where a species has several potential hydrogen abstraction sites, the most thermodynamically favorable site is that which results in the radical having the highest RSE. While kinetic factors can sometimes favor alternative reaction pathways, the consideration of RSE values can in the very least establish whether or not certain reactions are potentially feasible, and this in turn has important practical applications in chemistry and biology.

One such application of RSEs, or their constituent bond dissociation energies, is in the study of radical-mediated degradation mechanisms. For example, based on an examination of the relevant C–H and S–H bond dissociation energies in model peptides, Rauk et al. postulated a mechanism for generating and propagating oxidative damage via a Met residue of the Aβ peptide of Alzheimer's disease or the prion peptide of Creutzfeldt–Jakob disease.[31] In a similar manner, Li et al. used C–H BDE calculations to identify the most vulnerable sites for radical-mediated damage in

TABLE 5.5 RSEs for Disubstituted Methyl Radicals at 298.15K Calculated According to Equation 5.3 Together with BDE(C–H) Energies for Selected Systems (All in kJ/mol)

Radical	RSE(3) (G3(MP2)-RAD)	RSE(3) (Other)	RSE(3) (Exp.)	BDE (C–H) (Exp.)[a]
•CH(CH$_2$)$_2$	−20.0	−13.1 (G3)^c	−5.5	444.8 ± 1.0
•CH$_3$ (**4**)	0.0	0.0	0.0	439.3 ± 0.4
•CH(CH$_3$)-CF$_3$	+7.7			
•CHF$_2$	+8.9	+10.0 (G3)^c	+7.5	431.8 ± 4.2
•CH(CH$_3$)-CHF$_2$	+10.3			
•CH(CH$_3$)-CF$_2$CF$_3$	+11.5			
•CH(CH$_3$)-SO$_2$CH$_3$	+14.2			
•CH(CH$_2$)$_3$		+15.1 (G3)^c	+30.1	409.2 ± 1.3
•CH(CH$_3$)-CCl$_3$	+15.5			
•CH(CH$_3$)-CFH$_2$	+16.5			
•CH(CH$_3$)-CH$_2$OH	+17.4		+44.7	394.6 ± 8.4
•CH(CF$_3$)-Cl		+18.1 (G3)^c	+13.4	425.9 ± 6.3
•CH(CH$_3$)-CCl$_2$H	+19.3			
•CH(CH$_3$)-CH$_2$CH$_3$	+19.5	+21.2 (G3)^c	+28.2	411.1 ± 2.2
•CH(CH$_3$)-F	+19.7		+28.4	410.9 ± 8.4
•CH(CH$_2$)$_5$		+20.0 (G3)^c	+23.0	416.3
•CH(CH$_3$)-CH$_2$-C$_6$H$_5$	+20.1			
•CH(CH$_3$)-C(CH$_3$)$_3$	+20.1			
•CH(CH$_3$)-OCF$_3$	+20.6			
•CH(CH$_3$)-CH$_2$CH=CH$_2$	+20.9			
•CHFCl		+21.0 (G3)^c	+17.6	421.7 ± 10.0
•CH(CH$_3$)-CH$_2$CH (CH$_3$)COOCH$_3$	+22.4^{27}			
•CH(CH$_3$)-Br	+22.9		+32.6	406.7 ± 4.2
•CH(CH$_3$)$_2$	+23.0^{27}	+22.2 (G3)^c	+28.8	410.5 ± 2.9
•CH(CH$_3$)-CH$_2$C(CH$_3$)$_2$ COOCH$_3$	+23.0^{27}			
•CH(CH$_3$)-OCHO	+23.6			
•CH(CH$_3$)-OC(O)CH$_3$	+24.3	+24.3 (G3(MP2)-RAD(+))^{26}		
•CH(CH$_3$)-CH(CH$_3$)$_2$	+24.8^{27}			
•CH(CH$_3$)-CH$_2$Cl	+25.2		+30.0	409.3 ± 3.9
•CH(CH$_3$)- CH$_2$CH(CH$_3$)$_2$	+25.4			
•CH(CH$_2$)$_6$		+26.9 (G3)^c	+52.3	387.0 ± 4
•CH(CH$_3$)-Cl	+26.7	+27.0 (G3)^c +26.7 (G3(MP2)-RAD(+))^{26}	+32.7	406.6 ± 1.5
•CH(CH$_3$)-SiH$_3$	+28.8			
•CH(CH$_3$)-Si(CH$_3$)$_3$	+29.1			
•CH(CH$_3$)-CH(CH$_2$)$_2$	+29.1			
•CH(CH$_3$)-PH$_2$	+31.1			

(Continued)

Table 5.5 (*Continued*)

Radical	RSE(3) (G3(MP2)-RAD)	RSE(3) (Other)	RSE(3) (Exp.)	BDE (C–H) (Exp.)[a]
•CH(CH$_3$)-NO$_2$	+32.3		+28.8	410.5
•CH(CH$_2$)$_4$		+33.0 (G3)[c]	+39.3	400.0 ± 4.2
•CHCl$_2$	+32.2	+34.2 (G3)[c]	+32.2	407.1 ± 4.2
•CH(CH$_3$)-OCH$_3$	+36.5			
•CH(CH$_3$)-OCH$_2$CH$_3$	+36.5	+34.8 (G3)[c]	+50.2	389.1
•CH(CH$_3$)-OH	+38.3	+38.2 (G3)[c] +42.7 (CCSD(T)/ CBS)[c]	+42.7	396.6
•CH(CH$_3$)-P(CH$_3$)$_2$	+37.8			
•CH(CH$_3$)-CON(CH$_3$)$_2$	+38.6			
•CH(COOCH$_3$)- CH$_2$C(CH$_3$)$_2$COOCH$_3$	+39.2[27]			
•CH(COOCH$_3$)- CH$_2$CH(CH$_3$)COOCH$_3$	+39.4[27]			
•CH(CH$_3$)-CON(CH$_2$CH$_3$)$_2$	+40.1			
•CH(CH$_3$)-SCH$_2$COOCH$_3$	+40.8[23b]			
•CH(CH$_3$)-SH	+41.6			
•CH(CH$_3$)-COOC(CH$_3$)$_2$	+41.8			
•CH(CH$_3$)-COOH	+41.9		+40.5	398.8
•CH(CH$_3$)-COOCH$_3$	+41.9[27]			
•CH(CH$_3$)-COOCH$_2$CH$_3$	+42.0			
•CH(CH$_3$)-SC(CH$_3$)$_2$CN	+42.0[23b]			
•CH(CH$_3$)-CONH$_2$	+42.2			
•CH(CH$_3$)-SCH$_2$-C$_6$H$_5$	+43.7[23b]			
•CH(OCH$_2$CH$_2$CH$_2$)		+44.1 (G3)[c]	+54.0	385.3 ± 6.7
•CH(CH$_3$)-NHCOCH$_3$	+44.9			
•CH(COOCH$_3$)- CH$_2$CH(CH$_3$)$_2$	+45.4[27]			
•CH(CH$_3$)-NHCHO	+45.5			
•CH(CH$_3$)-SCH$_3$	+45.6[23b]			
•CH(CH$_3$)-CONHCH$_3$	+45.7			
•CH(CH$_3$)-SOCH$_3$	+46.6			
•CH(CH$_3$)-CN	+48.5	+49.0 (G3)[c] +48.5 (G3(MP2)- RAD(+))[26]	+46.0	393.3 ± 12.6
•CH(CH$_3$)-NH$_2$	+49.9		+62.3	377.0 ± 8.4
•CH(CH$_3$)-CO-C$_6$H$_5$	+50.0		+50.6	388.7
•CH(CH$_3$)-NHCH$_3$	+50.2			
•CH(CH$_3$)-COCH$_3$	+53.9	+52.2 (G3)[c]	+53.1	386.2 ± 7.1
•CH(CH$_3$)-N(CH$_3$)$_2$	+54.6			
•CH(CH$_3$)-CHO	+56.4			
•CH(CH$_3$)-CC-CH$_3$		+60.7 (G3)[c]	+74.0	365.3 ± 11.3
•CH(CH$_3$)-CCH	+63.9	+65.1 (G3)[c]	+66.3	373.0
•CH(CH$_3$)-BH$_2$	+65.2			

Table 5.5 (*Continued*)

Radical	RSE(3) (G3(MP2)-RAD)	RSE(3) (Other)	RSE(3) (Exp.)	BDE (C–H) (Exp.)[a]
•CH(CH$_3$)–C$_6$H$_5$	+68.3	+68.3 (G3(MP2)-RAD(+))[26]	+82.0	357.3±6.3
•CH(CH$_3$)–C$_6$H$_4$-pOH	+69.4			
•CH(CH$_3$)–C$_6$H$_4$-pOCH$_3$	+69.9			
•CH(CH$_3$)–C$_6$H$_4$-pCN	+71.0			
•CH(CH$_3$)–C$_6$H$_4$-pNO$_2$	+73.0			
•CH(Cl)–CH=CH$_2$		+83.0 (G3)[c]	+68.6	370.7±4.6
•CH(CH$_3$)–CH= CH–CH$_3$ (E)	+83.4		+97.5	341.8±6.3
•CH(CHCHCH$_2$CH$_2$)		+83.7 (G3)[c]	+92.6	346.7
•CH(CH$_3$)–CH=CH$_2$	+84.6	+81.7 (G3)[c]	+88.7	350.6
•CH(CH$_3$)–CH=C(CH$_3$)$_2$	+86.5		+107.3	332.0
•CH(C$_6$H$_5$)$_2$	+85.5	+72.4 (G3B3)	+85.8	353.5±2.1
•CH(CH)$_4$	+87.3			
•CH(OH)–CH=CH$_2$		+103.6 (G3)[c]	+97.9	341.4±7.5
•CH(CHCHCHCHCH$_2$)		+118.6 (G3)[c]	+134.3	305±21
•CH(CHCHCH$_2$CHCH)		+124.2 (G3)[c]	+119.6	319.7

[a]All BDE data taken from Ref. 3, if not specified otherwise.

[b]Reference data given at 0K and corrected to 298.15K by thermal corrections calculated at (U)B3LYP/6-31G(d) level using a scaling factor of 0.9806.

[c]Calculated from BDE(C–H) data at 298.15 K in Ref. 21.

RNA and DNA, and to study the effects of phosphorylation state and charge on the results.[24] In the polymer field, Berry et al. used BDE calculations for model alkyl urethanes to identify vulnerable sites to radical attack, and to examine the effect of branching on the photostability of urethane coatings.[32]

An advantage of using RSEs is that they provide a convenient framework for ranking and analyzing the effects of substituents on radical stability, and this information has predictive value, allowing one to choose substituents that target the optimum radical stability for a given application. For example, studies of substituent effects on the RSEs of glycyl and related radicals[33] have been exploited in the development of successful inhibitors of peptidylglycine α-amidating mono-oxygenase (PAM).[34] In this work, analogs of the substrates of PAM were designed so as to competitively bind to the enzyme, without being processed. To this end, the first step of the process, in which a copper-bound superoxide radical abstracts the pro-*S* hydrogen from the glycine residue, to give a glycyl radical was targeted. By replacing the glycine moiety with glycolate, this simple substitution (of the glycine NH for O) significantly reduced the RSE of the resulting glycolyl radical compared with the corresponding glycyl due to reduced lone pair donation and increased sigma withdrawal. This reduction in stability was sufficient to stop catalysis by monooxygenase, without affecting binding to the enzyme.

TABLE 5.6   RSEs for Trisubstituted Methyl Radicals at 298.15 K Calculated According to Equation 5.3 Together with BDE(C–H) Energies for Selected Systems (All in kJ/mol)

Radical	RSE(3) (G3(MP2)-RAD)	RSE(3) (Other)	RSE(3) (Exp.)	BDE (C–H) (Exp.)[a]
•CF$_3$	−13.1	−11.6 (G3)[c]	−10.1	449.4
•CF$_2$CF$_3$	−1.2	+0.2 (G3)[c]	+9.6	429.7 ± 2.1
•CH$_3$ (**4**)	0.0	0.0	0.0	439.3 ± 0.4
•CF$_2$Cl		+8.9 (G3)[c]	+18.0	421.3 ± 8.3
•C(CH$_3$)$_2$-OCHO	+17.2			
•C(CH$_3$)$_2$-CF$_3$	+17.3			
•C(CH$_3$)$_2$-OC(O)CH$_3$	+17.5		+47.0	392.3
•C(CH$_3$)$_2$-CHF$_2$	+17.6			
•C(CH$_3$)$_2$-OCF$_3$	+21.0			
•C(CH$_3$)$_2$-CH$_2$OH	+23.3			
•C(CH$_3$)$_2$-CH$_2$F	+23.4			
•C(CH$_3$)$_2$-F	+23.6			
•C(CH$_3$)$_2$-CCl$_3$	+24.5			
•C(CH$_3$)$_2$-CH$_2$CH$_3$	+25.8	+10.0 (G3)[c]	+38.5	400.8
•C(CH$_3$)$_2$-CH$_2$CH=CH$_2$	+26.3			
•CCl$_2$F		+26.7 (G3)[c]	+25.5	413.8 ± 5.0
•C(CH$_3$)$_2$-SO$_2$CH$_3$	+27.2			
•C(CH$_3$)$_2$-CF$_2$CF$_3$	+27.8			
•C(CH$_3$)$_2$-CH$_2$-C$_6$H$_5$	+27.9			
•C(CH$_3$)$_2$-Br	+27.9			
•C(CH$_3$)$_2$-C(CH$_3$)$_3$	+28.4			
•C(CH$_3$)$_3$	+28.5	+28.4 (G3)[c]	+28.8	410.5 ± 2.9
•C(CH$_3$)$_2$-CH(CH$_3$)$_2$	+29.0		+40.1	399.2 ± 13.0
•C(CH$_3$)$_2$-CHCl$_2$	+30.2			
•C(CH$_3$)$_2$-Cl	+30.6			
•C(CH$_3$)$_2$-CH(CH$_2$)$_2$	+31.6			
•C(CH$_3$)$_2$-NHC(O)CH$_3$	+35.4		+49.8	389.5
•C(CH$_3$)$_2$-CClH$_2$	+35.5			
•C(CH$_3$)$_2$-OCH$_3$	+36.3			
•C(CH$_3$)$_2$-OCH$_2$CH$_3$	+36.5			
•C(CH$_3$)Cl$_2$		+36.9 (G3)[c]	+48.7	390.6 ± 1.5
•C(CH$_3$)$_2$-NHCHO	+37.8			
•C(CH$_3$)$_2$-P(CH$_3$)$_2$	+38.6			
•C(CH$_3$)$_2$-C(O)N(CH$_2$CH$_3$)$_2$	+39.3			
•C(CH$_3$)$_2$-PH$_2$	+40.0			
•C(CH$_3$)$_2$-SCH$_2$-C$_6$H$_5$	+40.0			
•C(CH$_3$)$_2$-OH	+41.1	+40.8 (G3)[c]	+42.8	396.5
•C(CH$_3$)$_2$-SiH$_3$	+41.7			
•C(CH$_3$)$_2$-Si(CH$_3$)$_3$	+42.0			
•C(CH$_3$)$_2$-NO$_2$	+42.2		+44.4	394.9
•CCl$_2$-CHCl$_2$		+42.3 (G3)[c]	+46.3	393 ± 8

Table 5.6 (*Continued*)

Radical	RSE(3) (G3(MP2)-RAD)	RSE(3) (Other)	RSE(3) (Exp.)	BDE (C–H) (Exp.)[a]
•CCl$_3$	+42.5		+46.8	392.5 ± 2.5
•C(CH$_3$)$_2$–SC(CH$_3$)$_2$–CN	+43.3			
•C(CH$_3$)$_2$–SH	+44.3			
•C(CH$_3$)$_2$–C(O)N(CH$_3$)$_2$	+44.6			
•C(CH$_3$)$_2$–SCH$_3$	+45.6			
•C(CH$_3$)$_2$–SCH$_2$–COOCH$_3$	+46.9			
•C(CH$_3$)$_2$–C(O)NHCH$_3$	+49.6			
•C(CH$_3$)$_2$–N(CH$_3$)$_2$	+49.6			
•C(CH$_3$)$_2$–CONH$_2$	+50.8			
•C(CH$_3$)$_2$–NHCH$_3$	+50.9			
•C(CH$_3$)$_2$–NH$_2$	+51.5		+67.3	372.0 ± 8.4
•C(CH$_3$)(COOCH$_3$)–CH$_2$C(CH$_3$)$_2$–COOCH$_3$	+52.3			
•C(CH$_3$)$_2$–COOC(CH$_3$)$_3$	+53.9			
•C(CH$_3$)$_2$–SOCH$_3$	+54.0			
•C(CH$_3$)$_2$–COOCH$_3$	+54.6[27]			
•C(CH$_3$)$_2$–COOCH$_2$CH$_3$	+54.7		+51.9	387.4
•C(CH$_3$)(COOCH$_3$)–CH$_2$CH(CH$_3$)–COOCH$_3$	+55.2			
•C(CH$_3$)$_2$–COOH	+55.7		+50.3	389.0
•C(CH$_3$)(COOCH$_3$)–CH$_2$CH(CH$_3$)$_2$	+57.4			
•C(CH$_3$)$_2$–CN	+58.3[23b]	+58.3 (G3(MP2)-RAD(+))[26] +60.3 (G3)[c]	+54.8	384.5
•C(CH$_3$)$_2$–CO–C$_6$H$_5$	+58.8		+63.2	376.1
•C(CH$_3$)$_2$–CO–CH$_3$	+64.4			
•C(CH$_3$)$_2$–C$_6$H$_5$	+69.7			
•C(CH$_3$)$_2$–C$_6$H$_4$–pOH	+70.1			
•C(CH$_3$)$_2$–C$_6$H$_4$–pOCH$_3$	+70.8			
•C(CH$_3$)$_2$–CC–CH$_3$		+71.3 (G3)[c]	+95.0	344.3 ± 11.3
•C(CH$_3$)$_2$–CC–H	+72.0	+71.9 (G3)[c]	+94.1	345.2 ± 8.4
•C(CH$_3$)$_2$–C$_6$H$_4$–pNO$_2$	+73.2			
•C(CH$_3$)$_2$–C$_6$H$_4$–pCN	+73.6			
•C(CH$_3$)$_2$–C$_6$H$_4$–pCHO	+73.7			
•C(CH$_3$)$_2$–CH=C(CH$_3$)$_2$	+77.8			
•C(CH$_3$)$_2$–C(CH$_3$)=CH$_2$		+80.0 (G3)[c]	+88.2	351.1
•C(CH$_3$)$_2$–CH=CHCH$_3$	+87.2			
•C(CH$_3$)$_2$–CHCH$_2$	+88.7	+89.0 (G3)[c]	+106.7	332.6 ± 7.1
•C(CH$_3$)$_2$–BH$_2$	+91.0			

[a]All BDE data taken from Ref. 3 if not specified otherwise.

[b]Reference data given at 0K and corrected to 298.15K by thermal corrections calculated at (U)B3LYP/6-31G(d) level using a scaling factor of 0.9806.

[c]Calculated from BDE(C–H) data at 298.15K in Ref. 7.

TABLE 5.7   RSEs for Vinyl Radicals at 298.15K Calculated According to Equation 5.3 Together with BDE(C-H) Energies for Selected Systems (All in kJ/mol)

Radical	RSE(3) (G3(MP2)-RAD)	RSE(3) (Other)	RSE(3) (Exp.)	BDE(C-H) (Exp.)[a]
•CCH	−117.6	−122.6 (G3)[b]	−117.3	556.6 ± 2.9
•CN	−93.8	−100.9 (G3)[b]	−88.3	527.6 ± 1.7
•CF=CF$_2$		−58.4 (G3)[b]		
•CF=CFCl		−52.0 (G3)[b]		
•CF=CFH		−47.4 (G3)[b]		
•C$_6$H$_4$-pOH	−41.9			
•C$_6$H$_4$-pOCH$_3$	−41.6			
•C$_6$H$_4$-pNO$_2$	−40.4			
•C$_6$H$_4$-pCN	−39.4			
•C$_6$H$_5$	−37.0	−48.4 (G3)[b]	−32.9	472.2 ± 2.2
•CCl=CFCl		−38.3 (G3)[b]		
•CH=C(CH$_3$)$_2$	−34.3			
•CH=CHCH$_3$ (E)	−32.0		−25.5	464.8
•CCl=CHCl		−29.5 (G3)[b]		
•CH=CH$_2$	−26.6	−25.3 (G3)[b]	−26.0	465.3 ± 3.3
•CH=C=O		−12.8 (G3)[b]		
•CH$_3$ (**4**)	0.0	0.0	0.0	439.3 ± 0.4
•COOH	+14.2		+35.1	404.2
•COOCH$_3$	+16.2	+18.0 (G3)[b]	+51.4	387.9 ± 4.2
•COOCH$_2$CH$_3$	+18.1			
•COOC(CH$_3$)$_3$	+24.6			
•C(O)N(CH$_3$)$_2$	+37.8			
•C(O)NHCH$_3$	+38.1			
•C(O)N(CH$_2$CH$_3$)$_2$	+38.6			
•C(O)NH$_2$	+39.6			
•C(O)CF$_3$		+46.9 (G3)[b]		
•C(O)C$_6$H$_5$	+52.4		+68.2	371.1 ± 10.9
•C(O)CH=CH$_2$		+54.5 (G3)[b]		
•C(O)CH$_3$	+59.8	+62.0 (G3)[b]	+65.3	374.0 ± 1.3
•C(O)CH$_2$CH$_3$		+61.1 (G3)[b]		
•C(O)H	+64.6	+66.3 (G3)[b]	+70.9	368.4 ± 0.7

[a]All BDE data taken from Ref. 3, if not specified otherwise.
[b]Calculated from BDE(C–H) data at 298.15K in Ref. 7.

5.4.2   Assessment of Radical Stability in Other Types of Reactions

Although RSEs specifically measure radical stability toward hydrogen atom abstraction, they often provide a qualitative guide to radical stability and reactivity in other types of reactions. For example, one generally finds that radicals that have high RSEs such as benzyl radicals tend to have higher barriers and lower exothermicities in radical addition to alkenes, when compared with less stabilized radicals such as

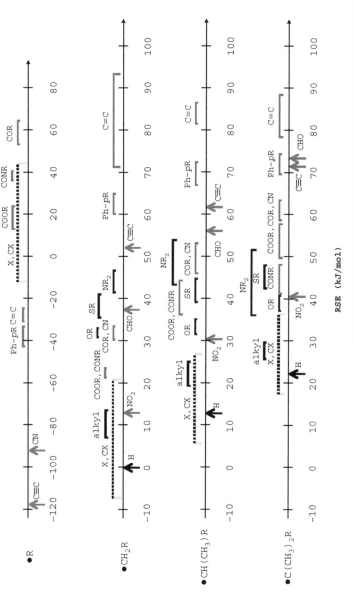

FIGURE 5.1 RSEs for primary, secondary, and tertiary carbon-centered radicals at 298.15 K, calculated according to Equation 5.3 (all in kJ/mol).

methyl.[35] Likewise in a competitive addition of a specific radical to two different alkenes, the favored reaction is usually that which results in the most stabilized product radical. Of course other factors, such as polar and steric effects are important; nonetheless, RSEs play an important role in interpreting structure–reactivity trends in these reactions, and this in turn has predictive value. For example, in a free-radical copolymerization of styrene with vinyl acetate, propagating radicals with styryl terminal units have significantly higher RSEs than those terminated by vinyl acetate units, due to the greater resonance stabilization of the unpaired electron in the former case. As a result, both types of propagating radicals tend to add to styrene rather than vinyl acetate, resulting in copolymers that are rich in styrene, even at low monomer feed ratios.[36]

RSEs have proven to be particularly useful in controlled radical polymerization processes such as reversible addition fragmentation chain transfer (RAFT) polymerization,[37] where a consideration of the effects of substituents on radical stability has contributed to the design of new and improved control agents.[38] Controlled radical polymerization processes allow for the synthesis of polymers with narrow molecular weight distributions, designer endgroups and special architectures, by minimizing the occurrence of bimolecular termination processes with respect to chain growth. This is achieved by establishing a dynamic equilibrium between the propagating radical and a dormant or protected species. However, their success rests upon a delicate balance of the rates of several competing reactions such that the concentration of the active species is orders of magnitude lower than that of the dormant species and the exchange between the two forms is rapid. Choosing optimal control agents for various types of monomer can be difficult. We have shown that RSEs, together with related isodesmic measures of the stabilities of the other reagents present, can provide a simple practical method for ranking and analyzing the effects of substituents on the controlling equilibrium in RAFT polymerization, and thereby aid in the design and selection of optimal control reagents for the various classes of monomer.[39] Moreover, on the basis of this analysis we were able to design first multipurpose RAFT agent, which was subsequently verified experimentally.[38] A similar approach, albeit based on the Rüchardt radical stability parameter,[40] has been used successfully in nitroxide-mediated polymerization (NMP),[41] where it aided the design of the first effective NMP agent for methyl methacrylate polymerization.[42]

5.5 CONCLUSIONS

Radical stabilization energies for a wide variety of carbon-centered radicals have been calculated at G3(MP2)-RAD or better level. While the interpretation of these values as the result of substituent effects on radical stability is not without problems, the use of these values in rationalizing radical reactions is straight forward. This is not only true for reactions involving hydrogen atom transfer steps but also for other reactions involving "typical" elementary reactions such as the addition to alkene double bonds and thiocarbonyl compounds.

REFERENCES

1. Zipse, H. *Topics Curr. Chem.* **2006**, *263*, 163–189.

2. Zavitsas, A. A. *J. Chem. Edu.* **2001**, *78*, 417–419.

3. Luo, Y. -R, *Comprehensive Handbook of Chemical Bond Energies*, CRC Press, 2007; Standard bond dissociation energies (in kJ/mol) for the systems in Tables 5.1 and 5.2 are listed as follows: CH_3-H +439.3 ±0.4 kJ/mol; CH_3CH_2-H +420.5 ± 1.3 kJ/mol; $(CH_3)_2CH$-H +410.5 ± 2.9 kJ/mol; $(CH_3)_3C$-H +400.4 ± 2.9 kJ/mol; CH_2CHCH_2-H +369 ± 3 kJ/mol; $C_6H_5CH_2$-H +375.5 ± 5 kJ/mol; CH_2F-H +423.8 ± 4.2 kJ/mol; CHF_2-H +431.8 ± 4.2 kJ/mol; CH_3-CH_3 +377.4 ± 0.8 kJ/mol; CH_3-CH_2CH_3 +370.3 ± 2.1 kJ/mol; CH_3CH_2-CH_2CH_3 +363.2 ± 2.5 kJ/mol; CH_3-CH_2F +388.3± 8.4 kJ/mol; CH_3-CHF_2 +405.0 ± 8.4 kJ/mol; CH_3-CF_3 +429.3 ± 5.0 kJ/mol.

4. (a) Henry, D. J.; Parkinson, C. J.; Radom, L. *J. Phys. Chem. A* **2002**, *106*, 7927. (b) Henry, D. J.; Sullivan, M. B.; Radom, L. *J. Chem. Phys.* **2003**, *118*, 4849–4860.

5. Baboul, A. G.; Curtiss, L. A.; Redfern, P. C.; Raghavachari, K. *J. Chem. Phys.* **1999**, *110*, 7650–7657.

6. Standard bond dissociation energies (in kJ/mol) as obtained from *CRC Handbook of Chemistry and Physics*, 88th ed.; David, R. L. Ed.; CRC publishers, 2007, are as follows: CFH_2-CFH_2 +368.2 ± 8.4 kJ/mol; H-CF_3 +445.2 ± 2.9 kJ/mol.

7. Feng, Y.; Liu, L.; Wang, J. T.; Huang, H.; Guo, Q. X. *J. Chem. Inf. Comput. Sci.* **2003**, *43*, 2005–2013.

8. Schwabe, T.; Grimme, S. *Acc. Chem. Res.* **2008**, *41*, 569–579.

9. Tarnopolsky, A.; Karton, A.; Sertchook, R.; Vuzman, D.; Martin, J. M. L. *J. Phys. Chem. A* **2008**, *112*, 3–8.

10. Zhao, Y.; Truhlar, D. G. *J. Phys. Chem. A* **2008**, *112*, 1095–1099.

11. Scott, A. P.; Radom, L. *J. Phys. Chem.* **1996**, *100*, 16502–16513.

12. Coote, M. L. *J. Phys. Chem. A* **2004**, *108*, 3865–3872.

13. Martin, J. M. L.; de Oliveira, G. *J. Chem. Phys.* **1999**, *111*, 1843–1843; Martin, J. M. L. *Chem. Phys. Lett.* **1999**, *310*, 271–276; Parthiban, S.; Martin, J. M. L. *J. Chem. Phys.* **2001**, *114*, 6014–6029.

14. Menon, A. S.; Wood, G. P. F.; Moran, D.; Radom, L. *J. Phys. Chem. A* **2007**, *111*, 13638–13644; correction: Menon, A. S.; Wood, G. P. F.; Moran, D.; Radom, L. *J. Phys. Chem. A* **2008**, *112*, 5554.

15. Berkowitz, J.; Ellison, G. B.; Gutman, D. *J. Phys. Chem.* **1994**, *98*, 2744–2765.

16. Ellison, G. B.; Davico, G. E.; Bierbaum, V. M.; DePuy, C. H. *Int. J. Mass. Spectrom. Ion Proc.* **1996**, *156*, 109–131.

17. Muralha, V. S. F.; Borges dos Santos, R. M.; Martinho Simoes, J. A. *J. Phys. Chem. A* **2004**, *108*, 936–942.

18. Blanksby, S. J.; Ellison, G. B. *Acc. Chem. Res.* **2003**, *36*, 255–263.

19. Wenthold, P. G.; Polak, M. L.; Lineberger, W. C. *J. Phys. Chem.* **1996**, *100*, 6920–6926.

20. Bernardes, C. E. S.; Minas da Piedade, M. E.; Amaral, L. M. P. F.; Ferreira, A. I. M. C. L.; Ribeiro da Silva, M. A. V.; Diogo, H. P.; Costa Cabral, B. J. *J. Phys. Chem. A* **2007**, *111*, 1713–1720.

21. Matus, M. H.; Nguyen, M. T.; Dixon, D. A. *J. Phys. Chem. A* **2007**, *111*, 113–126.

22. Henry, D. J.; Parkinson, C. J.; Mayer, P. M.; Radom, L. *J. Phys. Chem. A* **2001**, *105*, 6750–6756.

23. Coote, M. L. *Macromolecules* **2004**, *37*, 5023–5031.

24. Li, M. J.; Liu, L.; Wei, K.; Fu, Y.; Guo, Q. X. *J. Phys. Chem. B* **2006**, *110*, 13582–13589.

25. Wood, G. P. F.; Moran, D.; Jacob, R.; Radom, L. *J. Phys. Chem. A* **2005**, *109*, 6318–6325.

26. Lin, C. Y.; Coote, M. L.; Gennaro, A.; Matyjaszewski, K. *J. Am. Chem. Soc.* **2008**, *130*, 12762–12774.

27. Lin, C. Y.; Coote, M. L.; Petit, A.; Richard, P.; Poli, R.; Matyjaszewski, K. *Macromolecules* **2007**, *40*, 5985–5994.

28. (a) Vreven, T.; Morokuma, K. *J. Chem. Phys.* **1999**, *111*, 8799–8803. (b) Vreven, T.; Morokuma, K. *J. Comp. Chem.* **2000**, *21*, 1419–1432.

29. Izgorodina, E. I.; Brittain, D. R. B.; Hodgson, J. L.; Krenske, E. H.; Lin, C. Y.; Namazian, M.; Coote, M. L. *J. Phys. Chem. A* **2007**, *111*, 10754–10768.

30. Li, M.-J.; Liu, L.; Fu, Y.; Guo, Q.-X. *J. Phys. Chem. B* **2005**, *109*, 13818–13826.

31. Rauk, A.; Armstrong, D. A.; Fairlie, D. P. *J. Am. Chem. Soc.* **2000**, *122*, 9761–9767.

32. Berry, R. J.; Wilson, A. L.; Schwartz, M. *J. Mol. Struct. (Theochem)* **2000**, *496*, 121–129.

33. Croft, A. K.; Easton, C. J.; Radom, L. *J. Am. Chem. Soc.* **2003**, *125*, 4119–4124.

34. Barratt, B. J. W.; Easton, C. J.; Henry, D. J.; Li, I. H. W.; Radom, L.; Simpson, J. S. *J. Am. Chem. Soc.* **2004**, *126*, 13306–13311.

35. Fischer, H.; Radom, L. *Angew. Chem., Int. Ed.* **2001**, *40*, 1340–1371.

36. Coote, M. L.; Davis, T. P. *Prog. Polym. Sci.* **1999**, *24*, 1217.

37. Moad, G.; Rizzardo, E.; Thang, S. H. *Aust. J. Chem.* **2005**, *58*, 379.

38. Coote, M. L.; Henry, D. J. *Macromolecules* **2005**, *38*, 5774–5779.

39. Krenske, E. H.; Izgorodina, E. I.; Coote, M. L. *ACS Symp. Ser.* **2006**, *944*, 406–420.

40. (a) Rüchardt, C.; Beckhaus, H.-D. *Top. Curr. Chem.* **1985**, *130*, 1–22. (b) Welle, F. M.; Beckhaus, H. -D.; Rüchardt, C. *J. Org. Chem.* **1997**, *62*, 552–558. (c) Brocks, J. J.; Beckhaus, H. -D.; Beckwith, A. L. J.; Rüchardt, C. *J. Org. Chem.* **1998**, *63*, 1935–1943.

41. Bertin, D.; Gigmes, D.; Marque, S. R. A.; Tordo, P. *Macromolecules* **2005**, *38*, 2638.

42. Guillaneuf, Y.; Gigmes, D.; Marque, S. R. A.; Astolfi, P.; Greci, L.; Tordo, P.; Bertin, D. *Macromolecules* **2007**, *40*, 3108–3114.

6

INTERPLAY OF STEREOELECTRONIC VIBRATIONAL AND ENVIRONMENTAL EFFECTS IN TUNING PHYSICOCHEMICAL PROPERTIES OF CARBON-CENTERED RADICALS

VINCENZO BARONE[1], MALGORZATA BICZYSKO[2], AND PAOLA CIMINO[3]

[1]*Scuola Normale Superiore di Pisa, Pisa, Italy*

[2]*Dipartimento di Chimica, Università Federico II, Napoli, Italy*

[3]*Dipartimento di Scienze Farmaceutiche, Università di Salerno, Fisciano, Italy*

6.1 INTRODUCTION

Carbon-centered radicals play a significant role in a number of processes of technological and biological relevance, but as most of the radicals are unstable and very reactive species, difficult to study by experimental techniques. Many available experimental studies focused on the physicochemical properties of radicals come from the spectroscopic techniques, not only electronic spin resonance (ESR) but also IR/Raman and UV–vis, often obtained in matrices at low temperature. Unfortunately, in most cases, experimental results are extremely difficult to interpret, even in terms of appropriate identification of molecular species. For example, the entire UV–vis

Carbon-Centered Free Radicals and Radical Cations, Edited by Malcolm D. E. Forbes
Copyright © 2010 John Wiley & Sons, Inc.

spectrum recently reported and assigned to the phenyl radical[1] should be rather attributed to the phenoxy radical.[2] In fact, quite elaborate integrated experimental procedures involving the combined use of several complementary spectroscopic techniques and possibly few precursors are usually necessary to unequivocally identify radical species. On the other hand, recent developments in computational techniques, in particular the high accuracy of the results achievable for small systems,[3–7] clearly demonstrates the potentiality of computational chemistry experiments to become key tools for the prediction and understanding of spectroscopic properties of all kinds of molecular systems: this reflects a more general trend, since nowadays the experimental characterization of new systems relies more and more on computational approaches, for example, for the evaluation and rationalization of structural, energetic, electronic, and dynamic features.[8–10] Recently integrated approaches capable of accurately simulating vibronic spectra, at the same time easily accessible to nonspecialists, have been introduced[11]. As such, a fully automatic computation and visualization of vibrationally resolved UV–vis spectra complements tools already available for other spectroscopic ranges (e.g., IR/Raman), highly increasing the overall predictive/interpretative power of computational spectroscopy. Extension of such approaches to larger flexible systems in condensed phases involves a number of additional problems ranging from the development of cheaper, yet reliable, electronic methods to the implementation of effective vibrational approaches beyond the harmonic approximations and to the proper account of environmental effects by effective discrete/continuum models. Furthermore, procedures validated for closed-shell systems and/or for properties ruled by valence electrons need further elaboration and validation for open-shell systems and for properties (such as hyperfine couplings) receiving significant contributions from core electrons.

Here last-generation approaches rooted into the density functional theory (DFT) and its time-dependent extension (TD-DFT) have revolutionized the situation because they are able to couple in a very effective way computational speed and reliability. There are, of course, situations in which such approaches need further improvements (e.g., charge transfer electronic transitions or van der Waals interactions), but the present situation is, in general, already satisfactory and ongoing work in several research groups promises further significant progresses.

Direct comparison with experimental results requires that dynamic effects are properly taken into account. At present, the most common way to evaluate molecular properties is their computation at the "bottom of the well" corresponding to the global minimum. Nevertheless, this approach does not take the vibrational effects into account that can be nonnegligible even at very low temperatures. The molecular vibrations imply that molecular geometry is never fixed and might vary in a way defined by the zero point vibrational energy levels. In an ideal case, when harmonic approximation is valid and a perfectly symmetric potential energy profile represents correctly the true shape of the potential energy surface (PES) close to a minimum, the vibrationally averaged properties and those calculated at the "bottom of the well" are coincident. But such a condition is not generally fulfilled and the harmonic potential can only be considered as an approximation valid in a region close to the minimum. This model is useful and allows determining molecular properties in a simple manner;

however, in many cases it is crucial to take into proper account the effects of nuclear motions for the direct comparison between experimental and theoretical results. In particular, the ESR parameters often show a strong dependence on the molecular geometry. A number of studies have confirmed that the effects of vibrational motion on some parameters can change their values up to 25%, as, for example, in the case of a (H2) of cyclopropyl.[12]

6.2 EPR SPECTROSCOPY

6.2.1 Theoretical Background

The *effective* spin Hamiltonian (SH) that includes Zeeman and hyperfine interaction for a single unpaired electron and N nuclei is:

$$\hat{H} = \hat{H}_e + \hat{H}_{eN} = \frac{\beta_e}{\hbar} \mathbf{B}_0 \cdot \mathbf{g} \cdot \hat{\mathbf{S}} + \gamma_e \sum_n \hat{\mathbf{I}}_n \cdot \mathbf{A}_n \cdot \hat{\mathbf{S}} \qquad (6.1)$$

where the first term is the Zeeman interaction depending upon the $\mathbf{g}$ tensor, in the presence of a magnetic field $\mathbf{B}_0$; the second term is the hyperfine interaction of each couple nth nucleus/unpaired electron, defined with respect to hyperfine tensor $\mathbf{A}_i$.

The hyperfine coupling tensor (A_X), which describes the interaction between the electronic spin density and the nuclear magnetic momentum of nucleus X, can be split into three terms: $A_X = a_X I_3 + T_X + \Lambda_X$, where I_3 is the 3×3 unit matrix. The first term (a_X), usually referred to as Fermi-contact interaction, is an isotropic contribution, also known as hyperfine coupling constant (hcc), and is related to the spin density at the corresponding nucleus X. The second contribution (T_X) is anisotropic and can be derived from the classical expression of interacting dipoles. The last term, Λ_X, is due to second-order spin-orbit coupling and can be determined by methods similar to those used for the $\mathbf{g}$ tensor.[13] In the present case, because of the strong localization of spin density on the studied atoms and of their small spin-orbit coupling constants, its contribution can be safely neglected and will not be discussed in the following. Of course, upon complete averaging by rotational motions, only the isotropic part survives.

The gyromagnetic tensor can be written: $g = g_e I_3 + \Delta g_{RM} + \Delta g_G + \Delta g_{OZ/SOC}$ where g_e is the free-electron value ($g_e = 2.0023193$). Computation of the relativistic mass (RM) and gauge (G) corrections is quite straightforward because they are first-order contributions.[14] The last term arises from the coupling of the orbital Zeeman (OZ) and the spin-orbit coupling (SOC) operator. The OZ contribution is computed using the gauge-including atomic orbital (GIAO) approach,[15] whereas for light atoms, the two electron SOC operator can be reliably approximated by a one electron operator involving adjusted effective nuclear charges.[16] Although those charges were optimized for MCSCF/HF wavefunctions, a number of test computations showed that they are nearly optimal for DFT computations too. Upon complete averaging by rotational motions, only the isotropic part of the $\mathbf{g}$ tensor survives, which is given by $g_{iso} = 1/3\mathrm{Tr}(g)$. Of course, the corresponding shift from the free electron value is $\Delta g_{iso} = g_{iso} - g_e$.

6.2.2 Environmental Effects

The most promising general approach to the problem of environmental (e.g., solvent) effects can be based, in our opinion, on a system-bath decomposition. The system includes the part of the solute where the essential of the process to be investigated is localized together with, possibly, the few solvent molecules strongly (and specifically) interacting with it. This part is treated at the electronic level of resolution, and is immersed in a polarizable continuum, mimicking the macroscopic properties of the solvent. The solution process can then be dissected into the creation of a cavity in the solute (spending energy E_{cav}), and the successive switching on of dispersion–repulsion (with energy $E_{dis-rep}$) and electrostatic (with energy E_{el}) interactions with surrounding solvent molecules.

The so-called polarizable continuum model (PCM)[17] offers a unified and well sound framework for the evaluation of all these contributions both for isotropic and anisotropic solutions. In PCM, the solute molecule (possibly supplemented by some strongly bound solvent molecules, to include short-range effects such as hydrogen bonds) is embedded in a cavity formed by the envelope of spheres centered on the solute atoms. The procedures to assign the atomic radii[18] and to form the cavity[17] have been described in detail together with effective classical approaches for evaluating E_{cav} and $E_{dis-rep}$[17,19]. Here we recall only that the cavity surface is finely subdivided in small tiles (tesserae), and that the solvent reaction field determining the electrostatic contribution is described in terms of apparent point charges appearing in tesserae and self-consistently adjusted with the solute electron density.[17,20] The solvation charges ($\mathbf{q}$) depend, in turn, on the electrostatic potential ($\mathbf{V}$) on tesserae through a geometrical matrix $\mathbf{Q}$ ($\mathbf{q} = \mathbf{QV}$), related to the position and size of the surface tesserae, so that the free energy in solution G can be written as

$$G = E[\rho] + V_{NN} + \frac{1}{2}\mathbf{V}\,\mathbf{QV} \qquad (6.2)$$

where $E[\rho]$ is the free solute energy, but with the electron density polarized by the solvent, and V_{NN} is the repulsion between solute nuclei.

The core of the model is then the definition of the $\mathbf{Q}$ matrix, which in the most recent implementations of PCM depends only on the electrostatic potentials, takes into the proper account the part of the solute electron density outside the molecular cavity, and allows the treatment of conventional, isotropic solutions, ionic strengths, and anisotropic media such as liquid crystals. Furthermore, analytical first and second derivatives with respect to geometrical, electric, and magnetic parameters have been coded, thus giving access to proper evaluation of structural, thermodynamic, kinetic, and spectroscopic solvent shifts.

6.2.3 Vibrational Effects

In the framework of the Born–Oppenheimer approximation, we can speak of a PES and of a "property surface" (PS), which can be obtained from the quantum mechanical computations at different nuclear configurations. In this scheme, the expectation values of observables (e.g., isotropic hcc values) are obtained by averaging the different properties on the nuclear wave functions. Semirigid molecules are quite well described in terms of a harmonic model, but a second-order perturbative inclusion of principal anharmonicities provides much improved results at a reasonable cost[21–25] (see Ref. 25 for extensive review on the more computationally demanding variational approaches). In perturbative model, the vibrational energy (in wavenumbers) of asymmetric tops is given by

$$E_n = \xi_0 + \sum_i \omega_i \left(n_i + \frac{1}{2} \right) + \sum_i \sum_{j<i} \xi_{ij} \left(n_i + \frac{1}{1} \right) \left(n_j + \frac{1}{2} \right) \qquad (6.3)$$

where the ω values are the harmonic wave-numbers and the ξ values are simple functions of third (F_{ijk}) and semidiagonal fourth (F_{iijj}) energy derivatives with respect to normal modes Q^{22}. Next, the fundamental vibrational frequencies (ν_i) and zero point vibrational energy (E_0) are given by

$$\nu_i = \omega_i + 2\xi_i + \frac{1}{2} \sum_{j \neq i} \xi_{ij} \qquad (6.4)$$

$$E_n = \xi_0 + \frac{1}{2} \sum_i \left(\omega_i \frac{1}{2} \xi_{ii} \sum_{j>i} \frac{1}{2} \xi_{ij} \right) \qquad (6.5)$$

At the same level, the vibrationally averaged value of a property Ω is given by

$$\langle \Omega \rangle_n = \Omega_e + \sum_i A_i \left(n_i + \frac{1}{2} \right) \qquad (6.6)$$

where Ω_e is the value at the equilibrium geometry and

$$A_i = \frac{\beta_{ii}}{\omega_i} - \sum_j \frac{\alpha_j F_{iij}}{\omega_i \omega_j^2} \qquad (6.7)$$

α_i and β_i are the first and second derivatives of the property with respect to the ith normal mode. The first term of the right hand side of Equation 6.7 will be referred to in the following as harmonic and the second one as anharmonic. Also in this case, methods rooted in the density functional theory perform well provided that numerical problems are properly taken into account.[26,27]

6.2.4 Dynamical Effects

Large amplitude motions and solvent librations cannot be described by the perturbative approach sketched above, but a classical treatment is usually sufficient. Then, the computational strategy involves two independent steps: first, MD simulations are run for sampling with one or more trajectories the general features of the solute–solvent configurational space; then, observables are computed exploiting the discrete/continuum approach for supramolecular clusters, made by the solute and its closest solvent molecules, as averages over a suitable number of snapshots. It is customary to carry out the same steps also for the molecule in the gas phase, just to have a comparison term for quantifying solvent effects. The *a posteriori* calculation of spectroscopic properties, compared to other *on-the-fly* approaches, allows us to exploit different electronic structure methods for the MD simulations and the calculation of physical–chemical properties. In this way, a more accurate treatment for the more demanding molecular parameters, of both first (e.g., hyperfine coupling constants) and second (e.g., electronic **g** tensor shifts) order, could be achieved independently of structural sampling methods: first principles, semiempirical force fields, as well as combined quantum mechanics/molecular-mechanics approaches could be all exploited to the same extent, once the accuracy in reproducing reliable structures and statistics is proven.

6.3   CALCULATION OF EPR PARAMETERS

First, in this section we illustrate the performance of our methods on selected carbon radicals. In particular, we choose to use the new N07D[28–30] (Table 6.1) basis set (with PBE0[31] and B3LYP[32] functionals) because it seems to be the best choice for a number of properties: geometrical parameters, dipole moments, and magnetic properties. To demonstrate the accuracy that can be obtained in **A** and **g** tensor calculation, we selected 27 radicals (aliphatic and aromatic), including σ and π species (Fig. 6.1). The selected molecules are neutral, cationic, anionic; doublet, triplet, quartet; and localized and conjugated radicals.

TABLE 6.1   Functions to be Added to 6-31G Set to Obtain the N07D Basis Set for B3LYP and PBE0 Functionals

	s(B3LYP)	s(PBE0)	P	d	d
H		0.4	0.750		
B		7.0	0.035	0.343	
C	7.5	8.7	0.050	0.820	
N	12.6	13.7	0.053	1.015	
O	15.1	16.0	0.065	1.190	0.180
F	18.3	17.7	0.083	1.370	0.230
Al	3.1	3.1	0.015	0.189	
Si	3.6	5.0	0.033	0.275	
P	5.5	6.3	0.035	0.373	
S	8.0	7.5	0.041	0.479	
Cl	8.5	8.5	0.048	0.600	0.196

For He, Li, Be, Na, and Mg atoms N07D and 6-31 + G(d,p) are identical.

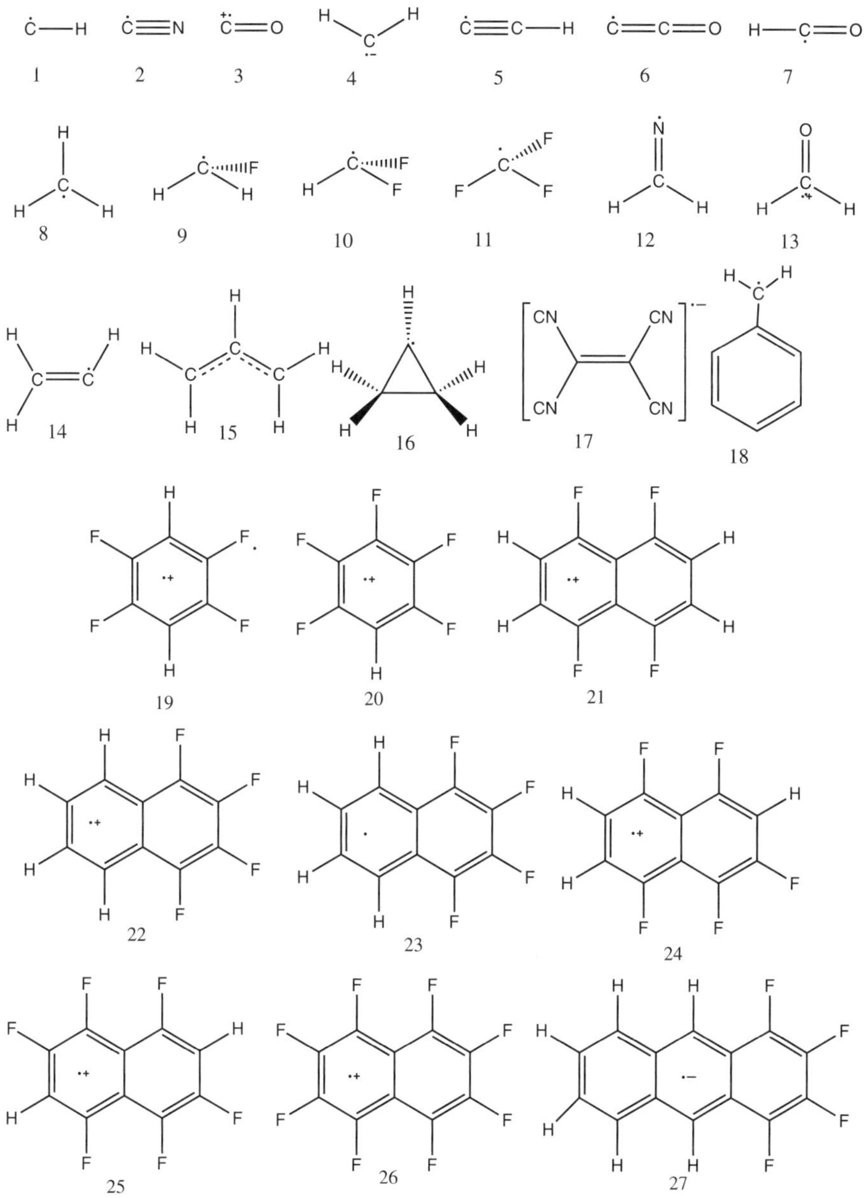

FIGURE 6.1   Structures of the radicals studied.

Then, as case study, we consider the glycine[33] and glycyl[30] radicals (Fig. 6.2) in solution. As mentioned above, the calculation of magnetic tensors needs to take into account the several factors such as the geometries, environmental effects, and dynamical effects (vibrational averaging from intramolecular vibrations and/or solvent librations). We use an integrated computational approach where the molecular

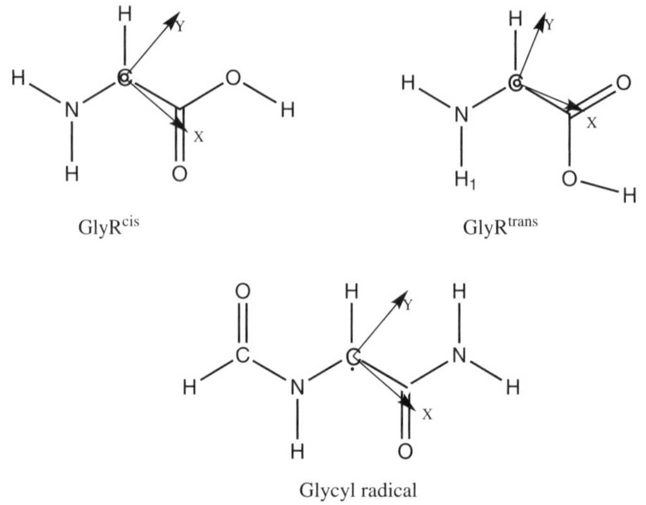

FIGURE 6.2 Structures of glycine and glycyl radicals. The orientations of the principal axes of the **g** tensor are also shown.

structure is accurately described with last generation models rooted in the density functional theory and the treatment of solvent effects is able to give full account of bulk (PCM) and specific interaction (solute–solvent H bond) with a discrete/continuum model. Finally, with molecular dynamic (MD) simulations we have a dynamical description of solute–solvent systems.

6.3.1 Geometric Parameters

The magnetic properties of a radical are tuned by small modifications of its geometry. Indeed, small geometry changes can be interpreted in terms of shape of the SOMO (single occupied molecular orbital), the orbital that directly affects the hyperfine coupling constants. For example, the deviation from planarity is the most influential geometric effect for π-like radicals (for example, the methyl radical); the bond distance between specific pair of atoms in the radical is another one. In fact, the variation of the bond distance contributing to the π-system can modify the shape of the SOMO, and, thus the magnetic properties. At the same way, in a σ-radical (vinyl) the distance between two atoms modifies the localization of the SOMO.

In summary, a good geometry is necessary to evaluate in the right way different effects, such as the direct and spin polarization contributions to the hyperfine coupling constant, for example. In this connection, the performance of the N07D basis set coupled with PBE0 and B3LYP functional are comparable to those of aug-cc-pVDZ, with increased computational efficiency.[29] For the purpose of illustration, we report in Fig. 6.3 some significant parameters of vinyl, propyl, and phenyl radicals. It is quite apparent that B3LYP/N07D results show remarkable agreement with highly accurate computational studies at CCSD(T)[34,35] and multireference[35] levels with extended basis sets. For C–C bond lengths maximum deviations do not exceed 0.01 Å, and in

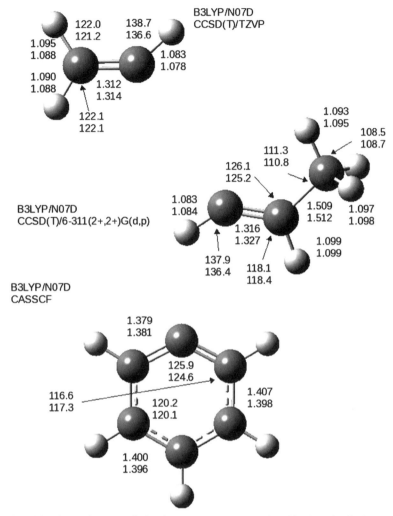

FIGURE 6.3   Ground-state optimized geometry structures (bond in Å angles in degrees) for vinyl, propanyl and phenyl radicals.

many cases are as small as 0.002 Å. For C–H bond lengths even better agreement has been found with a maximum discrepancy of 0.05 Å. The same stands for angles that are predicted by B3LYP/N07D model with accuracy of two degrees. These findings are particularly encouraging having in mind larger systems for which expensive coupled cluster studies are still unfeasible.

6.3.2   EPR Parameters

The N07D basis set with PBE0 and B3LYP functional provides a good description of the magnetic properties of carbon radicals in comparison with some standard basis set at a very convenient computational time.[28-30] The results are collected in Tables 6.2–6.6.

We report the number of data (N), mean absolute deviation (MAD), data range, and average absolute error and average percent error between calculated and experimental values. Next, we give the correlation coefficient (R^2), slope, and intercept of the least square line.

Hydrogen atoms require some specific considerations in view of the lack of inner shells and of the overwhelming role of small hcc values in an unbiased statistics. We have thus selected a specific set of data in which the presence of σ radicals (characterized by large hcc values) has been overemphasized. The results collected in Tables 6.2 and 6.3 show that different basis sets are nearly equivalent in this connection leading to a percent error approximately 10%, which is close to that of second row atoms. For the radicals containing carbon atoms, we compare (Table 6.4) our data with the theoretical ones calculated with four different basis sets [6-31G(d),

TABLE 6.2   Theoretical and Experimental Hyperfine Coupling Constants (Gauss) of Hydrogen Nuclei of the Radicals Studied. All the Theoretical Values have been Obtained with B3LYP Functional

		6-31G(d)	EPR-II	EPR-III	N07D	Exp[a]
1	H	−18.2	−17.8	−17.1	−17.0	20.6
4	2H	−17.1	−16.8	−13.8	−14.2	16.0
5	H	21.8	21.9	21.1	21.8	18.0
8	3H	−25.2	−23.8	−23.2	−23.0	25.0
9	2H	−18.3	−17.8	−17.6	−17.4	21.1
10	H	−23.1	−24.5	−25.3	−22.1	22.2
12	H	−78.8	−83.4	−83.0	−79.4	83.2
13	2H	−121.5	−128.2	−128.7	−120.0	132.7
14	H*cis.*	59.8	63.8	63.8	59.5	68.5
14	H*trans*	36.6	40.2	40.2	36.0	34.2
14	H(CH)	14.0	17.1	17.5	13.9	13.3
15	2H*cis*	−15.8	−14.9	−14.5	−14.3	13.5
15	2H*trans*	−16.5	−15.7	−15.4	−15.0	14.8
15	H(CH)	4.9	4.7	4.5	4.4	4.2
18	H(orto)	−6.0	−5.8	−5.6	−5.9	5.2
18	H(*meta*)	2.6	2.5	2.4	2.5	1.8
18	H(*para*)	−6.8	−6.6	−6.5	−6.7	6.2
18	H(CH2)	−18.3	−17.4	−16.8	−17.6	16.3
21	4H	−2.0	−2.0	−1.9	−2.0	4.0
22	2H	−5.1	−4.9	−4.9	−5.1	2.4
22	2H	−1.2	−1.2	−1.2	−1.2	0.6
23	2H	−7.3	−6.8	−6.7	−7.1	5.6
23	2H	−2.1	−2.0	−2.0	−2.1	1.4
24	H	−1.1	−1.1	−1.1	−1.1	2.1
24	H	−3.1	−2.9	−2.9	−3.0	4.2
25	2H	−0.1	−0.1	−0.1	−0.1	0.3

[a]Experimental data for 1, 4, 5, 8–10,12–15, 18 are from Ref. 36; for 21–25 are from Ref. 37.

TABLE 6.3   Data Analysis for Hydrogen Nuclei. MAD (Mean Absolute Deviation in Gauss) $= \Sigma |a_{calc} - a_{exp}|/N$; **E% (Percent Error)** $= |a_{calc} - a_{exp}|/a_{exp}$

	B3LYF/ 6-31G(d)	B3LYP/ EPR-II	B3LYP/ EPR-III	B3LYP/ NO7D	Exp.
Hydrogen: N = 29					
MAD	2.1	1.9	1.9	1.8	
Max absolute error	11.3	6.0	6.0	9.0	
Average E%	27.2%	27.1%	26.9%	27.4%	
Max E%	113.1%	108.4%	104.6%	113.9%	
R2	0.9946	0.9938	0.9934	0.9943	
Intercept	1.4588	0.9353	0.6423	0.9796	
Slope	0.9104	0.9670	0.9696	0.9060	
Max	121.5	128.2	128.7	120.0	132.7
Min	0.1	0.1	0.1	0.1	1.8

TABLE 6.4   Theoretical (B3LYP) and Experimental Hyperfine Coupling Constants (Gauss) of Carbon Nuclei

Structure	Carbon (^{13}C)	6-31G(d)[a]	EPR-III[a]	TZVP[a]	cc-pVQZ[a]	N07D[b]	Exp[a]
1	C	16.8	19.0	14.8	9.0	15.5	16.8
2	C	183.9	207.7	214.6	201.6	200.9	209.8
3	C	495.3	569.3	587.8	566.5	555.3	561.3
4	C	33.0	29.4	20.7	16.0	27.3	21.0
5	C	348.2	378.7	390.7	374.3	371.3	362.0
5	C(H)	72.9	81.2	82.9	83.5	78.9	76.0
6	C	16.7	14.2	12.0	8.1	13.3	15.7
6	C(O)	−2.7	−7.1	−8.5	−8.1	−6.5	10.7
7	C	152.2	138.5	142.7	136.8	137.1	134.7
8	C	44.4	28.6	27.0	19.9	28.7	27.0
9	C	62.3	56.4	50.9	40.0	56.3	54.8
10	C	152.8	143.4	146.8	137.6	145.4	148.8
11	C	258.9	264.5	274.0	261.3	266.9	271.6
12	C	−23.6	−24.5	−25.7	−24.4	−24.4	28.9
13	C	−30.1	−33.5	−35.1	−33.2	−33.2	38.9
14	C	121.9	107.7	109.6	101.5	109.5	107.6
14	CH2	−7.0	−4.9	−5.3	−4.0	−4.8	8.6
15	CH	−17.5	−16.0	−16.3	−14.9	−15.9	17.2
15	CH2	28.2	18.3	17.0	13.2	18.6	21.9
16	CH	108.8	93.6	95.5	87.7	95.5	95.9
17	CN	−8.3	−9.2	−9.7	−8.9	−9.1	9.5
18	CH2	32.0	20.4	19.0	15.1	21.0	24.5
18	Cl(3)	−14.4	−13.7	−14.1	−13.2	−13.6	14.5

[a] From Ref. 36.
[b] From Ref 29.

TABLE 6.5   Data Analysis for Carbon Nuclei. MAD (Mean Absolute Deviation in Gauss) $= \Sigma \, |a_{\text{calc}} - a_{\text{exp}}|/N_{(\text{totalnuclei})}$**;   E% (Percent Error)** $= |a_{\text{calc}} - a_{\text{exp}}|/a_{\text{exp}}$

	B3LYP/ 6-31G(d)	B3LYP/ EPR-III	B3LYP/ TVPZ	B3LYP/ cc-pVQZ	B3LYP/ NO7D	PBE0/ NO7D	Exp.
Carbon: $N = 23$							
MAD	10.7	4.0	5.0	6.6	3.5	2.9	
Average E%	18.6%	10.9%	9.3%	18.7%	10.5%	9.9%	
R2	0.9915	0.9988	0.9991	0.9983	0.9991	0.9994	
Intercept	7.5938	−1.7012	−3.6841	−6.2305	−1.1153	−0.9734	
Slope	0.9033	1.0182	1.0562	1.0196	0.9987	0.9968	
Max	495.3	569.3	587.8	566.5	555.3	555.3	561.3
Min	2.7	4.9	5.3	4.0	4.8	4.8	8.6

EPR-III, TVPZ, and cc-pVQZ]. In general, all DFT methods yield a_X values close to the experimental ones, and the best results are consistently delivered by the N07D basis set.

The performances of the different basis sets are compared in Table 6.5 (statistical analysis data). The N07D results for carbon atoms are much better than those delivered by other (even significantly larger) basis sets both in terms of MADs and closeness of the slope of the linear regression to the theoretical value of 1.0. The PBE0 and B3LYP/ N07D models give by far the lowest MAD (Table V) for C hcc values.

The N07D basis set has been assessed by comparison with EPR-III[33] basis set for the **g** tensors. The results collected in Table 6.6 show a very good agreement between computed and experimental values.

Furthermore, the availability of effective discrete/continuum solvent models and of different dynamical approaches, together with the N07D basis set, allow to perform comprehensive analyses aimed at evaluating the roles of stereoelectronic, vibrational,

TABLE 6.6   Theoretical and Experimental g_{iso} Tensor (in ppm)

Stucture	PBE0/EPR-III[a]	B3LYP/EPR-III[a]	PBE0/N07D	B3LYP/N07D	Exp.[b]
1	2.0008	2.0008	2.0010	2.0010	2.0007
2	2.0005	2.0005	2.0006	2.0006	2.0000
3	2.0002	2.0003	2.0004	2.0003	2.0003
4	2.0027	2.0027	2.0027	2.0027	2.0027
5	2.0040	2.0041	2.0039	2.0039	2.045
6	2.0040	2.0040	2.0033	2.0039	2.0041
7	2.0033	2.0034	2.0032	2.0033	2.0031
8	2.0028	2.0028	2,0027	2.0028	2.0028
9	2.0042	2.0042	2.0041	2.0041	2.0036
10	2.0027	2.0027	2.0026	2.0027	2.0027
11	2.0028	2.0028	2.0027	2.0028	2.0028
12	2.0029	2.0029	2.0028	2.0028	2.0026

[a] Single point calculations on geometries optimized at the PBE0/N07D level.
[b] From Ref. 30.

TABLE 6.7 Hyperfine Coupling Constants (in Gauss) of Glycine Radical

	Exp. pH:1–10	B3LYP/EPR-II		B3LYP/N07 D	
		cis	trans	cis	trans
N	6.4	5.5	5.5	6.3	6.3
H1	5.6	−5.4	−5.6	−5.3	−5.1
H2	5.6	−5.3	−5.4	−5.3	−5.1
Hα	11.8	−11.9	−11.7	−11.6	−11.7
	Exp. pH:1–10	PBE0/EPR-II		PBE0/N07D	
		cis	trans	cis	trans
N	6.4	5.5	5.5	6.4	6.1
H1	5.6	−5.4	−5.6	−5.0	−4.8
H2	5.6	−5.3	−5.4	−5.2	−4.9
Hα	11.8	−11.9	−11.7	–	–
				11.7	11.8

and environmental effects in determining the overall properties of radicals as shown with the following examples.

6.3.3 Case Studies: Glycine and Glycyl Radicals

The performances of the PBE0 and B3LYP/N07D models for a typical problem involving at the same time stereoelectronic, vibrational, and environmental effects can be judged by the results reported in Tables 6.7–6.10 for the glycine[28,29,33,38] and glycyl[30,39] radicals (Fig. 6.2) in aqueous solution.

6.3.3.1 Glycine Radical Since the hcc values computed for the minimum energy structure in vacuum are significantly tuned by both intramolecular vibrations and by solvent librations, the reported results (Table 6.7) are obtained by averaging more than 100 frames extracted at regular time steps from the *ab initio* dynamics described in Ref. 38. From a general point of view, all the computations provide, as expected, positive values for the C^α and N hcc values, and negative values for the hydrogen atoms. Moreover, dynamical effects reduce the differences between the pairs H_1, H_2, and C^α, H^α. Polar solvents increase delocalization along the GlyR backbone, due to an

TABLE 6.8 Hyperfine Coupling Constants (in Gauss) of Glycyl Radical

	Exp.	PBE0		B3LYP		PBE0 Best Estimate		B3LYP Best Estimate	
		EPR-II	N07D	EPR-II	N07D	EPR-II	N07D	EPR-II	N07D
C	16–21	18.3	17.5	18.3	17.5	13.3	12.5	12.9	13.3
Hα	14-15-17	−18.7	−19.9	−18.7	−19.9	−14.8	−15.8	−13.0	−13.7

TABLE 6.9 Calculated and Experimental g Tensors (ppm) of Glycine Radical in Aqueous Solution

	Exp.		pbe0/EPR-III		pbe0/N07D		b3lyp/EPR-III		b3lyp/N07D	
	cis	*trans*	*cis*	*trans*	*cis*	*trans*	*cis*	*trans*	*cis*	*trans*
Δg_{xx}			250.3	2620.4	2276.2	2388.2	2593.1	2702.6	2362.9	2468.5
Δg_{yy}			1386.3	1994.9	1259.1	1814.9	1437.8	2059.2	1295.9	1857.5
Δg_{zz}			−177.4	−175.7	−171.9	−172.0	−179.4	−177.6	−175.3	−173.2
g_{iso}	2.00340	2.00340	2.00356	2.00380	2.00344	2.00366	2.00360	2.00385	2.00348	2.00370
				with $\Delta_{\text{PCM-gas}}$						
Δg_{xx}			2586.3	2565.7	2359.2	2333.5	2676.1	2647.9	2445.9	2413.8
Δg_{yy}			1413.7	1928.2	1286.5	1748.2	1465.2	1992.5	1323.3	1790.8
Δg_{zz}			−191.8	−187.1	−186.3	−183.4	−193.8	−189.0	−189.7	−184.6
g_{iso}	2.00340	2.00340	2.00360	2.00375	2.00348	2.00361	2.00364	2.00380	2.00352	2.00365
				with $\Delta_{\text{GLOB/ADMP-gas}}$						
g_{iso}	2.00340	2.00340	2.00356	2.00371	2.00344	2.00357	2.00360	2.00376	2.00348	2.00361

increased importance of ionic resonance structures characterized by double N–C and C–C' bonds, and to the concomitant reduction of H^α hcc and of the pyramidalization of the aminic moiety. This last structural effect induces both a significant reduction of the H_2 hcc and an increased delocalization of the SOMO, with the consequent reduction of the C^α and H^α hcc values. After averaging by over MD snapshots in aqueous solution, the computed values are in general good agreement with experiment. It is, however, quite apparent that hydrogen atoms are described in a nearly equivalent way by the EPR-II and N07D basis set, whereas the nitrogen hcc is much improved by the new basis set, which is able to deliver quantitative agreement with experiment.

In Table 6.9 are reported the principal components of the **g** tensors (δg_{xx}, δg_{yy}, and δg_{zz}) and the g_{iso} values obtained using two functionals (PBE0 and B3LYP) and two basis sets (EPR-III and N07D) together with the g_{iso} value measured in aqueous solution (2.00340 ± 0.00005). As shown, the **g** tensors are sensitive to the conformation of the radical, for example, the **g** tensors of $GlyR^{cis}$ and $GlyR^{trans}$ are significantly different (Table 6.9). In particular, the results obtained with the N07D basis set for $GlyR^{cis}$ are in remarkable agreement with experiment, confirming the prevalence of this isomer in aqueous solution, as suggested by direct energetic evaluations.

6.3.3.2 Glycyl Radical   Also the magnetic parameters of glycyl radical are sensitive to a number of different physicochemical effects such as vibrational averaging and solvent shifts. In Tables 6.8 and 6.9 are shown the hyperfine coupling constants calculated with and without solvent and/or vibrational contributions. In Table 6.10, we compare our **g** tensors values obtained with N07D basis set for the glycyl radical with the theoretical ones issuing from PBE0/EPR-III computations and with experimental data obtained for different enzymes and for N-acetylglycyl radical.[30] In general, the **g** tensors calculated with the N07D basis set are slightly smaller than their EPR-III

TABLE 6.10   Calculated and Experimental g Tensors (ppm) of Glycyl Radical in Solution (Water). The Experimental Data are Given With an Error of ± 400 ppm

	$\langle \Delta g_{298} \rangle$	Δg_{solv}	Δg_{gas} N07D[a]	Δg_{gas} EPR-III[a]	Δg_{gas} EPR-III[b]
Δg_{xx}	−26.4	320.9	2190.1	2423.5	2363.9
Δg_{yy}	−28.9	−489.4	1402.1	1546.4	1524.9
Δg_{zz}	4.4	−5.4	−164.9	−170.3	−173.3

	Best Estimate				Exp[c]			N-Acetyl Glycyl
	N07D[a]	EPR-III[a]	EPR-III[b]	RNR	RNR	PFI	BSS	
Δg_{xx}	2484.6	2718.0	2658.4	1900	2000	2400	2200	2200/1900
Δg_{yy}	883.8	1028.1	1006.6	1000	1000	1600	1300	800/900
Δg_{zz}	−165.9	−171.3	−174.3	0	0	200	−100	−300/400

[a]Geometry optimized at the PBE0/N07D level.
[b]Geometry optimized at the PBE0/6-31 + G(d,p) level.
[c]From Ref. 30.

counterparts. Moreover, the **g** tensors are affected by both direct and indirect solvent effects as well as by intramolecular motions. Comparison with experiment confirms the remarkable performances of B3LYP/N07D and PBE0/N07D computational models when taking into the proper account dynamical and environmental effects.

6.3.4 Case Studies: Vibrationally Averaged Properties of Vinyl and Methyl Radicals

In the following section, we will see that vibrational averaging plays a significant role in the computation of reliable hcc values for a number of interesting systems. Here, we just discuss vibrational averaging effects related to inversion at the radical center of typical π (methyl) and σ (vinyl) radicals.

The results shown in Table 6.11 point out the effect of harmonic and anharmonic terms (see Eq. 6.7) on the total hyperfine coupling constants computed for the vinyl radical.[40] The harmonic contribution is the most important (although anharmonic terms are not negligible) except for H^{α} and, especially, C^{α}. As a matter of fact for this latter atom, there is a nearly exact compensation between harmonic contributions by in-plane and out-of-plane bendings, which have opposite signs. On the other hand, anharmonic terms are particularly large for C^{α} and H^{α} since they are not negligible for in-plane bending at the radical center, whereas they vanish (due to symmetry) for out-of-plane bendings.

The methyl radical has a planar equilibrium structure with a low frequency out-of-plane motion. The behavior of hcc values as a function of the out-of-plane angle (θ) is the following: a_C is always positive and increases with θ due to the progressive contribution of carbon's orbitals to the SOMO (see Fig. 6.4). The effect is similar for a_H, but since a_H is negative for the planar structure (due to spin polarization) the absolute value of a_H decreases up to $\theta = 10°$ and next increases due to the direct contribution (see Fig. 6.4). The ground vibrational function is peaked at the planar structure, thus vibrational averaging changes the coupling constant toward values that would have been obtained for pyramidal structures in a static description. The wave function of the ground vibrational state being symmetrically spread around the planar reference configuration introduces contributions of pyramidal configurations. The effect is even more pronounced in the first excited vibrational state, whose wave function has a node at the planar structure and is more delocalized than the fundamental one, thus giving increased weight to pyramidal structures.

TABLE 6.11   Harmonic and Anharmonic Contributions to hcc of Vinyl Radical Calculated at the PBE0/EPR-II//PBE/6-311 + G(d,p) Level

Atom	Min	ΔAnh	ΔHarm	Tot	Exp.
C^{α}	112.67	−2.41	0.05	110.31	107.6
C^{β}	−5.89	−0.61	−1.01	−7.51	−8.6
H^{α}	16.90	−1.61	0.52	15.81	13.4
H^{β}	64.58	0.32	1.65	66.55	65
H^{β}	41.46	0.33	1.59	43.38	37

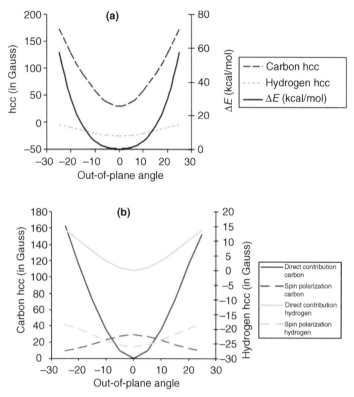

FIGURE 6.4 (a) Energy variation (in kcal/mol) and hcc for carbon and hydrogen atoms as a function of the hydrogen out-of-plane angle in methyl radical. PBE0/EPR-II computations. (b) Decomposition of the total hcc in its direct and spin polarization contribution.

As a first approximation, the vibrationally averaged value of a property can be written as

$$\langle \Omega \rangle = \Omega_{\text{ref}} + \left(\frac{\partial \Omega}{\partial s}\right)_{\text{ref}} \bullet \langle s \rangle + \frac{1}{2}\left(\frac{\partial^2 \Omega}{\partial s^2}\right)_{\text{ref}} \bullet \langle s^2 \rangle \tag{6.8}$$

For planar reference structures, the linear term is absent for symmetry reasons and the key role is played by mean square amplitudes, which, however, can be quite large and badly described at the harmonic level. From a quantitative point of view, vibrational averaging changes the equilibrium value of a_{H} by about 2 G (10%) and that of a_{C} by about 10 G (30%): thus quantitative (and even semiquantitative) agreement with experiment cannot be obtained by static approaches, irrespective of the quality of the electronic model.

The low frequency motions of vinyl radical correspond to out-of-plane vibrations (wagging and torsion) and in-plane inversion at the radical center. The out-of-plane motions have the same effect as the methyl inversion, albeit with a significantly smaller strength. On the other hand, in-plane inversion is characterized by a double-well potential with a significant barrier. Vibrational averaging now acts in an opposite direction, bringing the coupling constants to values that would have been obtained for less bent structures in a static description. The ground-state vibrational wave function is more localized inside the potential well, even under the barrier, than outside. So it introduces more contributions of "nearly linear" structures. Vibrational effects, while still operative, are less apparent in this case since high energy barriers imply high vibrational frequencies with the consequent negligible population of excited vibrational states and smaller displacements around the equilibrium positions. Unless Boltzmann averaging gives significant weight to states above the barrier, this kind of vibration is effectively governed by a single-well potential unsymmetrically rising on the two sides of the reference configuration. Now, $\partial\Omega/\partial s$ and $<s>$ do not vanish and usually have opposite signs, thus counterbalancing the positive harmonic term. The resulting correction to Ω_c is small and can be treated by perturbative methods.

Let us point out, as a final comment, that also large amplitude internal rotations can have a significant effect on hcc values: however, in most cases they can be treated by a simple average of the hcc values of different substituents.

6.4   VIBRATIONAL PROPERTIES BEYOND THE HARMONIC APPROXIMATION

Computations of harmonic frequencies have become nowadays key tools necessary to assist the studies of IR/Raman spectra and identification or assignment of vibrational bands. Usually scaling of harmonic frequencies by a uniform or frequency-dependent factor is exploited to obtain better agreement with experimental data. This procedure has been extensively applied to the calculations of vibrational frequencies of closed-shell molecules in the ground state, and in most cases adequate scaling factors obtained mainly by comparison of extensive data sets of well-defined experimental results are proposed in the literature (see Ref. 41 and references therein). However, such simple methodologies should not be applied to the studies of radicals for which experimental data is rather scarce and often quite error prone. On the other side, it is necessary to account for the anharmonicity to fully assist the analysis of experimental results. This goal can be achieved by performing the computation of the vibrational level beyond the harmonic approximation: in particular second-order perturbative approach[23,24](PT2), can be effectively applied for studies of relatively large semirigid molecules. It leads to simple expressions of vibrational levels in terms of third and semidiagonal quartic force constants (see Eq. 6.3–6.5) and the range of application can be further extended by taking the resonances into proper account. Of course, large amplitude modes require different methods based, for example, on direct dynamic simulations.

6.4.1   Case Studies: Anharmonic Frequencies of Phenyl and Naphtyl Cation Radicals

Several well-established experimental studies are available for the phenyl radical[42–44] and naphtyl cation radical,[45] thus these systems have been chosen to demonstrate the accuracy achievable by the above described theoretical approaches. Tables 6.12 and 6.13 compare the most reliable experimental frequencies usually obtained in low-temperature matrices with PT2 anharmonic frequencies obtained by with the B3LYP/N07D computational model. It is evident that a remarkable agreement has been obtained in both cases. In particular, a root mean square deviation of $\sim 10\,\mathrm{cm}^{-1}$ between computed and experimental frequencies is achieved, and in the case of phenyl radical even for strongly anharmonic C–H stretching modes, the maximum deviation does not exceed $30\,\mathrm{cm}^{-1}$. For naphtyl radical cation the maximum deviations has been observed for C–H in and out-plane deformation modes, but also in this case these discrepancies do not exceed $50\,\mathrm{cm}^{-1}$.

TABLE 6.12   Computed and Experimental Vibrational Frequencies (in cm^{-1}) of the Phenyl Radical

Mode	Symm	Exp Ref [42]	B3LYP/N07D Harm	B3LYP/N07D Anh
1	a2		400	393
2	b1	416	426	418
3	b2	587	598	591
4	a1	605	618	611
5	b1	657	663	654
6	b1	706	721	709
7	a2	816[a]	815	801
8	b1	874	895	873
9	a2	945[a]	968	946
10	a1	976	984	970
11	b1	972	994	970
12	a1	997	1017	1001
13	a1	1027	1052	1027
14	b2	1063	1074	1064
15	a1	1154	1176	1159
16	b2	1159	1177	1163
17	b2	1283	1306	1283
18	b2	1321	1337	1309
19	b2	1432	1463	1435
20	a1	1441	1473	1440
21	a1	1581	1578	1556
22	b2	1624	1633	1603
23	a1	3037	3174	3053
24	b2	3060	3182	3037
25	a1	3072	3195	3058
26	b2	3072[b]	3197	3059
27	a1	3086	3207	3067

[a] Observed only in the Raman spectrum[43] tentative assignment.
[b] From a high-resolution gas-phase spectrum[44].

TABLE 6.13   Computed and Experimental Vibrational Frequencies (in cm^{-1}) of the Naphtyl Radical Cation

Mode	Symmetry	Exp.[45]	B3LYP/N07D harm	anh	Delta (Calc.-Exp.)
1	Ag		3209	3084	
2		1584	1629	1575	9
3		1463	1504	1472	9
4		1394	1417	1379	15
5		1187	1204	1192	5
6		1043	1068	1048	5
7		763	775	762	1
8		507	514	508	1
9	B1u		3230	3091	
10			1019	1006	
11		848	874	856	8
12		550	557	554	4
13		176	183	178	2
14	B1g	940	982	961	21
15		705	755	737	32
16		365	376	368	3
17	B1u		3219	3077	
18			3205	3085	
19		1525	1555	1528	3
20			1433	1404	
21			1305	1282	
22			1124	1107	
23			807	797	
24		365	357	357	8
25	B2g	1016	1020	1010	6
26			933	927	
27			722	765	
28			440	432	
29	B2u		3230	3088	
30			3206	3077	
31		1518.8	1573	1535	17
32		1400.9	1427	1396	4
33		1215	1240	1217	2
34			1193	1180	
35		1023.2	1041	1029	6
36			608	602	
37	B3g		3219	3096	
38			3203	3084	
39		1468	1515	1490	25
40		1401	1469	1445	44
41		1218	1261	1242	24
42		1098	1120	1107	9
43		927	939	926	1

TABLE 6.13   (*Continued*)

Mode	Symmetry	Exp.[45]	B3LYP/N07D harm	anh	Delta (Calc.-Exp.)
44		455	473	469	14
45	B3u	965	1004	982	17
46		759	779	759	0
47	B1u	423	427	421	2
48		125	158	155	30

6.4.2 Case Studies: Gas and Matrix Isolated IR Spectra of the Vinyl Radical

It can be postulated that the overall accuracy achievable by good quality computational studies in cases of huge discrepancies with experimental findings casts serious doubts on the latter. As an example, we would like to recall recent findings on the vinyl infrared spectrum. The most recent study[46] on infrared absorption spectra of vinyl radical isolated in low temperature neon matrices reports all fundamental vibrations of C_2H_3. In this work, band assignments have been assisted by experiments performed for the ^{13}C and D isotopically substituted species and computation of harmonic frequencies. The obtained results show exceptionally large deviations between the solid Neon and gas-phase[47] vibrational wavenumbers putting severe doubts on the assignment of band lines from the earlier gas-phase experiments.[47] The experimental frequencies in gas-phase and in Argon[48] and Neon[46] low temperature matrices are compared in Table 6.14 with computational results. Harmonic and PT2[23] anharmonic frequencies have been computed by B3LYP/N07D and CAM-B3LYP/N07D computational models. Since low-temperature experimental frequencies in Argon and Neon are very close, matrix effects have been taken into account by continuum medium

TABLE 6.14   **Computed Anharmonic and Experimental Vibrational Frequencies (in cm^{-1}) of the Vinyl Radical in the Gas-Phase and Low-Temperature Matices**

	Computed gas phase		Exp. Gas phase Ref. (47)	Computed Ar matrix (CPCM)		Exp. Ne Ref. (46)	Exp. Ar Ref. (48)
	B3LYP	CAM-B3LYP		B3LYP	CAM-B3LYP		
$\nu 1$	3108	3146	3235	3107	3144	3141.0	
$\nu 2$	3010	3049	3164	3009	3048	2953.6	
$\nu 3$	2899	2943	3103	2898	2941	2911.5	
$\nu 4$	1611	1650	1700	1609	1649		
$\nu 5$	1364	1372	1277	1361	1369	1357.4	1356.7
$\nu 6$	1004	1011	1099	1003	1010		
$\nu 7$	678	683	758	678	683	677.1	677.0
$\nu 8$	917	936	955	916	935	895.3	900.8
$\nu 9$	797	823	895	797	822	857.0	

CPCM[49] model, with the dielectric constant corresponding to Ar. It is immediately visible that an inert gas environment has an almost negligible effect on the computed frequencies, as it was expected. Moreover, the anharmonic frequencies computed with both functionals agree very well with experimental data from solid-state studies, while quite large discrepancies (average of $\sim 110\,\mathrm{cm^{-1}}$) can be observed by comparison to the gas-phase experimental data. Such a disagreement is far off the usual error associated to PT2 anharmonic frequencies for small molecules[50] and strengthens the need for a reinvestigation of the gas-phase IR absorption spectrum of vinyl radical, possibly accompanied by accurate computational studies.

6.5 ELECTRONIC PROPERTIES: VERTICAL EXCITATION ENERGIES, STRUCTURE, AND FREQUENCIES IN EXCITED ELECTRONIC STATES

6.5.1 Theoretical Background

The latest developments of time-dependent methods rooted in the density functional theory, especially by the so-called range separated functionals like LC-ωPBE or LC-TPSS[51,52] are allowing computation of accurate electronic spectra even for quite large systems. Moreover, the recent availability of analytical gradients for TD-DFT[53,54] allows an efficient computation of geometry structures and harmonic frequencies (through the numerical differentiation of analytical gradients) also for excited electronic states.

6.5.2 Case Studies: Vertical Excitation Energies of the Vinyl Radical

Vertical excitation energies for the first eight doublet electronic states of vinyl radical computed by highly accurate *ab initio* methodologies like equation of motion coupled cluster single and double excitations (EOM-CCSD)[55] and multireference configuration interaction (MRCI)[56], are compared to the results obtained by time-dependent density functional theory. For TD-DFT several functionals have been tested starting from the standard B3LYP or PBE0 to the functionals corrected for the long-range interactions CAM-B3LYP, LC-ωPBE, and LC-TPSS. In all cases, the appropriate N07D basis set has been applied. All vertical excitation energies have been computed for a ground-state geometry computed with the related functional. The results are gathered in Table 6.15 together with a short description of the nature of electronic states and an indication of the molecular orbitals involved in the transition.

The plots of molecular orbitals involved in all reported electronic transitions are collected in Fig. 6.5. The available experimental energies[57–60] are also listed, and assigned to the electronic states as proposed by Krylov et al.[34] In general, a good overall agreement between TD-DFT and high-level computations[34,35] has been achieved. TD-B3LYP provides accurate results for the lowest excited state, and only slightly overestimates the VE of higher electronic states, while the LC functionals

TABLE 6.15 Vertical Excitation (VE) Energies, (in eV) and Oscillator Strength of the First Eight Doublet Excited Electronic States of the Vinyl Radical

State	Transition	MO[d]	Exp.	EOM[a]		MRCI[b]		B3LYP[c]		CAM-B3LYP[c]		PBE0[c]		LC-ωPBC[c]		LC-TPSS[c]	
				VE	f	VE	f	VE	f	VE	f	VE	f	VE	f	VE	f
2A''	n→π	7b→8b	3.08[e]	3.31	0.001	3.24	0.001	3.29	0.001	3.34	0.001	3.51	0.001	3.65	0.001	3.58	0.001
2A''	π*←n	8a→9a		4.93	0.003	4.78	0.001	4.48	0.003	4.54	0.003	4.64	0.003	4.69	0.003	4.64	0.003
2A'	π*←π	7a,b→9a,b		5.60	0.000	5.58	0.000	4.40	0.000	4.32	0.000	4.24	0.000	4.21	0.000	4.25	0.000
2A'	3s←n	8a→10a		6.31	0.005	6.25	0.007	6.65	0.027	6.97	0.027	6.92	0.026	7.61	0.035	7.55	0.034
2A'	3px←n	8a→11a		6.88	0.013	6.80	0.017	7.11	0.081	7.46	0.035	7.36	0.101	7.74	0.058	7.65	0.018
2A'	3py←n	7a,b→9a,b		7.09	0.058	7.31	0.019	7.21	0.049	7.39	0.138	7.44	0.045	7.95	0.188	7.89	0.201
2A'	3pz←n	8a→12a	7.37[f]	7.38	0.010	7.48	0.020	7.63	0.001	7.87	0.001	7.94	0.001	8.57	0.002	8.54	0.002
2A''	3s←π	6,5b→8,9b	7.53[f]	7.47	0.059	8.08	0.003	7.62	0.081	7.75	0.068	7.80	0.080	8.15	0.022	8.12	0.029

[a] EOM-CCSD/6-311(2 + ,2 +)G(d,p)[34] .

[b] MRCI + D/ANO(2 +)**[35].

[c] excitation energies computed by TD-DFT with N07D basis set for the structures optimised at the respective DFT/N07D level.

[d] for plots of Molecular Orbitals see figure 5.

[e] Experimental results from Ref. [57–59].

[f] Ref. [60].

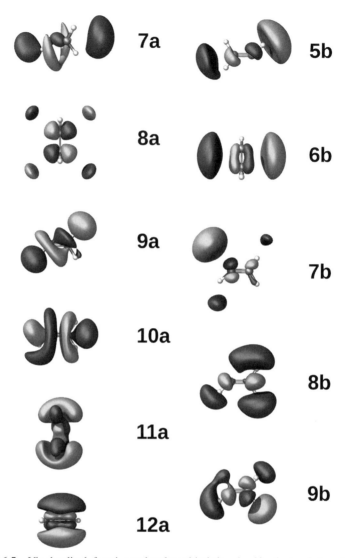

FIGURE 6.5 Vinyl radical, frontier molecular orbitals involved in electronic transitions into the first eight doublet excited electronic states.

systematically overestimates vertical excitation energies. The main discrepancy is the underestimation of excitation energy for $\pi^* \leftarrow \pi$ transition, large enough to reverse the ordering of second and third electronic states. Nevertheless, it should be mentioned that accurate computation of electronic excitations is still a challenging task even for the most elaborate (and expensive) *ab initio* methodologies. In summary, TD-DFT is normally able to provide quite reliable information about the nature and properties of highly excited electronic states at a reasonable cost.

6.5.3   Case Studies: Structures and Frequencies of Vinyl Radical in First Three Doublet Excited Electronic States

Electronic excitation can lead to significant changes of the geometry structure and vibrational properties of molecular systems. Vinyl radical stands as an example where such modifications are particularly enhanced. Computational studies of geometry changes induced by electronic excitations require geometry optimization in excited electronic states. This can be accomplished at TD-DFT level in a quite easy manner, in particular without the need of reducing the number of molecular orbitals considered as often necessary in multireference approaches (CASSCF).[61] It is also possible to compute TD-DFT harmonic frequencies from numerical differentiation of analytical gradients. Moreover, computational studies provide information on the electron density in both electronic states in a way that it is possible to analyze in detail its changes upon excitation, and describe appropriately their influence on radical properties. Such approach has been chosen to study first three doublet electronic states of vinyl radical with the TD-CAMB3LYP model. Changes in geometry structure and frequencies upon excitation are reported in Tables 6.16 and 6.17, while the plots of electron density difference between excited and ground states are shown in Fig. 6.6.

The first excited electronic state is related to the transfer of an electron from a π orbital into the SOMO leading to a lone pair. This induces strong repulsive interaction and leads to the strong deviation of HC1C2 angle, and slight elongation of C–C bond. The TD-CAMB3LYP results are in remarkable agreement with their CASSCF and EOM-CCSD counterparts. Such transition influences mostly the frequency related to the $\nu7$ mode, the C–H out of plane bending (see Fig. 6.7). Next excited state of $^2A''$ symmetry can be described as the $\pi^* \leftarrow n$ transition, and leads to changes of opposite tendency than for $1^2A''$ state, increase of HC1C2 angle, and smaller frequency for $\nu7$ mode. The lowest state of $2A'$ symmetry is related to the $\pi^* \leftarrow \pi$ transition thus weakening of C–C bond. Indeed the most pronounced change in the structure is a C–C bond elongation by about 0.2 Å, while all other geometry parameters remain almost unchanged from their ground-state values. Conversely, this state is characterized by most pronounced changes in frequency values mainly for modes $\nu7$ and $\nu8$ related to out-of plane bending.

6.6   VIBRONIC SPECTRA

Valuable information on the physical–chemical properties of radicals can be often obtained by photoelectron studies in which the electron is detached, so that open-shell systems can be created. Moreover, excited electronic states of radicals can be studied by absorption spectroscopy in the UV–vis regions. An analysis of the resulting experimental spectra can be even more difficult than for ground-state IR or Raman ones. The additional factors can be related to the often not trivial identification of electronic band origin, possible overlap of several electronic transitions and nonadiabatic effects. Although such complications are challenging also for the theoretical approaches, some[5,62] examples show already their interpretative efficiency.

TABLE 6.16 Structures, and Adiabatic Excitation (AE) Energies (in eV) of the First Three Doublet Electronic Excited States of the Vinyl Radical. Bond Lengths are in Å and Angles in Degrees

State transition	2A' ground			2A'' n←π				2A'' π*←n		2A' π*←π	
MO				7b→8b				8a→9a		7a,b→9a,b	
	MRCI/CAS[a]	CCSD[b]	CAM-B3LYP	Exp.[c]	MRCI/CAS[a]	EOM CCSD[b]	TD-CAM-B3LYP	MRCI[a]	TD-CAM-B3LYP	MRCI[a]	TD-CAM-B3LYP
AE				2.49	2.37	2.47	2.69	4.16	4.07	4.67	3.64
	Geometry Structure						Geometry Changes Upon Excitation				
Bonds [Å]											
CC	1.326	1.325	1.301		0.14	0.13	0.10	0.06	0.05	0.21	0.17
HC1	1.073	1.084	1.079		0.05	0.03	0.03	0.00	−0.01	0.02	0.00
$H_{cis}C2$	1.079	1.095	1.093		0.03	0.00	0.00	0.03	0.00	0.02	−0.01
$H_{trans}C2$	1.076	1.090	1.088		0.02	0.00	0.00	0.02	0.00	0.01	−0.01
Angles [degrees]											
HC2H	117.0	116.7	115.7		−2	−1	−2	−1	−2	3	3
HC1C2	133.4	136.6	138.8		−28	−31	−30	36	22	2	−1
$H_{cis}C2C1$	121.5	121.4	122.0		4	4	4	0	1	−2	−2
$H_{trans}C2C1$	121.5	121.9	122.2		−2	−3	−2	1	1	−1	−1

TABLE 6.17   Vibrational frequencies (in cm^{-1}) of Vinyl Radical in its Ground and First Three Doublet Excited Electronic States

State	2A′	2A″		2A″		2A′	
Transition	Ground	$n \leftarrow \pi i$		$\pi^* \leftarrow n$		$\pi^* \leftarrow \pi$	
Mode	CAM-B3LYP	7b- > 8b	$\Delta\nu$	7a,b- > 12a,b	$\Delta\nu$	8a- > 12a	$\Delta\nu$
$\nu 1$	3284	3001	−284	3437	153	3302	18
$\nu 2$	3202	3191	−11	3164	−38	3296	94
$\nu 3$	3102	3078	−24	3087	−15	3176	74
$\nu 4$	1680	1621	−59	1537	−143	1845	165
$\nu 5$	1404	1448	44	1645	241	1406	2
$\nu 6$	1061	1282	221	1268	207	1057	−5
$\nu 7$	951	1059	108	646	−305	111	−840
$\nu 8$	842	1479	637	1146	304	5154	4312
$\nu 9$	720	944	224	756	36	711	−9

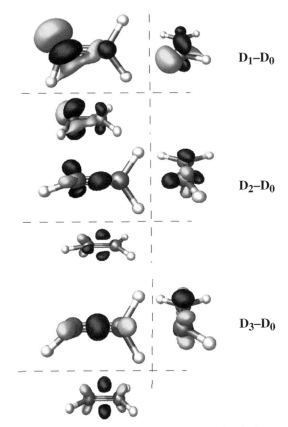

FIGURE 6.6   Vinyl radical: plots of the difference in electron density between the ground and the first three doublet excited electronic states. The regions that have lost electron density as a result of the transition are shown in dark blue, and the bright yellow regions gained electron density.

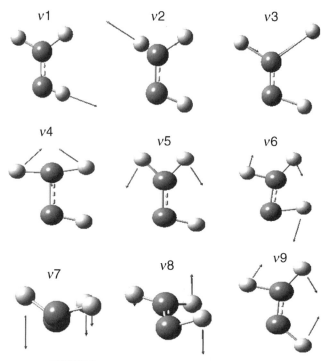

FIGURE 6.7   Normal modes of vinyl radical.

6.6.1   Theoretical Background

A vibronic stick spectrum can be obtained by summing the intensities of the lines of absorption or emission. For a given incident energy, these intensities are proportional to the square of the transition dipole moment integral between the electronic states. Using the Born–Oppenheimer approximation and the Eckart conditions,[63] this integral can be obtained through the analysis of the transitions between vibronic states. This, is however, insufficient due to the lack of an analytic solution for the electronic transition dipole moment. An approximation derived from the Franck–Condon principle[64–66] allows to expand the electronic transition dipole moment in a Taylor series whenever the electronic transition is fast enough that the relative positions and velocities of the nuclei are nearly unaltered by the molecular vibrations. This concept is sketched in Fig. 6.8. Replacing the electronic transition dipole moment by its Taylor expansion about the equilibrium geometry of the final state, it is possible to compute the probability of transition, and so the intensity of a line of absorption or emission, by calculating the overlap integrals between the vibrational states of the initial and final electronic states. The so-called Franck–Condon (FC) approximation assumes that the transition dipole moment is unchanged during the transition. The Herzberg–Teller (HT) approximation takes into account a linear variation of the transition dipole moment along normal coordinates. For most systems, the FC and HT approximations are sufficient to correctly describe both absorption and emission spectra.

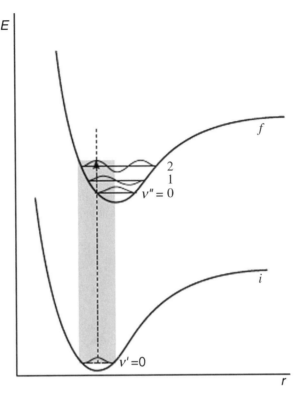

FIGURE 6.8 The Franck–Condon principle is shown by the vertical dotted line. The clear rectangle shows schematically the broader possibilities of transitions when the Herzberg–Teller approximation is used.

In the framework of the Franck–Condon principle,[64–66] time-independent *ab initio* approaches to simulate vibronic spectra are based on the computation of overlap integrals (known as FC integrals), between the vibrational wave functions of the electronic states involved in the transition. The computation of FC integrals requires a detailed knowledge of the multidimensional PES of both electronic states or, within the harmonic approximation, at least computation of equilibrium geometry structures and vibrational properties. For Herzberg–Teller calculations also the transition dipole moment and its first derivatives are required. Moreover, it is necessary to take into account mixing between the normal modes of the initial and the final states, using the linear transformation proposed by Duschinsky[67]:

$$\mathbf{Q} = \mathbf{J}\mathbf{Q}' + \mathbf{K} \tag{6.9}$$

where $\mathbf{Q}$ and $\mathbf{Q}'$ represent the mass-weighted normal coordinates of the initial and final electronic states, respectively. The Duschinsky matrix $\mathbf{J}$ describes the projection of the normal coordinate basis vectors of the initial state on those of the final state and represents the rotation of the normal modes upon the transition. The displacement vector K represents the displacements of the normal modes between the initial-state and the final-state structures.

6.6.2 Computational Strategy

Till recently, computations of vibronic spectra have been limited to small systems or approximated approaches, mainly as a consequence of the difficulties to obtain accurate descriptions of excited electronic states of polyatomic molecules and to computational cost of full dimensional vibronic treatment. Recent developments in electronic structure theory for excited states within the time-dependent density functional theory (TD-DFT)[53,54] and resolution-of-the-identity approximation of coupled cluster theory (RI-CC2)[68] and in effective approaches to simulate electronic spectra[11,69–74] have paved the route toward the simulation of spectra for significantly larger systems.

Recently integrated approaches, capable of accurately simulating one-photon absorbtion (OPA) or one-photon emission (OPE) vibronic spectra and at the same time easily accessible to nonspecialists, have been introduced.[11] The computational strategy is based on an effective evaluation method[72,73] able to select *a priori* the relevant transitions to be computed. The details of the procedure used to compute the spectrum can be found in Refs11,72–74. In brief, simulation of vibrationally resolved electronic spectra starts with the computation of the equilibrium geometries, frequencies, and normal modes for both electronic states involved in the transition. The computational tool has been set within the harmonic approximation, but a simple correction scheme to derive excited state's anharmonic frequencies from ground-state data has been implemented.[74] The simplest computation of Franck–Condon spectrum requires the following data: Cartesian coordinates of the atoms, atomic masses, energy of the ground and excited states, frequencies for the two electronic states involved in the transition, and normal modes for the two electronic states, expressed by the atom displacements.

6.6.3 Case Studies: Electronic Absorption Spectrum of Phenyl Radical

The electronic absorption spectrum of phenyl radical is an interesting example where computational approaches can be compared to the experimental spectrum assigned to the above-mentioned radical on the basis of the strict correlation of intensity evolution in simultaneously measured IR and UV–vis spectra,[2] for which several independent precursors gave consistent results.

In general, the accuracy of a simulated spectrum depends on the quality of the description of both the initial and the final electronic states of the transition. This is obviously related to the proper choice of a well-suited computational model: a reliable description of equilibrium structures, harmonic frequencies, normal modes, and electronic transition energy is necessary. In the study of the $A^2B_1 \leftarrow \tilde{X}^2A_1$ electronic transition of phenyl radical[75] the structural and vibrational properties have been obtained with the B3LYP/TDB3LYP//N07D model, designed for computational studies of free radicals.[28,29] Unconstrained geometry optimizations lead to planar

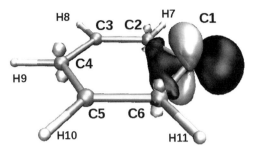

FIGURE 6.9 Phenyl radical, atom numbering scheme and plot of the electron density difference between the A^2B_1 and $\tilde{X}^2A_1$ electronic states. The regions that have lost electron density as a result of the transition are shown in bright yellow, and the darker blue regions gained electron density.

structures for both ground and first excited electronic states. The geometry parameters are compared in Table 6.18 with the results reported by Kim et al.[76] while the atom numbering can be found in Fig. 6.9. It can be observed that the DFT results are in good agreement with their multireference (CASSCF) counterparts.[76] Both computational models predict the same trend in the geometry changes upon electronic excitation and agree also on their magnitude. The main geometry changes are related to the increase of C1–C2 and C1–C6 bond lengths and decrease of C2–C1–C6 angle, in line with the $n \leftarrow \pi$ transfer of electron density from the aromatic ring to the carbon orbital. Indeed,

TABLE 6.18 Geometry Structure of Phenyl Radical in the Ground $\tilde{X}^2A_1$ and First Excited A^2B_1 Electronic States. Bond Lengths are in Å and Angles in Degrees

	CASSCF Ref. [76]			TD-B3LYP/N07D		
	X^2A_1	A^2B_1	Δ	X^2A_1	A^2B_1	Δ
Bonds [Å]						
C1–C2	1.381	1.468	0.087	1.379	1.463	0.084
C2–C3	1.398	1.373	−0.025	1.407	1.381	−0.026
C3–C4	1.396	1.415	0.019	1.400	1.415	0.016
C2–H7	1.074	1.077	0.003	1.086	1.089	0.003
C3–H8	1.076	1.076	0.000	1.087	1.088	0.000
C4–H9	1.075	1.075	0.000	1.086	1.087	0.001
Angles [degrees]						
C6–C1–C2	124.6	112.4	−12.2	125.9	112.3	−13.6
C6–C2–C3	117.3	124.1	6.8	116.5	124.5	7.9
C2–C3–C4	120.1	119.5	−0.6	120.2	119.0	−1.2
C3–C4–C5	120.7	120.3	−0.4	120.6	120.7	0.1
C1–C2–H7	121.5	117.4	−4.1	122.4	117.3	−5.1
C3–C2–H7	121.2	118.5	−2.7	121.0	118.2	−2.8
C4–C3–H8	120.0	119.2	−0.8	120.2	119.4	−0.8
C2–C3–H8	119.9	121.3	1.4	119.6	121.5	1.9
C3–C4–H9	119.7	119.8	0.1	119.7	119.6	0.0

the excited electronic state is characterized by an electron lone pair on the carbon atom, as confirmed by the difference density plot of Fig. 6.9.

The simulated Franck–Condon Hertzberg–Teller (FC-HT) spectra (Fig. 6.10) computed taking into account changes in structures, normal modes, and vibrational frequencies between both electronic states closely resemble their experimental counterparts. The most striking difference is a relative shift of both spectra. It is worth to recall that this part of the spectrum is very weak and the measurements have been close to the performance limit of spectrometer even by application of a multiple-pass technique, as described in detail in Ref. 2. The reason for the discrepancy can be the weak intensity of the 0–0 transition: as a matter of fact, an analysis of the experimental spectrum from Ref. 2 shows a weak progression preceding the first intense band assigned to the spectrum origin. For the weakly allowed transitions the unequivocal assignment of the 0–0 transition may be cumbersome, and the theoretical spectrum suggests that transition assigned as 0–0 can be already a result of the progression to the excited vibrational state of A^2B_1. The comparison of theoretical spectra with the experimental spectrum shifted by $\sim 850\,\text{cm}^{-1}$ (as to match the transition origin reported in Ref. 77) shows a very good agreement, also for the band position of the most intense transitions, suggesting possible revision of the experimental data. It is interesting to recall that earlier theoretical results have not been able to reproduce correctly the spectrum shape, discrepancies being attributed to nonadiabatic couplings.[76,78] Nevertheless, this seems to be due to the limited dimensionality models both studies have been performed with. It is thus strongly advisable to exploit full-dimensional vibronic models prior to analyze the possible role of nonadiabatic effects.

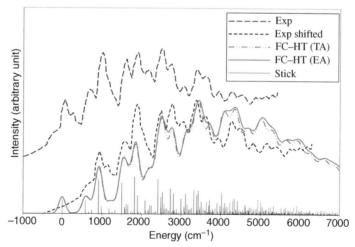

FIGURE 6.10 Theoretical, convoluted and stick FC–HT spectrum of the $A^2B_1 \leftarrow \tilde{X}^2A_1$ electronic transition of phenyl. The experimental1 spectrum is shown for comparison. Spectra have been arbitrarily shifted along energy axis to achieve best match, see text for details.

6.7 CONCLUDING REMARKS

The present paper summarizes the results of systematic computational studies devoted to the calculation of several properties of carbon-centered radicals using DFT and TD-DFT approaches and the new N07D basis set. The results for a representative set of organic free radicals seem accurate enough to allow for quantitative studies. This finding together with the computational efficiency of the approach suggests that we dispose of a quite powerful tool for the study of free radicals, especially taking into account that the same density functional and basis set can be used for different properties and for second and third row atoms. Furthermore, the availability of effective discrete/continuum solvent models and of different dynamical approaches, together with the reduced dimensions of the N07D basis set allow to perform comprehensive analyses aimed at evaluating the roles of stereoelectronic, vibrational, and environmental effects in determining the overall properties of large flexible radicals of current biological and/or technological interest.

REFERENCES

1. Engert, J. M.; Dick, B. *Appl. Phys. B Lasers Opt.* **1996**, *63*, 531.
2. Radziszewski, J. G. *Chem. Phys. Lett.* **1999**, *301*, 565.
3. Jensen, P.; Bunker, P. R., Eds.; *Computational Molecular Spectroscopy*; Wiley: Chichester, 2000.
4. Carter, S.; Handy, N. C.; Puzzarini, C.; Tarroni, R.; Palmieri, P. *Mol. Phys.* **2000**, *98*, 1697.
5. Biczysko, M.; Tarroni, R.; Carter, S. *J. Chem. Phys.* **2003**, *119*, 4197.
6. Puzzarini, C.; Barone, V. *Chem. Phys. Lett.* **2008**, *462*, 49.
7. Puzzarini, C.; Barone, V. *Chem. Phys. Lett.* **2009**, *467*, 276.
8. Clary, D. C. *Science* **2006**, *314*, 5797.
9. Barone, V.; Polimeno, A. *Chem. Soc. Rev.* **2007**, *36*, 1724.
10. Barone, V.; Improta, R.; Rega, N. *Acc. Chem. Res.* **2008**, *41*, 605.
11. Barone, V.; Bloino, J.; Biczysko, M.; Santoro, F. *J. Chem. Theory Comp.* **2009**, *5*, 540.
12. Barone, V.; Adamo, C.; Brunel, Y.; Subra, R. *J. Chem. Phys.* **1996**, *105*, 3168.
13. Neese, F. *J. Chem. Phys.* **2003**, *118*, 3939.
14. Cheesman, J. R.; Trucks, G. W.; Keith, T. A.; Frisch, M. J. *J. Chem. Phys.* **1998**, *104*, 5497.
15. Koseki, S.; Schmidt, M. W.; Gordon, M. S. *J. Phys. Chem.* **1992**, *96*, 10768.
16. Rega, N.; Cossi, M.; Barone, V. *J. Chem. Phys.* **1996**, *105*, 11060.
17. Tomasi, J.; Mennucci, B.; Cammi, R. *Chem. Rev. (Washington, D. C.)* **2005**, *105*, 2999.
18. Barone, V.; Cossi, M.; Tomasi, J. *J. Chem. Phys.* **1997**, *107*, 3210.
19. Benzi, C.; Cossi, M.; Improta, R.; Barone, V. *J. Comput. Chem.* **2005**, *26*, 1096.
20. Cossi, M.; Scalmani, G.; Rega, N.; Barone, V. *J. Chem. Phys.* **2002**, *117*, 43.
21. Clabo, D. A. Jr.; Allen, W. D.; Remington, R. B.; Yamaguchi, Y.; Schaefer, H. F. III, *Chem. Phys.* **1988**, *123*, 187.
22. Barone, V.; Minichino, C. *THEOCHEM* **1995**, *330*, 325.

23. Barone, V. *J. Chem. Phys.* **2005**, *122*, 014108.

24. Ruud, K.; Astrand, P.-O.; Taylor, P. R. *J. Chem. Phys.* **2000**, *112*, 2668.

25. Christiansen, O. *PCCP* **2007**, *9*, 2942.

26. Barone, V. *J. Chem. Phys.* **1994**, *101*, 10666.

27. Dressler, S.; Thiel, W. *Chem. Phys. Lett.* **1977**, *273*, 71.

28. Barone, V.; Cimino, P. *Chem. Phys. Lett.* **2008**, *454*, 139.

29. Barone, V.; Cimino, P.; Stendardo, E. *J. Chem. Theory Comp.* **2008**, *4*, 751.

30. Barone, V.; Cimino, P. *J. Chem Theory Comp.* **2009**, *5*, 192.

31. Adamo, C.; Barone, V. *J. Chem. Phys.* **1996**, *110*, 158.

32. Hariharan, P. C.; Pople, J. A. *Theor. Chim. Acta* **1973**, *28*, 213.

33. (a) Rega N.; Cossi M.; Barone, V. *J. Am. Chem. Soc.* **1997**, *119*, 12962 (b) Rega, N.; Cossi; M. Barone, V. *J. Am. Chem. Soc.* **1998**, *120*, 5723

34. Koziol, L.; Levchenko, S. V.; Krylov, A. I. *J. Phys. Chem. A* **2006**, *110*, 2746.

35. Mebel, A. M.; Chen, Y.-T.; Lin, S.-H. *Chem. Phys. Lett.* **1997**, *275*, 19.

36. Hermosilla, L.; Calle, P.; Garcia de la Vega, J. M.; Sieiro, C. *J. Phys. Chem. A* **2005**, *109*, 114.

37. Rakitin, A. R.; Yff, D.; Trapp, C. *J. Phys. Chem.* **2003**, *107*, 6281.

38. Brancato, G.; Rega, N.; Barone, V. *J. Am. Chem. Soc.* **2007**, *129*, 15380.

39. Patchkovskii, S.; Ziegler, T. *J. Phys. Chem. A* **2001**, *105*, 5490.

40. Improta, R.; Barone, V. *Chem. Rev.* **2004**, *104*, 1231.

41. Borowsk, P.; Fernàndez-Gómez, M.; Fernàndez-Liencres, M.-P.; Penã Ruiz, T. *Chem. Phys. Lett.* **2007**, *446*, 191.

42. Friderichsen, A. V. et al. *J. Am. Chem. Soc.* **2001**, *123*, 1977.

43. Lapinski, A. et al. *J. Phys. Chem. A* **2001**, *105*, 10520.

44. Sharp, E. N.; Roberts, M. A.; Nesbitt, D. *J. Phys. Chem. Chem. Phys.* **2008**, *10*, 6592.

45. Negri, F.; Zgierski, M. Z. *J. Chem. Phys* **1997**, *107*, 4827.

46. Wu, Y.-J.; Lin, M.-Y.; Cheng, B.-M.; Chen, H.-F.; Lee, Y.-P. *J. Chem. Phys.* **2008**, *128*, 294509.

47. Letendre, L.; Liu, D.-K.; Pibel, C. D.; Halpern, J. B. *J. Chem. Phys.* **2000**, *112*, 9209.

48. Tanskanen, H.; Khriachtchev, L.; Räsänen, M.; Feldman, V. I. Sukhov, F. F.; Orlov, A. Y.; Tyurin, D. A. *J. Chem. Phys.* **2005**, *123*, 064318.

49. Cossi, M.; Scalmani, G.; Rega, N.; Barone, V. *J. Comp. Chem.* **2003**, *24*, 669.

50. Carbonniere, P.; Barone, V. *Chem. Phys. Lett.* **2004**, *399*, 226.

51. Jacquemin, D.; Perpète, E.; Scalmani; G.; Frisch, M. J.; Kobayashi, R.; Adamo, C. *J. Chem. Phys.* **2007**, *126*, 144105.

52. Barone, V. unpublished.

53. Scalmani, G.; Frish, M. J.; Menucci, B.; Tomasi, J.; Cammi, R.; Barone, V. *J. Chem. Phys.* **2006**, *124*, 094107.

54. Furche, F.; Ahlrichs, R. *J. Chem. Phys.* **004**, *121*, 12772.

55. Krylov, A. I. *Annu. Rev. Phys. Chem.* **2008**, *59*, 433.

56. Knowles, P. J.; Werner, H.-J. *Theor. Chim. Acta* **1992**, *84*, 95.

57. Can, H. R. *J. Chem.* **1983**, *61*, 993.

58. Pibel, C. D.; Mcllroy, A. Taatjes, C. A.; Alfred, S. Patrick, K.; Halpern, J. B. *J. Chem. Phys.* **1998**, *110*, 1841.

59. Shahu, M.; Yang, C.-H.; Pibel, C. D.; Mcllroy, A.; Taatjes, C. A.; Halpern, J. B. *J. Chem. Phys.* **2002**, *116*, 8343.

60. Fahr, A.; Laufer, A. H. *J. Phys. Chem.* **1988**, *92*, 7229.

61. Werner, H.-J.; Knowles, P. J. *J. Chem. Phys.* **1985**, *82*, 5053.

62. Tarroni, R.; Carter, S. *Mol. Phys.* **2004**, *102*, 2167.

63. Eckart, C. *Phys. Rev.* **1935**, *47*, 552.

64. Franck, J. *Trans. Faraday Soc.* **1926**, *21*, 536.

65. Condon, E. *Phys. Rev.* **1926**, *28*, 1182.

66. Condon, E. *Phys. Rev.* **1928**, *32*, 858.

67. Duschinsky, F. *Acta Physicochim. URSS* **1937**, *7*, 551.

68. Köhn, A.; Hättig, C. *J. Chem. Phys.* **2003**, *199*, 5021.

69. Berger, R.; Fischer, C.; Klessinger, M. *J. Phys. Chem. A* **1998**, *102*, 7157.

70. Dierksen, M.; Grimme, S. *J. Chem. Phys.* **2005**, *122*, 244101.

71. Jankowiak, H.-C.; Stuber, J. L.; Berger, R. *J. Chem. Phys.* **2007**, *127*, 234101.

72. Santoro, F.; Improta, R.; Lami, A.; Bloino, J.; Barone, V. *J. Chem. Phys.* **2007**, *126*, 084509.

73. Santoro, F.; Improta, R.; Lami, A.; Bloino, J.; Barone, V. *J. Chem. Phys.* **2007**, *126*, 184102.

74. Bloino, J.; Biczysko, M.; Crescenzi, O.; Barone, V. *J. Chem. Phys.* **2008**, *128*, 244105.

75. Biczysko, M.; Bloino, J.; Barone, V. *Chem. Phys. Lett.* **2009**, *471*, 143.

76. Kim, G. S.; Mebel, A.; Lin, S. *Chem. Phys. Lett.* **2002**, *361*, 421.

77. Miller, J. H. et al. *J. Chem. Phys.* **1980**, *73*, 4932.

78. Reddy, V. S.; Venkatesan, T. S.; Mahapatra, S. *J. Chem. Phys.* **2007**, *126*, 074306.

7

UNUSUAL STRUCTURES OF RADICAL IONS IN CARBON SKELETONS: NONSTANDARD CHEMICAL BONDING BY RESTRICTING GEOMETRIES

GEORG GESCHEIDT

Institute of Physical and Theoretical Chemistry, Graz University of Technology, Graz, Austria

7.1 INTRODUCTION

Interactions between formally nonbonded molecular moieties have been of interest in many fields of molecular research e.g., as theoretical models or for the study of π–π interactions in DNA. Significant discoveries comprise, for example, dimeric sulfur–sulfur[1,2] and nitrogen–nitrogen[3–6] two-center three-electron bonded radical cations (a whole variety of analogous systems has been reported[2,7–10]). In these cases, it was shown, particularly by optical and EPR spectroscopy, that the newly formed bond is caused by the interaction of lone pairs and the unpaired electron resides in a formally antibonding orbital. This is schematically illustrated in Fig. 7.1.

Generally, the persistence of these unusually bonded stages is particularly high when the geometry of the respective molecular system is confined facilitating unusual orbital interactions and if they are structurally hindered to undergo (stabilizing) transformations (e.g., "anti-Bredt protection"). Illustrative examples are the bicyclic diamines synthesized in the group of R. Alder (**1**) or 1,6:8,13-diimino[14]annulenes (**2**), or tricyclic tetramines (**3**).

Carbon-Centered Free Radicals and Radical Cations, Edited by Malcolm D. E. Forbes
Copyright © 2010 John Wiley & Sons, Inc.

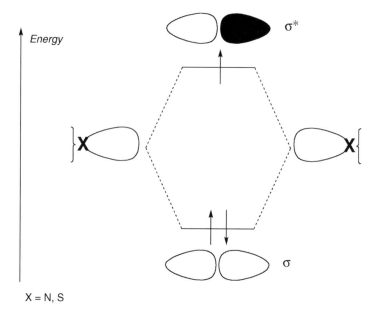

X = N, S

FIGURE 7.1   Two-center three-electron bond formed by the lone pairs of an arbitrary atom X.

All these lone pair interactions lead to a considerable thermodynamic stabilization of the one-electron oxidized stages mirrored by remarkably low oxidation potentials.

In the field of hydrocarbons, radical cations of extended π systems tend to form π dimers under appropriate reaction conditions. This was established for classical aromatics such as naphthalene or anthracene. In cyclophanes, the geometric restrictions force the aromatic constituents to interact through space.[11–15]

Much less persistent radical cations with remarkable electronic structures can be produced by γ-irradiation of small hydrocarbons[16–18] in freon matrices at low temperatures. Analogously, such short-lived species can also be detected by CIDNP spectroscopy.[19] Moreover, radical cations can be generated in zeolites.[20]

In such systems, rearrangements of the molecular skeletons, particularly of strained systems are often observed.

This chapter shows that unusual electronic structures of radical ions can be established in constraining carbon environments. The molecular features introduced in this chapter are derived from cyclovoltammetric measurements and, predominately, from EPR spectroscopy and theoretical calculations.

7.2 THE TOOLS

Before discussing the specific molecular skeletons and the properties of their radical ions a brief survey of the spectroscopic parameters is presented.

7.2.1 Cyclovoltammetry

Cyclovoltammetry provides redox potentials for a variety of molecules. By applying a well-defined (versus a reference electrode) voltage to a dissolved substrate, it can be determined, whether a molecule can be reversibly reduced or oxidized or whether a first electron transfer is followed up by chemical transformations. Accordingly, the thermodynamic stability of the radical ion, which is formed by a primary electron transfer between the substrate and the working electrode can be measured together with the detection of electron transfer-induced follow-up reactions (Scheme 7.1).

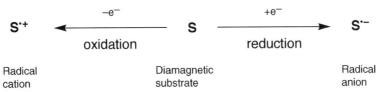

$$S^{\cdot+} \xleftarrow{\quad -e^- \quad} S \xrightarrow{\quad +e^- \quad} S^{\cdot-}$$

oxidation   reduction

Radical   Diamagnetic   Radical
cation   substrate   anion

SCHEME 7.1   Formation of radical ions by redox (electron transfer) reactions.

7.2.2 EPR Parameters: Experimental and Calculated

EPR spectra of organic radicals recorded in fluid solution present two principal parameters: The g factor and the isotropic hyperfine coupling constant, hfc. The g factor characterizes the center of the EPR signal and spin orbit coupling leads to specific deviations from the value of the free electron, 2.0023.

The hfc represents the orientation-independent interaction between the free electron and a magnetic nucleus (e.g., ^1H, ^{13}C, ^{19}F, etc.). It reflects the spin density and the spin population at the magnetic nucleus and the orbital character of the spin-carrying atom (Fermi contact).

In most cases, carbon-centered radicals carry the unpaired electron in p(π)-type orbitals. Since ^{13}C isotopes are present at a very low percentage (natural abundance, 1.11%), the spin distribution is monitored by adjacent ^1H nuclei. The spin from the p-type orbital of the C atom is transferred to the adjacent H atom (H_α) via π–σ spin polarization (Fig. 7.2a); however, spin transfer to more distant protons follows the model of a hyperconjugation mechanism illustrated in Fig. 7.2b. The closer the C_β–H_β is oriented toward the z-axis of the C_α p_z orbital, ($\theta = 0°$) the bigger becomes the $^1H_\beta$ hfc. When θ is equal to 90°, the $^1H_\beta$ hfc reaches its minimum value. This behavior is described by an empirical formula[21]:

$$^1H_\beta \text{ hfc} = \rho(C_\alpha) \cdot (A + B \cos^2(\theta))$$

with $\rho(C_\alpha)$ being the spin population at C_α and A and B being empirical parameters.

A third arrangement for a rather efficient long-range spin transfer is a W-plane-like arrangement between the z-axis of the C_α p_z orbital and a C_γ–H_γ bond (Fig. 7.2c).

Obviously, the interactions sketched in Fig. 7.2 are very helpful, yet rough, empirical models for efficient spin transfer. Substantially more precise hfc values

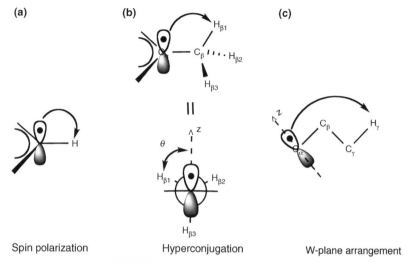

(a) Spin polarization   **(b)** Hyperconjugation   **(c)** W-plane arrangement

FIGURE 7.2   Models for spin transfer.

can be obtained by density functional theory calculations and, therefore, this type of calculation is utilized throughout for rationalizing the experimental results.

7.3 PAGODANE AND ITS DERIVATIVES

This remarkable class of hydrocarbons has been developed by the Prinzbach group. In C_{20} cages, possessing a geometry resembling a (double) pagoda, particularly in the case of [1.1.1.1]pagodane (Fig. 7.3), a notable feature is the central peralkylated cyclobutane ring. This highly strained C_{20} carbon cage had originally been synthesized as potential precursor of the pentagonal dodecahedrane. The rather surprising properties of the [1.1.1.1]isopagodane radical cation (and dication), first observed along the routes taken for the pagodane → dodecahedrane conversion motivated the construction of the homologous [2.2.1.1]/[2.2.2.2](iso)pagodanes and the respective pagodadienes. Ultimately, also two unsaturated dodecahedranes with their extreme olefinic pyramdalization could be added (Chart 7.1).

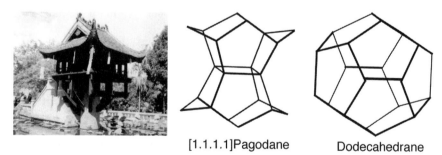

[1.1.1.1]Pagodane   Dodecahedrane

FIGURE 7.3   Pagoda and [1.1.1.1]pagodane and dodecahedrane.

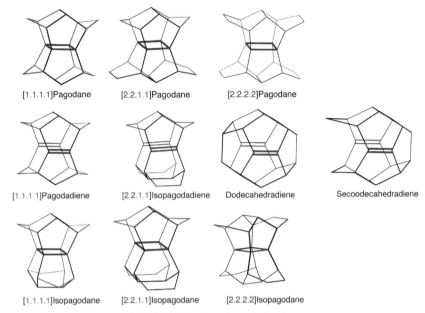

CHART 7.1   Pagodane derivatives.

The parent pagodane molecule or more precisely, [1.1.1.1]pagodane with "[1.1.1.1]" possesses four symmetrically equivalent bridging methylene groups. In [1.1.1.1] isopagodane, two facing methylene groups are twisted by 90°. Elongation of two facing methylene bridges leads to [2.2.1.1]pagodane and its iso derivative. Moreover, dodecahedraene-type molecules dodecahedraene and diene could be synthesized. All these hydrocarbons are displayed in Chart 7.1.

Unexpectedly, several of the pagodane derivatives possess rather low oxidation potentials, indicating a considerable thermodynamic stability of the one-electron oxidized stages, the radical cations. The oxidation potentials of pagodane derivatives are listed in Table 7.1.

TABLE 7.1   One-Electron Oxidation Potentials of Pagodane Derivatives[22]

Pagodane Derivative	First Oxidation Potential V versus Ag/AgCl
[1. 1. 1. 1]Pagodane	$+1.20$ (E_p)
[1.1.1.1]Pagodadiene	$+0.66$ ($E_{1/2}$)
[1. 1. 1. 1]Isopagodane	$+1.72$ (E_p)
[2.2.2.2]Pagodane	$+1.33$ (E_p)
[2.2.2.2]Pagodadienee	$+0.77$ ($E_{1/2}$)
[2.2.1.1]Pagodane	$+1.46$ (E_p)
[2.2.1.1]Isopagodane	$+1.36$ (E_p)
[2.2.1.1]Isopagodadiene	$+0.80$ ($E_{1/2}$)
Dodecahedra-1,6-diene	$+0.99$ (E_p)

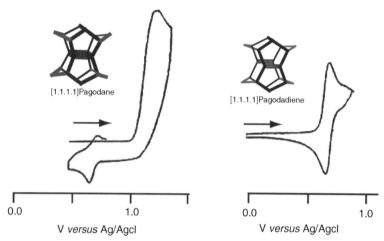

FIGURE 7.4 Cyclovoltammograms of [1.1.1.1]pagodane and [1.1.1.1]pagodadiene versus Ag/AgCl (solvent, CH_2Cl_2; supporting salt, tetrabutylammonium perchlorate).

According to cyclovoltammetric investigations some of the derivatives possess even quasi-reversible oxidation waves although no extended π systems are present. This is illustrated in Fig. 7.4 for [1.1.1.1]pagodadiene. A first quasi-reversible redox process can be detected at $E_{1/2} = 0.66$ V versus Ag/AgCl resembling the formation of a persistent radical cation.

Indeed, the EPR spectrum recorded after oxidation of [1.1.1.1]pagodadiene is in very good agreement with the formation of a well-defined radical cation. The splitting of the EPR spectrum into nine equidistant line groups (Fig. 7.5) that indicates the hyperfine interaction of 8 equivalent protons with the unpaired electron can be attributed to the eight protons in β-position relative to the four sp^2 alkene C atoms. The considerable size of the proton hyperfine coupling constant (hfc) that is equal to 1.54 mT (Table 7.2) mirrors the spin and the charge being predominately located in the central C_4 fragment. The splittings within the nine line groups stem from the remaining protons.

These hyperfine data can be rationalized by an in-plane interaction between the two ethene moieties embedded in the polycyclic skeleton representing a four-center three-electron radical cation (Fig. 7.6).

The hyperfine data presented in Table 7.2 are in favorable agreement with their calculated counterparts. Particularly, the dominant ^1H hfc values attributed to the β-hydrogen atoms serve as well-suited reporters of the bonding situation within the "inner" cyclobutanoid fragment.

When parent [1.1.1.1]pagodane is oxidized in the same way as the isomeric diene above, an identical EPR spectrum is detected. Accordingly, the oxidation leads to a formal ring opening of the cyclobutane fragment producing the identical radical cation as the diene. Such release of strained bonds has often been observed upon oxidation and is in straightforward agreement with cyclovoltammetric measurements displayed in Fig. 7.4. Whereas the oxidation of the diene leads to a quasi-reversible oxidation

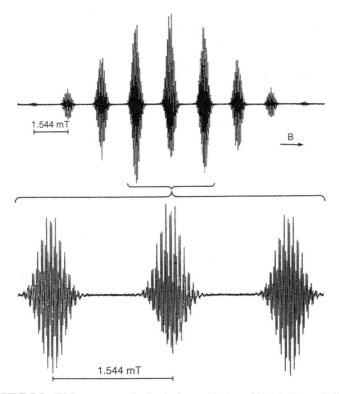

FIGURE 7.5   EPR spectrum obtained after oxidation of [1.1.1.1]pagodadiene.

wave at 0.66 V versus Ag/AgCl, oxidation of the parent [1.1.1.1]pagodane shows an irreversible wave at 1.2 V versus Ag/AgCl with a rereduction at 0.65 V, the identical value as the reoxidation of [1.1.1.1]pagodadiene. The follow-up voltammogram shows a newly emerging wave being identical to that of the diene.

Hence, the radical cation derived directly from the parent [1.1.1.1]pagodane has to have rather a short lifetime. Fortunately, the application of a time-resolved technique, fluorescence-detected magnetic resonance[23] revealed a radical cation possessing 8 equivalent protons with a ^{1}H hfc of 0.96 mT being substantially smaller than

TABLE 7.2   Selected Hyperfine Data for Pagodane Derivatives

	^{1}H hfc(β protons)/mT		
		Calculated[a]	
Molecule	Experimental	"Tight"	"Extended"
[1.1.1.1]Pagodadiene	1.54		1.48
[1. 1. 1. 1]Pagodane	1.05	1.01	1.48
[2.2.1.1]Pagodane	0.96/1.76	0.97/1.73	1.11/1.54
[2.2.2.2]Pagodane	0.06	−0.03	0.0

[a] UB3LYP/6-31G(d)//UB3LYP6-31G(d)

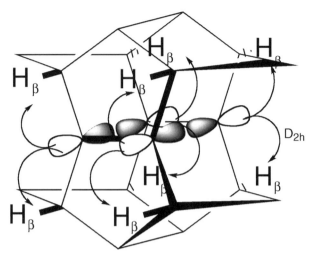

FIGURE 7.6   Sketch of the structure of a four-center three-electron radical cation embedded in [1.1.1.1]pagodadiene and the interaction of the π orbitals with the β-hydrogen atoms.

that obtained by conventional EPR. This short-lived radical cation represents the ring-closed ("tight") inner radical cation. The markedly smaller ^{1}H hfc clearly resembles the more pronounced pyramidalization of the cyclobutanoid carbon atoms. The thus lowered π character leads to the decrease of the ^{1}H hfc of the β-hydrogens.

What happens, if the upper and the lower part of [1.1.1.1]pagodane are twisted by 90° yielding D_{2d} symmetric [1.1.1.1]isopagodane where no preferred ring opening can occur?

The radical cation of [1.1.1.1]isopagodane could be generated by γ-irradiation in a freon matrix. The corresponding EPR spectrum indicated two ^{1}H hfc values, one of 0.95 (4 equivalent H) and one of 0.11 mT (4 equivalent H) showing that the D_{2d} symmetry of the parent molecule was reduced to C_{2v}. This mirrors a slight ring extension but still a cyclobutanoid configuration as confirmed by quantum mechanical calculations.

Up to now, the valence isomers of [1.1.1.1]pagodane were regarded. The next aspect is extending the methylene bridges connecting the two symmetry-equivalent molecular moieties by one methylene group leading to [2.2.2.2]pagodane.

Remarkably, different EPR spectra are obtained upon oxidation of [2.2.2.2] pagodane depending on the experimental conditions. The EPR spectra obtained upon oxidation by γ-irradiation in a freon matrix and with Tl(CF$_3$COO)$_3$ in fluid solution are distinctly different but both are substantially narrower than that of the [1.1.1.1]pagodanes above (Fig. 7.7). This is due to markedly smaller ^{1}H hfc values of the latter two radical cations. In contrast to the "[1.1.1.1] cases", the largest ^{1}H hfc of 0.562 mT (freon matrix) is attributed to the γ′ protons (*exo* γ-hydrogens in the ethylene bridge, see Fig. 7.8) and *not* to the β hydrogens (these hfc values are only 0.060 mT here). An even narrower EPR spectrum is recorded after chemical oxidation with the prominent ^{1}H hfc of only −0.167 mT (γ″ protons, Fig. 7.8). This, in first respect astonishing finding can be rationalized by the formation of a

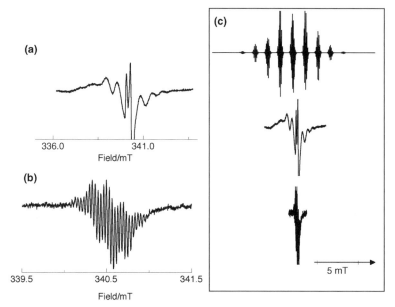

FIGURE 7.7 EPR spectra obtained upon oxidation of [2.2.2.2]pagodane. (a) γ-Irradiation, freon matrix ($CFCl_3$, 77K), (b) oxidation with $Tl(CF_3COO)_3$ in CH_2Cl_2, 253K, and (c) comparison of the EPR widths with the radical cation of [1.1.1.1]pagodadiene (cf. Fig. 7.4).

cyclobutanoid, tight [2.2.2.2] radical cation in the freon matrix and of an extended structure in CH_2Cl_2. However, in both cases, the orientation of the extended bonds is perpendicular to that in the [1.1.1.1]pagodane radical cation (Fig. 7.8).

As for the lower homolog, the "twisted" derivative [2.2.2.2]isopagodane could be prepared,[24] yet, various attempts of oxidation did not provide EPR spectra.

More promising were oxidations of derivatives possessing a "mixed geometry," such as the [2.2.1.1]pagodane family. Both derivatives, [2.2.1.1]pagodane and [2.2.1.1]

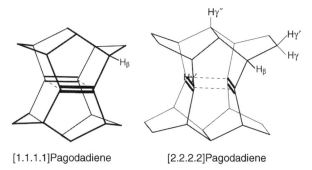

[1.1.1.1]Pagodadiene   [2.2.2.2]Pagodadiene

FIGURE 7.8 Preferred ring opening in the radical cations of [1.1.1.1]pagodane and [2.2.2.2] pagodane.

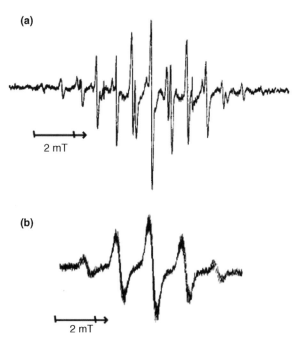

FIGURE 7.9   EPR spectra obtained after oxidation of (a) [2.2.1.1]pagodane and (b) [2.2.1.1] iospagodane (solvent, CH_2Cl_2; oxidant, $AlCl_3$).

isopagodane possess C_{2v} symmetry and can be oxidized to species yielding well-distinguishable EPR spectra at 213K (Fig. 7.9).[25]

The dominating splittings in the EPR spectra displayed in Fig. 7.9 (1.76 mT for [2.2.1.1]pagodane and 1.63 mT for [2.2.1.1]iospagodane) are clearly attributable to β hydrogens. Again, the assignments of the experimental data can be straightforwardly performed by DFT calculations and the significant ^1H hfc values of the β and γ hydrogens serve as "reporters" for the bonding situation (analogous to the molecules displayed in Fig. 7.8). For the [2.2.1.1]pagodane radical cation, the tight geometry reveals an energy minimum, but for the isomer "iso" an extended form was established.

Lateral C–C bond formation in [1.1.1.1]pagodadiene leads to the (seco)dodecahe-dradienes. The distance between the two coplanar double bonds in the latter molecules is substantially longer than in [1.1.1.1]pagodadiene and amounts to 352 versus 262 pm.[26] Is a through-space delocalization still possible at such a long distance? Again the EPR spectrum attributed to the dodecahedradiene radical cation (detected after γ irradiation in freon matrices) shows a nonet pattern with a ^1H hfc of 1.45 mT reflecting 8 equivalent β protons. This value is essentially twice the value obtained for the reference compound dodecahedraene with just one double bond (^1H hfc of 3.10 mT, quintet, 4 equivalent β protons). Thus, even at low temperatures (77 K), the two π bond moieties in the docecahedradiene radical cation do communicate.

A related effect can be followed by regarding derivative secododecahedradiene (and reference secododecahedraene).[27] In the secodiene, the two C=C bonds are

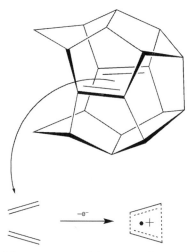

FIGURE 7.10   Trapezoidal arrangement of the two coplanar double bonds in secododeca-hedra-1,6-diene.

essentially coplanar but not parallel (Fig. 7.10). According to the calculations, the π–π distances between the formally nonbonded ethene moieties equal to 290/325 pm in the neutral precursor and are slightly reduced in the radical cation (277/311 pm). Here, two sets of β hydrogen atoms (4 equivalent H each) are present, one adjacent to the longer nonbonded distance and one related to the shorter one. The EPR spectrum obtained upon γ irradiation (CFCl$_3$ matrix, 77K) is split into a nonet spaced by 1.5 mT. This experimentally determined ^{1}H hfc is very close to the calculated counterparts of 1.49 and 1.59 mT for the β/β' protons and shows that the two symmetrically nonequivalent positions are not distinguishable in the experiment. Nevertheless, through-space delocalization in the 4c–3e system is retained even in a trapezoidal arrangement of the two double bonds. The reference compound secododecahedraene allows distinction between the two different β hydrogen atoms: Here, the experimental β ^{1}H hfc values are equal to 4.3 (H$_\beta$,calc.: 4.0 mT) and 2.9 mT (H$_\beta$,calc.: 2.7 mT).

7.4   DIFFERENT STAGES OF CYCLOADDITION/CYCLOREVERSION REACTIONS WITHIN CONFINED ENVIRONMENTS

The array of radical cations derived from pagodane/dodecahedrane-related cages introduced in the preceding paragraph showed that the spin and the charge delocalize in a "tight" or "extended" way.

Whereas a [2 + 2] pericyclic reaction is essentially forbidden in the ground state, a [2 + 1] open-shell reaction is feasible. In this respect, the radical cations detected in this context represent distinct stages of pericyclic, radical-cation catalyzed cycloadditions/cycloreversions.[28] In Fig. 7.11, three distinct stages, a "tight" (cyclobutane-like), an "extended" (bis ethene), and a trapezoid, of a hole- (or radical-cation) catalyzed cycloaddition/cycloreversion are presented in a schematic way.[22,23,29,30]

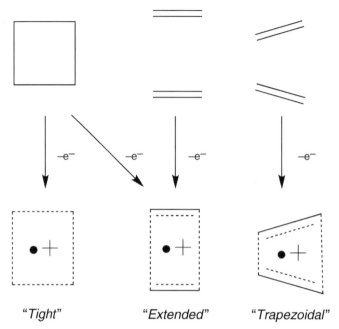

"Tight"   "Extended"   "Trapezoidal"

FIGURE 7.11   A schematic representation of 4c–3e system geometries established within hydrocarbon cages.

7.5   EXTENDING THE "CAGE CONCEPT"

In the above sections, it was shown that restricted carbon cages lead to unusual structures of one-electron oxidized stages. This concept is extendable to molecular skeletons comprising heteroatoms.

For, example, joining azo groups into are rigid carbon polycycle, such as in bis (diazenes) **N1** and **N2**. These proximate, parallel in-plane preoriented bis(diazenes) were synthesized by the Prinzbach group are candidates for lacking N=N/N=N photocycloadditions. The π π distances (d) are equal to approximately 280 pm and the nitrogen lone pairs are unable to interact because of steric reasons. On the other hand, an efficient overlap of the p_z orbitals is enforced.

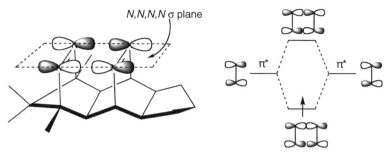

N,N,N,N σ plane

π*   π*

FIGURE 7.12   A schematic view of the "through-space" delocalization (as established for **N1**˙⁻ and **N2**˙⁻) and a scheme for a 4N/5e bonding. Only the interaction of the antibonding π-orbitals is displayed.

Oxidation of **N1** and **N2** cannot be established. However, one-electron reduction is feasible in a straightforward way. Exposure with alkali-metal mirrors in THF or dimethoxyethane under super dry conditions allows the detection of EPR spectra attributable to the radical anions **N1**˙⁻ and **N2**˙⁻. The ^{15}N hfc values of the pairwise equivalent nitrogen nuclei are 0.420/0.394 and 0.430/0.340 mT, respectively. This is approximately half the size of the corresponding values of (mono) diazenes and reveals that the spin is (almost) evenly distributed between the virtually equivalent nitrogen centers.

Are **N1**˙⁻ and **N2**˙⁻ really "through-space" delocalized radical anions as illustrated in Fig. 7.12? A rather clear indication can be derived from the conspicuously large ^{1}H hfc values of the γ protons in **N1**˙⁻ and **N2**˙⁻ of 0.625 and 0.842 mT. This size can only be rationalized by a dominating electron density between the formally nonbonded diazene units and is in perfect agreement with density functional theory calculations.

In summary, the specific arrangement of the two diazene moieties allows an in-plane delocalization of five (in-plane) π electrons within four almost coplanar nitrogen centers (4N/5e bonding). This type of stabilization can even be extended to the corresponding dianions (4N/6e bonding), which are remarkably persistent and can be characterized by NMR spectroscopy. The unusual feature of these bonds is the fact that they are formed by the interaction of antibonding π* orbitals (Fig. 7.12). The confinement of the additional charge(s) to only four atoms causes intense ion pairing.[31,32]

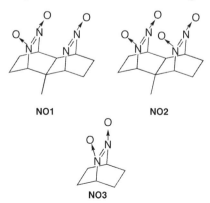

NO1   NO2

NO3

Oxidation of the four N atoms in **N1** leads to trinitroxide **NO1** and tetranitroxide **NO2**. By comparison with spectra of the bis(nitroxide) radical cation, **NO3**$^{\cdot+}$, it could be concluded that in **NO2**$^{\cdot+}$, a radical cation comprising all four NO moieties is formed. This can be anticipated from the EPR spectrum (narrow, g factor $= 2.0061$) and the substantially lower oxidation potential of **NO2** (1.37 V versus Ag/AgCl) in comparison to **NO1** and **NO3** (1.65 V versus Ag/AgCl, both).; moreover, a characteristic absorption at 1020 nm was found for **NO2**$^{\cdot+}$.[33]

Structurally related to bis(diazene) **N2** are tetrazolidine **N3** and **N4**, caged, proximate *syn* periplanar bishydrazines. Their oxidation led to novel highly persistent 4N/7e radical cations with dominant electron delocalization along the cage bonds mirrored by virtually identical EPR spectra dominated by a splitting by 4 equivalent nitrogen nuclei with a ^{14}N hfc of 0.98 mT.[34] A closely related 4c/7e bonding situation was established in 1,3,6,8-tetraazatricyclo[4.4.1.1^{3,8}]dodecane (**TTD**). In **TTD**, however, the nitrogen atoms are embedded in a more flexible skeleton and undergo a more pronounced planarization upon oxidation. This leads to an attenuated σ character at the nitrogen centers and, consequently, a distinct decrease of the ^{14}N hfc values (4 equivalent N atoms) to 0.343 mT.[35]

7.6 SUMMARY

Unusually persistent remarkable open-shell structures were discovered upon one-electron oxidation/reduction of C_4 and N_4 fragments embedded into rigid carbon skeletons. The thus generated radical ions reveal "electron deficient bonding."

For pagodane-related carbon skeletons 4C/3e radical cations with "tight" and "extended" geometries could be established by spectroscopy (predominately EPR) and quantum chemical calculations at the DFT level of theory. Such structures resemble frozen stages of cycloadditions/cycloreversions on the hyper energy surface of the hole-catalyzed cyclobutane formation.

Related unusual electron deficient bonds formed by the interaction of nonbonding orbitals (4N/5e, 4N/7e) can also be established between azo, nitroxide, and amino groups when they are appropriately arranged within a rather rigid molecular framework.

ACKNOWLEDGMENTS

The author thanks Professor Horst Prinzbach (University of Freiburg) for a long-standing very fruitful collaboration, a lot of fun, many beautiful molecules that we have been investigating over the past 25 years and his suggestions for the manuscript. The author is indebted to Professors Stephen Nelsen (Madison, Wisconsin) and Fred Brouver (Amsterdam) for the joint investigations on **TTD**. The author also appreciates the help of Professor Itzhak Bilkis (Hebrew University of Jerusalem) for his comments, and of Markus Griesser and Arnulf Rosspeintner (both Graz) for their keen eyes for details.

REFERENCES

1. Asmus, K. D. *NATO ASI Ser. A* **1990**, *197*, 155–172.

2. Wenska, G.; Filipiak, P.; Asmus, K.-D.; Bobrowski, K.; Koput, J.; Marciniak B. *J. Phys. Chem. B* **2008**, *112*, 10045–10053.

3. Kirste, B.; Alder, R. W.; Sessions, R. B.; Bock, M.; Kurreck, H.; Nelsen, S. F. *J. Am. Chem. Soc.* **1985**, *107*, 2635–2640.

4. Alder, R. W.; Sessions, R. B.; Symons, M. C. R. *J. Chem. Res. Synop.* **1981**, 82–83.

5. Nelsen, S. F.; Alder, R. W.; Sessions, R. B.; Asmus, K. D.; Hiller, K. O.; Goebl, M. *J. Am. Chem. Soc.* **1980**, *102*, 1429–1430.

6. Alder, R. W.; Sessions, R. B. *J. Am. Chem. Soc.* **1979**, *101*, 3651–3652.

7. Fourre, I.; Berges, J.; Braida, B.; Houee-Levin, C. *Chem. Phys. Lett.* **2008**, *467*, 164–169.

8. Joshi, R.; Ghanty, T. K.; Naumov, S.; Mukherjee, T. *J. Phys. Chem. A* **2007**, *111*, 2362–2367.

9. Asmus, K.-D. *Nukleonika* **2000**, *45*, 3–10.

10. Kishore, K.; Asmus, K. D. *J. Phys. Chem.* **1991**, *95*, 7233–7239.

11. Badger, B.; Brocklehurst, B. *Trans. Faraday Soc.* **1969**, *65*, 2582–2587.

12. Batsanov, A. S.; John, D. E.; Bryce, M. R.; Howard, J. A. K. *Adv. Mater.* **1998**, *10*, 1360–1363.

13. Fujitsuka, M.; Cho Dae, W.; Tojo, S.; Yamashiro, S.; Shinmyozu, T.; Majima, T. *J Phys Chem A* **2006**, *110*, 5735–5739.

14. Ohya-Nishiguchi, H.; Terahara, A.; Hirota, N.; Sakata, Y.; Misumi, S. *Bull. Chem. Soc. Jpn.* **1982**, *55*, 1782–1789.

15. Roth, H. D.; Schilling, M. L. M.; Hutton, R. S.; Truesdale, E. A. *J. Am. Chem. Soc.* **1983**, *105*, 153–157.

16. Gerson, F. *Acc. Chem. Res.* **1994**, *27*, 63–69.

17. Knolle, W.; Janovsky, I.; Naumov, S.; Williams, F. *J. Phys. Chem. A* **2006**, *110*, 13816–13826.

18. Rideout, J.; Symons, M. C. R.; Swarts, S.; Besler, B.; Sevilla, M. D. *J. Phys. Chem.* **1985**, *89*, 5251–5255.

19. Roth, H. D. *Electron Transfer in Chemistry*; Balzani, V., Ed.; Wiley-VCH, Weinheim, **2001**; Vol 2, 55–132.

20. Garcia, H.; Roth, H. D. *Chem. Rev.* **2002**, *102*, 3947–4007.

21. Heller, C.; McConnell, H. M. *J. Chem. Phys.* **1960**, *32*, 1535.

22. Prinzbach, H.; Gescheidt, G.; Martin, H. D.; Herges, R.; Heinze, J.; Surya Prakash, G. K.; Olah, G. A. *Pure Appl. Chem.* **1995**, *67*, 673–682.

23. Trifunac, A. D.; Werst, D. W.; Herges, R.; Neumann, H.; Prinzbach, H.; Etzkorn, M. *J. Am. Chem. Soc.* **1996**, *118*, 9444–9445.

24. Wollenweber, M.; Etzkorn, M.; Reinbold, J.; Wahl, F.; Voss, T.; Melder, J.-P.; Grund, C.; Pinkos, R.; Hunkler, D.; Keller, M.; Worth, J.; Knothe, L.; Prinzbach, H. *Eur. J. Org. Chem.* **2000**, 3855–3886.

25. Gescheidt, G.; Herges, R.; Neumann, H.; Heinze, J.; Wollenweber, M.; Etzkorn, M.; Prinzbach, H. *Angew. Chem., Int. Ed.* **1995**, *34*, 1016–1019.

26. Weber, K.; Prinzbach, H.; Schmidin, R.; Gerson, F.; Gescheidt, G. *Angew. Chem., Int. Ed* **1993**, *32*, 875–877.

27. Prinzbach, H.; Reinbold, J.; Bertau, M.; Voss, T.; Martin, H.-D.; Mayer, B.; Heinze, J.; Neschchadin, D.; Gescheidt, G.; Prakash, G. K. S.; Olah, G. A. *Angew. Chem., Int. Ed.* **2001**, *40*, 911–914.

28. Fagnoni, M.; Dondi, D.; Ravelli, D.; Albini, A. *Chem. Rev.* **2007**, *107*, 2725–2756.

29. Etzkorn, M.; Wahl, F.; Keller, M.; Prinzbach, H.; Barbosa, F.; Peron, V.; Gescheidt, G.; Heinze, J.; Herges, R. *J. Org. Chem.* **1998**, *63*, 6080–6081.

30. Gescheidt, G.; Prinzbach, H.; Davies, A. G.; Herges, R. *Acta Chem. Scand.* **1997**, *51*, 174–180.

31. Exner, K.; Hunkler, D.; Gescheidt, G.; Prinzbach, H. *Angew. Chem., Int. Ed.* **1998**, *37*, 1910–1913.

32. Exner, K.; Cullmann, O.; Voegtle, M.; Prinzbach, H.; Grossmann, B.; Heinze, J.; Liesum, L.; Bachmann, R.; Schweiger, A.; Gescheidt, G. *J. Am Chem Soc.* **2000**, *122*, 10650–10660.

33. Exner, K.; Prinzbach, H.; Gescheidt, G.; Grossmann, B.; Heinze, J. *J. Am. Chem. Soc.* **1999**, *121*, 1964–1965.

34. Exner, K.; Gescheidt, G.; Grossmann, B.; Heinze, J.; Bednarek, P.; Bally, T.; Prinzbach, H. *Tetrahedron Lett.* **2000**, *41*, 9595–9600.

35. Zwier, J. M.; Brouwer, A. M.; Keszthelyi, T.; Balakrishnan, G.; Offersgaard, J. F.; Wilbrandt, R.; Barbosa, F.; Buser, U.; Amaudrut, J.; Gescheidt, G.; Nelsen, S. F.; Little, C. D. *J. Am Chem Soc.* **2002**, *124*, 159–167.

8

MAGNETIC FIELD EFFECTS ON RADICAL PAIRS IN HOMOGENEOUS SOLUTION

JONATHAN. R. WOODWARD

Chemical Resources Laboratory, Tokyo Institute of Technology, Midori-ku, Yokohama, Japan

8.1 INTRODUCTION

The idea that the application of a magnetic field might alter the course of a chemical reaction is a tantalizing one that was, however, considered by physicists for a long time to be unlikely due to the very small magnitude of the interaction of molecules with magnetic fields relative to the thermal energy and typical reaction barriers of most reactions. However, the development of the radical pair mechanism (RPM) in the 1960s[1-3] led to the prediction that chemical reactions proceeding through radical pair (RP) intermediates might show sensitivity to externally applied magnetic fields. The first experimental verifications of this prediction followed in the 1970s. Buchachenko *et al.* demonstrated a magnetic field effect (MFE) on the reaction of substituted benzyl chlorides with *n*-butyl lithium,[4,*] and Brocklehurst *et al.* showed MFEs on the fluorescence and absorption intensities in the pulse radiolysis of fluorene in squalene.[7] Research in this field flourished and soon overwhelming evidence was amassed confirming that indeed both the rate and yield of RP reactions could be controlled by the application of magnetic fields easily generated by common permanent and

---

* This experiment was carefully repeated more recently by Hayashi *et al.*, [5,6] who were unable to reproduce the effect, although the experimental conditions varied slightly from the original work.

Carbon-Centered Free Radicals and Radical Cations, Edited by Malcolm D. E. Forbes
Copyright © 2010 John Wiley & Sons, Inc.

electromagnets. Alongside theoretical treatments,[8–10] these experiments showed that RP reactions show a complex dependence on magnetic field strength that could be used to gain insight into the dynamic processes involved. A comprehensive and seminal review of the first 15 years of MFE studies was written by Steiner and Ulrich[11] and is essential reading for those with an interest in this field.

8.2   THE SPIN-CORRELATED RADICAL PAIR

Central to the magnetic field sensitivity of chemical reactions is the spin-correlated radical pair (SCRP).[12,13] The creation of free radicals from neutral molecules requires the separation of two electrons and thus always results in the formation of a pair of radicals. Furthermore, the spin states of the unpaired electrons on the two radicals are correlated and defined by the multiplicity of the precursor molecule (Fig. 8.1). Typically, RPs generated in thermal reactions will be born from singlet-state (S) precursors and thus are generated in a pure singlet state, whereas for photochemical reactions, both singlet and triplet (T) RPs can be prepared. Generation of a triplet RP represents an unusual chemical scenario. The excess energy supplied to generate the original triplet molecule through absorption of a photon is rapidly removed from the newly born RP through collisions with the surrounding solvent. This now leaves two reactive radical species next to one another in solution, but unable to react together as the spin states of the two electrons prevent them from entering the same molecular orbital due to the restrictions imposed by the Pauli principle. As the first triplet excited state of the molecule is usually energetically inaccessible, this means that for neutral free radicals in solution, triplet-correlated RPs are nonreactive. For reaction to take place between the two radicals, the RP must first undergo spin-state interconversion to

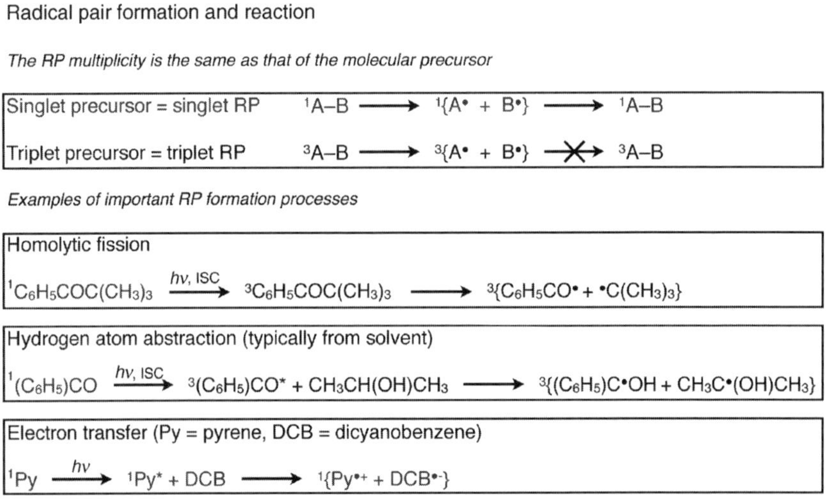

Radical pair formation and reaction

The RP multiplicity is the same as that of the molecular precursor

Singlet precursor = singlet RP	$^1A\text{–}B \longrightarrow {}^1\{A^\bullet + B^\bullet\} \longrightarrow {}^1A\text{–}B$
Triplet precursor = triplet RP	$^3A\text{–}B \longrightarrow {}^3\{A^\bullet + B^\bullet\} \xrightarrow{\quad\times\quad} {}^3A\text{–}B$

Examples of important RP formation processes

Homolytic fission

$$^1C_6H_5COC(CH_3)_3 \xrightarrow{hv,\ ISC} {}^3C_6H_5COC(CH_3)_3 \longrightarrow {}^3\{C_6H_5CO^\bullet + {}^\bullet C(CH_3)_3\}$$

Hydrogen atom abstraction (typically from solvent)

$$^1(C_6H_5)CO \xrightarrow{hv,\ ISC} {}^3(C_6H_5)CO^* + CH_3CH(OH)CH_3 \longrightarrow {}^3\{(C_6H_5)C^\bullet OH + CH_3C^\bullet(OH)CH_3\}$$

Electron transfer (Py = pyrene, DCB = dicyanobenzene)

$$^1Py \xrightarrow{hv} {}^1Py^* + DCB \longrightarrow {}^1\{Py^{\bullet+} + DCB^{\bullet-}\}$$

FIGURE 8.1   The formation and subsequent recombination of S and T RPs from closed shell, neutral molecules with conservation of spin state. Examples of the three most common methods of RP formation are illustrated.

a singlet state. The key to the magnetic field sensitivity of RPs is that the process of conversion of a triplet RP to and from a singlet one (referred to as S–T state mixing) is driven by weak magnetic interactions in the radicals and can be influenced by the presence of an external magnetic field.

8.2.1 Radical Pair Interactions

RP reactions have been found to be well described by the application of a spin Hamiltonian,[14] a common approach used in the field of magnetic resonance, which reduces the full Hamiltonian to one that contains only spin-dependent terms. The interactions capable of influencing spin-state mixing processes in RPs are concisely introduced in the expression for the spin Hamiltonian of a RP, which can be written as a sum of interradical, intraradical, and external interactions.

$$\hat{H}_{RP} = \hat{H}_{inter} + \hat{H}_{intra} + \hat{H}_{ext} \tag{8.1}$$

The intraradical interactions provide the mechanism for coherent spin-state mixing and the interradical interactions act contrary to this process.

The spin state of a given radical is commonly and simply described by a spin vector operator

$$\hat{\mathbf{S}} = \hat{S}_x \mathbf{i} + \hat{S}_y \mathbf{j} + \hat{S}_z \mathbf{k} \tag{8.2}$$

Where $\mathbf{i}, \mathbf{j},$ and $\mathbf{k}$ are unit vectors along the $x, y,$ and z directions. The expectation value of this operator is the electron spin magnetization vector, which describes the bulk electron spin state for a given radical.

$$\left\langle \hat{\mathbf{S}} \right\rangle = \left\langle \hat{S}_x \right\rangle \mathbf{i} + \left\langle \hat{S}_y \right\rangle \mathbf{j} + \left\langle \hat{S}_z \right\rangle \mathbf{k} \tag{8.3}$$

8.2.2 Intraradical Interactions

Electrons possess the properties of charge and spin angular momentum and thus possess a magnetic moment (e.g., Ref. 15 and references therein). This magnetic moment is capable of interacting with other magnetic moments in its vicinity. For an isolated free radical, the only other source of magnetic moments is those generated by the nuclei in the molecule with nonzero spin quantum numbers. The coupling between electron and nuclear magnetic moments is known as the hyperfine interaction[15] and has two components. The first is the direct, through-space dipolar interaction between the electron and a given nucleus. This interaction is anisotropic and for radicals in a homogeneous solution, rapid tumbling serves to average this interaction to zero. An isotropic interaction between electron and nuclear magnetic moments exists only when the electron penetrates inside a nucleus. This is only possible for s-orbitals or orbitals that possess some s-character. Being isotropic, it is not influenced by the relative orientation of electron and nuclear spins. It is usually written as

$$\hat{H}_{hfi} = a_i \hat{\mathbf{S}} \cdot \hat{\mathbf{I}}_i \tag{8.4}$$

where a_i is the isotropic hyperfine coupling constant (HFC) for the interaction between electron spin S and nuclear spin I_i. Thus, for the RP, we must consider the total set of electron–nuclear spin interactions for each electron (i.e., one on each radical, labeled 1 and 2) with all the nuclear spins in the given radical.

$$\hat{H}_{\text{intra}} = \sum_i a_i \hat{\mathbf{S}}_1 \cdot \hat{\mathbf{I}}_{1i} + \sum_k a_k \hat{\mathbf{S}}_2 \cdot \hat{\mathbf{I}}_{2i} \tag{8.5}$$

Unlike a free electron, an electron in a molecule also experiences a complex interaction between spin and orbital angular momentum, spin–orbit coupling. These interactions are described in terms of a tensor, g.[15,16] EPR spectroscopy is the most common method for determining g-tensors. Indeed, g-tensor analysis in complex biomolecules can give important orientational information on paramagnetic centers.[16,17] For radicals free to tumble in isotropic solution, the EPR spectra reveal a reduction to an isotropic g-value. For typical small organic free radicals, this g-value is very similar to the value for a free electron ($g_e = 2.0023$) but can differ much more substantially (0–6) for species such as transition metal ions.

8.2.3 Interradical Interactions

The Hamiltonian for interradical interactions can be decomposed into two terms corresponding to the electron exchange interaction and the electron dipolar interaction

$$\hat{H}_{\text{inter}} = \hat{H}_{\text{exchange}} + \hat{H}_{\text{D}} \tag{8.6}$$

The electron exchange interaction is critical to the magnetic field sensitivity of reactions. It is a purely quantum mechanical effect arising from the fact that the wavefunction of indistinguishable particles (in this case electrons) is subject to exchange symmetry as defined by the Pauli principle. It results in an energy separation between S and T RP states as the radicals approach close enough for the electrons to become correlated and bonding begins to occur.

Figure 8.2a shows a diagrammatic representation of the orbital energies for a pair of hydrogen atoms as a function of the separation of these two radicals. For separations of greater than about 1 nm, the S and T states have equal energy, as the two electrons are uncorrelated. For shorter distances, bonding can occur for the singlet state but not for the triplet state. The energy separation between the two is the electron exchange interaction and rises very rapidly for close RPs. At these short RP separations, the exchange interaction dominates the spin Hamiltonian and serves to halt S–T state mixing. This is significant because it means that when radicals approach one another close enough to react, the ability to undergo spin-state interconversion is lost. Thus, for a RP originally born in a triplet state, no interconversion to singlet, and thus no reaction can occur until the radicals diffuse apart sufficiently that the exchange interaction no longer swamps the hyperfine interaction. The exchange interaction is typically included in the spin Hamiltonian in the following form, where J is the value of the

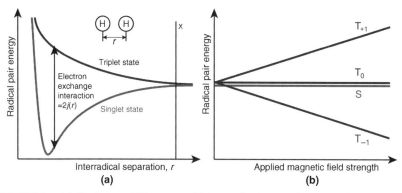

FIGURE 8.2 (a) Variation of RP energy with interradical separation exemplified by a pair of hydrogen atoms. The difference in energy between the S and T states is given by the electron exchange interaction and falls rapidly as the radicals become separated. (b) Variation of RP energy with applied external magnetic field strength due to the electron Zeeman effect. The S and T_0 states remain unchanged, their separation determined by the interradical separation (see (a)).

exchange integral between the two electron spins, r is the separation of the radicals, and r_J and J_0 are both empirically determined parameters.

$$\hat{H}_{\text{exchange}} = -J(r)\left(\frac{1}{2} + 2\mathbf{S}_1 \cdot \mathbf{S}_2\right) \tag{8.7}$$

$$J(r) = J_0 e^{-r/r_J} \tag{8.8}$$

The exchange interaction typically drops to magnitudes of the same order as hyperfine couplings within a single diffusive step.[18–20] For neutral RPs, the exchange interaction is always negative, but positive J has been proposed and observed in some radical ion pairs (RIPs).[21–24]

The dipolar interaction is a direct interaction between the magnetic dipoles of the electrons on the two radicals.

$$\hat{H}_{\text{D}} = \frac{\mu_0\mu_B^2 g_1 g_2}{4\pi\hbar^2 r^3}\left(\hat{\mathbf{S}}_1 \cdot \hat{\mathbf{S}}_2 - \frac{3}{r^2}(\mathbf{S}_1 \cdot \mathbf{r})(\mathbf{S}_2 \cdot \mathbf{r})\right) \tag{8.9}$$

where g_1 and g_2 are the isotropic g-values of radicals A and B and r is the vector separation of the two radicals, usually defined to be between the centers of the relevant electron orbitals. The dipolar interaction has a shallower distance dependence than the exchange interaction and is capable of retarding spin-state interconversion at larger radical separations. The significance of the dipolar interaction in differing field strengths is discussed later, but in general, for RP reactions in homogeneous solution, the dipolar interaction is neglected and the total spin Hamiltonian for a RP in zero magnetic field is usually given as

$$\hat{H}_{\mathrm{RP},\,B=0} = -J(r)\left(\frac{1}{2} + 2\mathbf{S}_1 \cdot \mathbf{S}_2\right) + \sum_i a_i \hat{\mathbf{S}}_1 \cdot \hat{\mathbf{I}}_{1i} + \sum_k a_k \hat{\mathbf{S}}_2 \cdot \hat{\mathbf{I}}_{2k} \qquad (8.10)$$

8.3 APPLICATION OF A MAGNETIC FIELD

Application of an external magnetic field alters the nature of some of the magnetic interactions in a RP and also leads to additional terms in the spin Hamiltonian through the electron Zeeman interaction.

8.3.1 The Zeeman Effect

The Zeeman effect is the name given to the interaction between an electron and an external magnetic field.[15] For a radical tumbling freely in solution, it can be written as

$$\hat{H}_{\mathrm{Zeeman}} = g\mu_\mathrm{B}\hat{\mathbf{S}} \cdot \mathbf{B} = g\mu_\mathrm{B} S_z B \qquad (8.11)$$

where g is the g value of the radical concerned and μ_B is the Bohr magneton. It causes the two possible spin states ($m_\mathrm{s} = +1$, α and $m_\mathrm{s} = -1$, β) to become nondegenerate; their energy separation increasing linearly with the strength of the applied magnetic field. EPR spectroscopy is based on using resonant microwave radiation to cause transitions between these two spin states.

For a pair of radicals, the Zeeman effect serves to energetically separate the triplet RP into three sublevels, written as T_{+1}, T_0, and T_{-1}. The energy of the T_{+1} state increases, while that of the T_{-1} state is reduced by an equal amount. The S and T_0 states possess no magnetic moment in the direction of the applied field and thus are unaffected. This is illustrated in Fig. 8.2b.

The effect of the application of an external field on an electron is well described in many EPR texts, for example, Ref. 15. A vector picture is often used that, while approximate, describes the interaction sufficiently for most situations. The electron magnetic moment experiences a torque that causes it to precess around the direction of the applied magnetic field at the *Larmor frequency*.

$$\omega = \frac{g\mu_\mathrm{B} B_{\mathrm{local}}}{h} \qquad (8.12)$$

Figure 8.3 shows vector pictures for the four RP spin states in an external magnetic field. A static image is insufficient to visualize this model, and we must remember that all the electron magnetic moments are in constant precession about the direction of the magnetic field at their respective Larmor frequencies.

The orientation of electron spins in this manner influences the electron–electron dipolar interaction described above. For strong magnetic fields, the diffusive

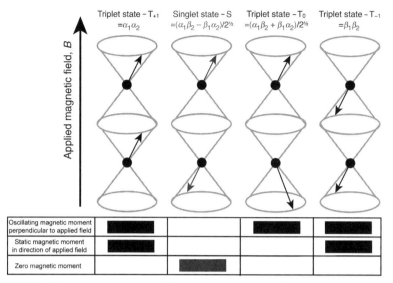

| | Triplet state – T_{+1}
$= \alpha_1 \alpha_2$ | Singlet state – S
$= (\alpha_1 \beta_2 - \beta_1 \alpha_2)/2^{1/2}$ | Triplet state – T_0
$= (\alpha_1 \beta_2 + \beta_1 \alpha_2)/2^{1/2}$ | Triplet state – T_{-1}
$= \beta_1 \beta_2$ |

Oscillating magnetic moment perpendicular to applied field	■■■■		■■■■	■■■■
Static magnetic moment in direction of applied field	■■■■			■■■■
Zero magnetic moment		■■■■		

FIGURE 8.3 Vector representation of RP spin states in an applied external magnetic field. The vectors are each in constant precession about the magnetic field axis at the Larmor frequency of the respective radical.

movement of radicals around one another causes an averaging of the dipolar interaction to zero. However, in zero and weak magnetic fields, the situation is more complicated. Detailed simulations have shown that, in general, the dipolar interaction serves to reduce the magnitude of MFEs in weak fields[25] except for the case in which the dipolar and exchange interaction energies are similar, where MFEs may still be manifest.[26]

8.4 SPIN-STATE MIXING

Critical to the observation of magnetic field effects in solution is the ability of the RP to interconvert between triplet (nonreactive) and singlet (reactive) spin states. Having established the various interactions present in the RP, we can now consider how such a mixing process might take place.

8.4.1 Coherent Spin-State Mixing

The complete spin Hamiltonian for the RP in zero and applied field can be written as follows:

$$\hat{H}_{\mathrm{RP}} = -J(r)\left(\frac{1}{2} + 2\mathbf{S_1} \cdot \mathbf{S_2}\right) + \sum_i a_i \hat{\mathbf{S}}_1 \cdot \hat{\mathbf{I}}_{1i} + \sum_k a_k \hat{\mathbf{S}}_2 \cdot \hat{\mathbf{I}}_{2k} + \mu_B B(g_1 S_{1z} + g_2 S_{2z})$$

$$(8.13)$$

The vector model introduced above provides a very simple visual representation of the RP spin states in an applied magnetic field and allows us a clear glimpse of how spin-state mixing occurs. Examination of Fig. 8.3 reveals that the S and T_0 states differ only in the phase of the precession of the two electron spins. The Larmor precession frequency depends on the strength of the applied magnetic field, the radical g-value, and the HFCs. Thus, if the two radicals in a RP possess different g-values or are in different hyperfine states, their Larmor precession frequencies will differ and the phase difference between their electron spins will change in time. Thus, they will coherently oscillate between S and T_0 states at a frequency dependent on the difference in their Larmor frequencies. Typically, the g-values of organic radicals are very similar. As the g-value-dependent term is magnetic field dependent, spin-state mixing brought about by a difference in g-values is generally only manifest at strong magnetic fields.

Thus, it is hyperfine couplings in the individual radicals that drive the spin-state mixing process. While in high field, hyperfine coupling is not capable of causing S–$T_{\pm 1}$-state mixing due to the separation in their energies, the case is different in zero magnetic field. It is useful to consider the spin Hamiltonian for a real RP. The simplest of RPs is one in which one member is an electron and the other has an electron with coupling to a single spin 1/2 nucleus. Figure 8.4 shows the spin Hamiltonian for such a pair, written in a basis of the high-field singlet/triplet states, expressed in the form of a table. It is clear that the diagonal terms in the Hamiltonian are dependent on J or B, the former giving rise to the energetic separation between S and T states and the latter to the separation between S, T_0, and $T_{\pm 1}$ states in an applied field. Spin-state mixing between particular states occurs

	$\lvert S_{\alpha N}\rangle$	$\lvert T_{0\alpha N}\rangle$	$\lvert T_{+1\alpha N}\rangle$	$\lvert T_{-1\alpha N}\rangle$	$\lvert S_{\beta N}\rangle$	$\lvert T_{0\beta N}\rangle$	$\lvert T_{+1\beta N}\rangle$	$\lvert T_{-1\beta N}\rangle$
$\lvert S_{\alpha N}\rangle$	$2J$	$q+a/4$	0	0	0	0	$-a/\sqrt{8}$	0
$\lvert T_{0\alpha N}\rangle$	$q+a/4$	0	0	0	0	0	$a/\sqrt{8}$	0
$\lvert T_{+1\alpha N}\rangle$	0	0	$\omega_0+a/4$	0	0	0	0	0
$\lvert T_{-1\alpha N}\rangle$	0	0	0	$-\omega_0\ a/4$	$a/\sqrt{8}$	$a/\sqrt{8}$	0	0
$\lvert S_{\beta N}\rangle$	0	0	0	$a/\sqrt{8}$	$2J$	$q-a/4$	0	0
$\lvert T_{0\beta N}\rangle$	0	0	0	$a/\sqrt{8}$	$q-a/4$	0	0	0
$\lvert T_{+1\beta N}\rangle$	$-a/\sqrt{8}$	$a/\sqrt{8}$	0	0	0	0	$\omega_0-a/4$	0
$\lvert T_{-1\beta N}\rangle$	0	0	0	0	0	0	0	$-\omega_0+a/4$

FIGURE 8.4 The spin Hamiltonian for the simplest RP consisting of one electron and a radical with a single spin-1/2 nucleus on the basis of the high-field S/T states. The coupling between individual RP spin states can be directly obtained from the off-diagonal elements.

through the terms that connect them in the Hamiltonian. We can see that there are two singlet states, with the nuclear spin in either the α state or the β state. If we consider the $S_{\alpha N}$ state, we can see that this has a diagonal element allowing its conversion to the $T_{0\alpha N}$ state (i.e., $S-T_0$) mixing and that the value of this term increases with magnetic field strength (thus remaining active in applied magnetic fields). This is the $S-T_0$ process described above. However, the $S_{\alpha N}$ state also has a term allowing its interconversion to the $T_{1\beta N}$ state that is dependent only on the HFCs and is magnetic field independent. However, no terms exist that allow conversion to any of the triplet states with α nuclear spin. This arises due to the requirement for conservation of total spin angular momentum. For $S-T_{\pm 1}$ interconversion, electron and nuclear spins must flip simultaneously. For the one nucleus RP, the available $S-T_{\pm 1}$ conversion channels are $S_{\alpha N} \leftrightarrow T_{+1\beta N}$ and $S_{\beta N} \leftrightarrow T_{-1\alpha N}$.

Thus, at zero field for a general RP, spin-state mixing can occur between the singlet and all three triplet states. The rate of this spin-state mixing depends on the strength of the HFCs in the radicals. If any other magnetic interactions become larger than this interaction, spin-state mixing can be interrupted. Examples include the exchange and dipolar couplings at short interradical separations and the Zeeman interaction at magnetic fields greater than the HFCs. Typical average hyperfine couplings for organic radicals lie in the range of 1–10 mT. By using the Zeeman resonance condition, we can estimate the typical mixing times to be about 3–30 ns. This means that for external magnetic fields to influence RP reactions, the RP must exist for periods longer than this under the condition of negligible exchange and dipolar interactions.

8.4.2   The Life Cycle of a Radical Pair

We are now in a position to consider the lifetime of a RP. For simplicity, we will consider a RP born in a triplet state. Figure 8.5 shows the formation and subsequent reaction processes of a RP born from a photoexcited triplet state.

A ground-state molecule is photoexcited, initially to an excited singlet state, and undergoes intersystem crossing to produce an excited triplet state. This state can then react (typically by homolytic bond fission, electron transfer, or hydrogen atom transfer; see Fig. 8.1) to produce a triplet-correlated RP. At the moment of formation, the two radicals are close together and the exchange energy is large and dominates all other interactions. No reaction between the two radicals can take place. The radicals diffuse apart and once the interradical separation becomes sufficiently large, the exchange interaction becomes essentially negligible. As the RPs continue their diffusive motion at such distances and beyond, coherent spin-state mixing gradually causes the RP to oscillate between S and T spin states (this is a coherent quantum mechanical process, and after birth in a pure spin state, the RP is best described as being in a mixed quantum state with a certain degree of singlet/triplet character). If the radicals continue to diffuse apart from one another, their probability of reencounter drops away and they eventually become the familiar solution "free radicals." Alternatively, the radicals may reencounter, now in a state with some singlet character.

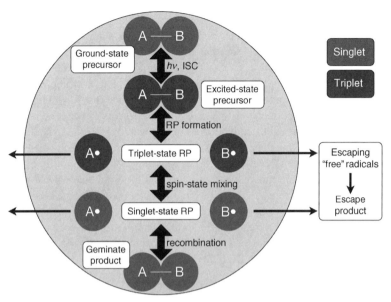

FIGURE 8.5   The life cycle of a geminate RP formed from a photoexcited triplet state.

Their probability of reaction is now dependent on the degree of this singlet character. If they do not react on this encounter, they diffuse apart again and the process repeats. This initial RP is known as a geminate ("born together") pair or g-pair, the stage of the reaction is referred to as the geminate phase, and any reaction that takes place produces geminate product. For those radicals that escape the geminate phase and diffuse into the bulk, the products subsequently formed (e.g., by reaction with solvent) are referred to as escape products.

If a magnetic field is applied, the rate of singlet–triplet mixing is altered. For increases in the rate of spin-state mixing, RPs develop singlet character more rapidly and so the rate of geminate recombination increases. As well as this change in reaction rate, this produces a change in the geminate product: escape product ratio. For decreases in the rate of spin-state mixing, singlet character is less rapidly established in the RP and so geminate reaction is slower and the ratio of geminate:escape product decreases. For a singlet-born RP, the formation process usually provides sufficient excess energy to propel the RPs apart, meaning that not all RPs instantly recombine. In this case, S–T mixing serves to reduce the rate of formation of geminate product, and so the effects of increases and decreases in spin-state mixing rates are reversed relative to a triplet-born pair.

The effect of spin-state mixing on the overall reaction rates and yields of RPs is strongly dependent on how long the RP is able to undergo spin-state mixing while still maintaining the possibility of reencounter and reaction. This means that the diffusion process is key to the observed RP dynamics. For nonviscous solutions, the escape rate of RPs from the solvent cage is more rapid than the spin-state mixing processes, meaning that field effects are seldom observed.[27] However, increasing the viscosity of the solvent allows the RP to live long enough for spin-state mixing to matter and MFEs become manifest. In the case of RIPs, the electrostatic attraction between the pair members means that they remain associated for longer and so MFEs can be observed even in nonviscous solution.[27] Studies on RIPs in mixed solvents have found evidence for the role of microheterogeneity in the solvent influencing the RP dynamics[28–31] and indeed MFEs have been used to measure solvent microviscosity in a range of solvents.[32]

8.4.3 Incoherent Spin-State Mixing

We have thus far considered coherent processes that take place in RPs (which in some cases been have been modulated by stochastic motion). However, the common spin-lattice and spin–spin relaxation processes familiar from magnetic resonance also come to bear on the dynamics of RPs. Typical values of T_1 and T_2 for small organic radicals in homogeneous solution are on the microsecond timescale and as such are rather slow relative to coherent mixing and RP diffusion. Thus, for the most part, effects of incoherent spin relaxation are not manifest in such reactions. However, for reactions in which the RP lifetime is substantially extended, for instance, by constraining the RP inside a microreactor such as a micelle (many examples in Ref. 14), relaxation effects become significant.

8.5 THE MAGNETIC FIELD DEPENDENCE OF RADICAL PAIR REACTIONS

The response of RPs to fields of increasing magnitude is distinctly nonlinear. Figure 8.6 shows how the geminate product yield typically changes with field strength for a triplet-born RP. Experimentally, the dependence of reaction yield on magnetic field strength is referred to as a MARY (magnetic affected reaction yield or magnetic effect on reaction yield have both been used) curve.

The curve exhibits three distinct regions of differing field sensitivity. For typical fields produced by static and electromagnets, the field dependence is dominated by the field saturation observed in the middle region of the curve and fields of this magnitude are most sensibly employed to determine if a reaction is field sensitive or not.

8.5.1 "Normal" Magnetic Fields

For applied magnetic fields that are larger than the HFCs present in the RP ("normal" strength magnetic fields), the effect of a magnetic field is generally referred to as a MFE. Here, the field dependence is due to a decrease in the rate of S–T mixing caused by the energetic separation of the $T_{\pm 1}$ states from the S and T_0 ones through the Zeeman interaction (Fig. 8.2b). As the energy gap increases, the S–$T_{\pm 1}$ mixing becomes less efficient until the energy separation becomes greater than the size of the hyperfine interaction. A plateau is observed once S–$T_{\pm 1}$ mixing ceases completely. The standard approach for analyzing the field dependence in this region is to determine

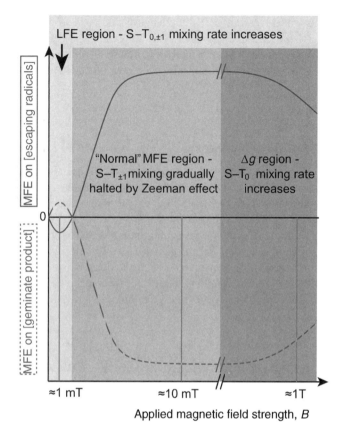

FIGURE 8.6 The effect of magnetic field strength on the yield of geminate product and radicals escaping the geminate cage for a triplet-born RP.

the size of the saturated field effect and then read off the magnetic field value where the field effect is half this value; this is known as the $B_{1/2}$ value. This value is usually compared with values predicted theoretically from average HFCs in the two radicals calculated as follows:

$$\bar{a}_r = \left[\sum_i^r I_i(I_i + 1)(a_i)^2 \right]^{\frac{1}{2}} \qquad (8.14)$$

$$B_{1/2,\text{theo}} = \frac{2(\bar{a}_1^2 + \bar{a}_2^2)}{\bar{a}_1 + \bar{a}_2} \qquad (8.15)$$

This analysis is based on the semiclassical approximation[33,34] (see Section 8.6). This analysis is highly useful as a rapid evaluation of whether the expected RP is responsible for the field dependence can be performed and deviations due to effects like spin relaxation or electron exchange can be rapidly discerned.

8.5.2 Weak Magnetic Fields

At very weak fields, it is sometimes possible to observe an increase in the rate of S–T mixing, which in Fig. 8.6 corresponds to an increase in geminate product yield. This is known as a low-field effect (LFE) and has been a topic of interest recently due to the interest in the effect of environmental electromagnetic fields on human health [35–37] and the recent resurgence of the proposal of the SCRP as a carrier of magnetic field sensing ability in some animals. [38–45] Despite the fact that at weak magnetic fields a level crossing between (typically) the S and T_{-1} states takes place, movement through this region is usually too rapid for this to serve as a mechanism for enhanced S–T mixing (although in constrained systems, this is an important mechanism; see examples in Ref. 15).

The LFE arises due to a purely quantum mechanical effect. Let us return to our one-proton RP, which can exist in one of the eight different spin states obtained by diagonalization of the matrix in Fig. 8.4. Figure 8.7 shows the zero-field wavefunctions of these eight states and the dependence of their energy on the strength of the applied magnetic field. Examination of these wavefunctions reveals that the different states have different degrees of singlet/triplet character. At zero field, the RPs exist as two degenerate groups of six and two states. Quantum coherences exist between some of the degenerate states, but have oscillations of zero amplitude and thus the states are unmixed. As a magnetic field is applied, the degeneracy of many of these states is broken and so they can now oscillate with nonzero amplitude. In other words, the application of a weak magnetic field unlocks the quantum coherences and provides additional pathways for S–T mixing due to the differing S/T character of the now mixed states. Figure 8.7 also shows that S–T state mixing is only enhanced for weak magnetic fields as the energetic separation of the states through the Zeeman interaction soon dominates and the total rate of S–T state mixing drops.

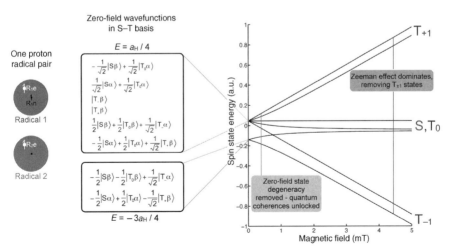

FIGURE 8.7 The zero-field wavefunctions of the eight spin states of a one-proton RP and the dependence of their energy on the strength of an applied magnetic field.

There have been a number of observations of the LFE in systems where the RP lifetime has been extended through the use of RIPs or microreactors (summarized in a recent review).[46] Of interest for neutral RPs in solution reactions is the hyperfine dependence of these LFEs. Early experimental and theoretical studies concluded that substantial LFEs are only observed for RPs in which there are substantial HFCs in one pair member and negligible couplings in the other.[47,48] In a more recent study, RPs with a wide range of combinations of HFCs and lifetime were screened for LFEs by simulation on a supercomputer.[49] This study concluded that indeed for short-lived RPs, the large:small HFC combination is required to produce substantial LFEs, but for long-lived RPs the restriction vanishes.

Until recently, however, there were virtually no observations of the LFE in homogeneous solution. A single data point was observed in an early paper by Fischer[50] and a LFE can be discerned in a study on the electronically excited radicals born in glycerol solution.[51] However, distinct LFEs have now been observed for a range of benzoyl radical containing neutral RPs in homogeneous solutions of moderately high viscosity.[52,53] The observed MARY dependence in these systems has been well reproduced using Monte Carlo simulations. By varying the RP partners, the size of the LFE can be substantially modified (and even removed) (see Fig. 8.8). In such systems, those pairs with large HFCs in one pair member and negligible couplings in the other produce the strongest LFEs in good agreement with previous prediction[47,48] and the short lifetime regime found in the theoretical screening work.[49] The presence of substantial LFEs also causes a problem for the useful MARY $B_{1/2}$

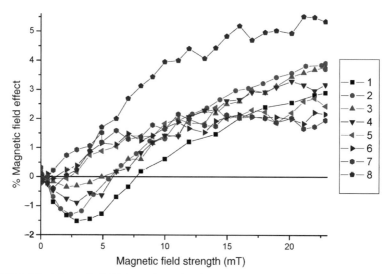

FIGURE 8.8 Magnetic field dependence for the photolysis of a series of benzoyl containing molecules in cyclohexanol solution obtained by time-resolved infrared spectroscopy of the carbonyl group of the resulting benzoyl radical. Key: $1 = \alpha,\alpha,\alpha$-trimethylacetophenone, $2 = $ 2-hydroxy-4-(2-hydroxyethoxy)-2-methylpropiophenone, $3 = $ 1-benzoylcyclohexanol, $4 = $ 2-hydroxy-2-methylpropiophenone, $5 = $ benzoin, $6 = $ methyl ether benzoin, $7 = $ dimethyl ether benzoin, $8 = $ 2-dimethylamino-2-(4-methyl-benzyl)-1-(4-morpholin-4-yl-phenyl)-butan-1-one.

analysis method. By analyzing a range of MARY curves with LFEs of varying magnitude, simple subtraction of the field value at which the field effect passes through zero was shown to produce $B_{1/2}$ values in good agreement with those calculated from average HFCs.[53]

An alternative approach to investigating the LFE is to choose RPs with extremely large HFCs to shift this field region to larger magnetic field strengths. Such studies have been undertaken using phosphorous-centered radicals[54,55] and although the fields are not true low fields, useful information about spin dynamics in the LFE region has been reported.

Thus, the shape of the magnetic field dependence for weak and normal strength fields can provide information on the HFCs present in the RP and some details about its lifetime. The shape of the curve is also affected by incoherent processes occurring in the RP. The effect of spin relaxation on long-lived RPs causes deviation from the usual field saturation characterized by average HFCs. The rate at which the MFE increases and saturates is reduced and the behavior is well described in terms of the relaxation mechanism [56] that results directly from spin-lattice relaxation in the pair members. As neutral RPs in free solution do not live long enough to be subject to such effects, readers are directed to Azumi, Nagakura and Hayashi's book that provides an excellent summary.[14] An incoherent relaxation process acting on a more rapid timescale has recently been observed and explained[57] that involves spin dephasing brought about by the modulation of exchange and dipolar interactions during RP diffusion. Thus far this effect has only been observed in micelles and seems to require the constrained motion and thus frequent movement into regions where D and J operate that they provide.

For RIPs, the shape of the field dependence is often modified by another spin-scrambling process, degenerate electron exchange (DEE). Here, electrons can hop between a donor or an acceptor and its uncharged form (i.e., another precursor molecule). This process serves to broaden and then narrow the MARY curve as the rate of DEE increases in a manner similar to that observed in chemical exchange in magnetic resonance.[58–62] One of the benefits of working with short-lived uncharged RPs is that both of these relaxation effects can be ignored.

MARY has been developed into a precise form of spectroscopy by Russian spin chemists.[63–67] For many RPs generated using radiolysis, crossings of energy levels during the move from zero to normal field can bring about specific resonances due to level crossings of the various RP spin states, which in ideal cases can be observed as peaks in the MARY spectrum. Early experiments only showed such resonances in RPs with degenerate HFCs,[63] but recent work has revealed experimentally observable resonances even in RPs with nondegenerate couplings.[67]

8.5.3 Strong Magnetic Fields

The third field region observed in the MARY curve shows a reversion of the field effect relative to the Zeeman effect saturation. We saw previously that the rate of S–T_0 mixing depends on the difference in Larmor precession frequencies of the two radicals and that this difference may arise from a difference in the g-values of the radicals. This frequency difference is proportional to the strength of the applied field. As the g-value difference is small for most organic radicals, the mixing term is normally dominated by the hyperfine-induced differences. However, at strong magnetic fields, this term can become significant and produces an increase in the rate of S–T_0 mixing. This effect is referred to as the Δg effect.[11–14] Magnetic field studies have been performed in magnetic fields of up to approximately 30 T.[68] Although the bulk of measurements have focused on reactions inside micelles,[14,69] high-field studies have been performed in homogeneous solution and the Δg effect can be observed. Theory predicts that the Δg effect should be saturated at fields of about 10^3 T. Hayashi and coworkers observed saturation of the Δg effect on the photoreduction of methoxybenzophenone with thiophenol in 2-methyl-1-propanol solution in fields of up to 30 T.[70] This was explained in terms of spin–spin relaxation competing with coherent S–T_0 mixing at high field.

8.6   THEORETICAL APPROACHES

The theoretical approaches for describing the RP and the influence of magnetic fields are well established. Let us first consider the case of a RP in which the members are sufficiently separated that the exchange and dipolar interactions are negligible.

8.6.1   General Approaches

The initial singlet or triplet RP is a nonstationary state that evolves under the influence of the interactions described above, primarily the hyperfine interaction. This process is readily simulated exactly, even for systems with many nondegenerate nuclei. To perform calculations, the electron spin vector operator is written in terms of Hilbert space matrices, typically

$$\hat{\mathbf{S}} = \begin{pmatrix} 0 & 1/2 \\ 1/2 & 0 \end{pmatrix}\mathbf{i} + \begin{pmatrix} 0 & -i/2 \\ i/2 & 0 \end{pmatrix}\mathbf{j} + \begin{pmatrix} 1/2 & 0 \\ 0 & -1/2 \end{pmatrix}\mathbf{k} \qquad (8.16)$$

The system is described by a spin density matrix, $\rho(t)$. The electron spins are then allowed to evolve under the spin Hamiltonian (Equation 8.13 with the exchange term removed) by application of the Liouville–Von Neumann equation.

$$\frac{d\hat{\rho}(t)}{dt} = i\left[\hat{H}, \hat{\rho}(t)\right] \qquad (8.17)$$

Thus, the precise time evolution of the population of the various hyperfine states can be calculated for any given magnetic field. For the determination of MFEs, the singlet or triplet character of the pair is readily calculated, for example, using the singlet projection operator.

$$\langle \hat{P}_S \rangle(t) = Tr(\hat{\rho}(t)\hat{P}_S)\hat{P}_S = \frac{1}{4} - \hat{\mathbf{S}}_1 \cdot \hat{\mathbf{S}}_2 \qquad (8.18)$$

The problem however, is that for a RP that can undergo diffusion, this spin-state mixing process becomes retarded every time the radicals become close enough for the exchange and dipolar interactions (where applicable) to become significant. Thus, a quantum mechanical approach alone is not sufficient to describe the RP, and the diffusion of the pair must be considered. The simplest theoretical approaches make the assumption that the spin and spatial dependences of the RP can be treated separately by assuming simply that beyond a certain separation, spin-state mixing occurs unhindered but is halted completely at separations less than a certain encounter distance. Thus the (for example) singlet product yield is determined by modifying the above spin evolution with the reencounter probability, $f(t)$, of the RP, which depends inherently on its diffusion.

$$\Phi_S = \int_0^\infty \langle \hat{P}_S \rangle (t) f(t) dt \tag{8.19}$$

8.6.2 Modeling Diffusion

For homogeneous solutions, two different models are usually employed to determine the reencounter probability. The simplest, and most often employed, is the "exponential model" [2,71] that assumes an exponential falloff in reencounter probability with time.

$$f(t) = ke^{-kt} \tag{8.20}$$

Here, k is a phenomenological rate constant. The great advantage of this model is its computational simplicity, allowing rapid calculations even for real RPs with multiple nondegenerate HFCs. A more realistic model of diffusion in free solution developed by Adrian [72-74] assumes a random walk of radicals and can be reduced to the following form:

$$f(t) = \frac{R_\sigma(R_0 - R_\sigma)}{R_0} \left(\frac{1}{4\pi Dt}\right)^{\frac{3}{2}} e^{-\frac{(R_0 - R_\sigma)^2}{4Dt}} \tag{8.21}$$

Here, spin-state mixing is possible only at distances beyond R_0, and R_σ is the separation at which the radicals can recombine. D is the mutual diffusion coefficient for the radicals in the given solvent.

Notably, recent studies have used some modern mathematical deconvolutions to try to extract reencounter probabilities from experimental MARY curves.[49] It is significant that experiments employing time-insensitive measurements can be used to deliver time-resolved information on rapid diffusional processes.

8.6.3 The Semiclassical Approach

A different approach to simplifying RP spin evolution for calculation is the semiclassical approach developed by Schulten and coworkers.[33,34] In this approximation, electron spins are treated quantum mechanically, but nuclear spins are treated classically and the model involves the precession of the electron spin about the magnetic field and the resultant of the nuclear spins. Thus, it is possible to define an "average" HFC for a given radical (Eq. 8.14, and the MARY curve shows a Lorentzian saturation characterized by the $B_{1/2}$ value (Eq. 8.15) as discussed previously. The semiclassical approach is also highly useful in calculations that try to model J and D more precisely.[57] In these cases, use of a full density matrix would make calculations impractically slow. Great care must be taken, however, when using the semiclassical approximation, as it is incapable of predicting and thus accounting for the presence of LFEs.

8.6.4   The Stochastic Liouville Equation

It is possible to perform more precise calculations that simultaneously account for the coherent quantum mechanical spin-state mixing and the diffusional motion of the RP. These employ the stochastic Liouville equation.[75] Here, the spin density matrix of the RP is transformed into Liouville space and acted on by a Liouville operator (the commutator of the spin Hamiltonian and density matrix), which is then modified by a stochastic superoperator, to account for the random diffusive motion. Application to a RP and inclusion of terms for chemical reaction, W, and relaxation, R, generates the equation in the form that typically employed

$$\frac{\partial}{\partial t}\hat{\rho}(r,t) = \left[-i\hat{\hat{L}}(r) + \hat{\hat{K}} + \hat{\hat{R}} + \hat{\hat{W}} \right]\hat{\rho}(r,t) \qquad (8.22)$$

The stochastic Liouville equation is highly useful when applied at high field, as techniques exist to reduce in size the typically large matrices it produces, and it has thus been used to simulate electron and nuclear spin polarizations in magnetic resonance experiments.[76–78] A relatively recent book describes the approach in detail.[79] However, for determining field dependences, such reductions are not possible, meaning that the sizes of the matrices are too large for even modern computers, and so this approach is seldom used for the simulation of field effects.

8.6.5   Monte Carlo Approaches

In order to better model the effects of J and D without the complexity of the stochastic Liouville equation, Monte Carlo approaches have recently been employed.[25,53,57,80] Such calculations have been used to discuss dephasing effects introduced by the dipolar interaction that can both influence the LFE in low fields and also cause changes to the time evolution of the RP in micellar cages.

8.7   EXPERIMENTAL APPROACHES

There are two basic strategies available for the experimental monitoring of MFEs: measurement of products and measurement of intermediate radicals. Much early work in the field focused on measuring product distribution (many examples in Ref. 11), that is, comparing the amount of products formed through the geminate and escape product channels. The advantage of such an approach is that a wide range of standard analytical techniques can be used in the product analysis. The disadvantage is that useful information about the reaction kinetics is not easily obtained.

8.7.1 Fluorescence Detection

One of the problems facing spin chemists performing these measurements is that the observed field effects can be rather small. Thus the method of detection should be as sensitive as possible. In some systems, it is possible to use the inherent fluorescence of one of the species involved in the reaction as a probe of RP activity. The most common of these approaches is the situation with the formation of RIPs that can often lead to spin-selective exciplex formation via the singlet RIP. Systems involving conjugated aromatic molecules, for example, anthracene and pyrene as electron donors/acceptors, amines as electron donors, and substituted benzenes (e.g., dicyanobenzenes) as electron acceptors, have been commonly employed and are now extremely well characterized. Grampp and coworkers have used these systems in recent years [59–62] to examine electron transfer processes in detail. These systems have been employed in many studies in Oxford, being successfully employed to study DEE,[58] the effects of radiofrequency oscillating magnetic fields,[81–84] and reaction yield detected magnetic resonance at low[85,86] and high fields.[87–90] Although fluorescence-based experiments are the most sensitive available, their great limitation is the small number of reactions that can be studied. In recent years, Basu and coworkers have been extending the range of systems of this type and revealing distinct solvent polarity dependences for different electron/donor pairs.[91–94]

Both time-resolved and continuous-wave experiments have been performed using fluorescence detection. For the former experiments, the sample is typically excited with a nanosecond pulsed laser (originally nitrogen lasers, later excimer lasers, and now typically solid-state Nd:YAG lasers) and the fluorescence is detected orthogonally with a PMT, solid-state detector, or photon counting system. In the latter experiments, an approach similar to CW EPR is employed where the magnetic field is modulated at an audiofrequency and phase-sensitive detection is used to measure only the component of the fluorescence that follows this oscillation, thus substantially improving the signal-to-noise ratio. In such experiments, RPs are continuously generated with light from a bright lamp source (typically a Xe arc lamp).

The alternative to generating RPs photochemically is to use radiolysis: excitation of a sample with high-energy (X- or gamma ray) photons. This is a very important technique and has generated much of the important data that informs our knowledge of RP field sensitivity. In analogous experiments to the photoinitiated reactions described above, CW and pulsed radiolysis experiments can both be readily performed. The high-energy photons initiate solvent ionization generating solvated electrons and holes that are then scavenged to produce RIPs. The reaction systems employed in radiolysis studies tend to be complementary to photochemical ones, and the two approaches together have allowed the study of a diverse range of RPs.

8.7.2   Optical Absorption Detection

Alongside fluorescence, the other most commonly employed approach is to monitor intermediate free radicals directly through their optical absorption in the UV/visible region of the spectrum, time-resolved optical absorption (TROA). This is a more general technique than the fluorescence methods described above, but is still limited to radicals that have appropriate optical absorptions, which in general means the presence of (mostly conjugated) aromatic groups. In ideal cases, the radical kinetics can be directly observed, but in many cases, there are absorptions from other species (typically excited triplet states and reaction products) in the same spectral region, which means that observed kinetic traces require deconvolution.

Recently, time-resolved experiments have been performed that employ molecular vibrations of the radicals to allow their detection.[52,53,95] The concept is similar to the TROA technique, but instead uses strong IR absorptions in a radical to monitor its concentration dependence. To date this technique has been employed to examine RPs containing benzoyl radicals using the carbon–oxygen double bond stretching frequency close to $1800\,cm^{-1}$. This technique has the potential to extend the range and type of RPs available for study. The technique relies on the use of a solid-state IR diode laser and a fast mercury cadmium telluride (MCT) detector.

8.7.3   Rapid Field Switching

A significant step in the measurement of MFEs is the development of rapid field switching apparatus.[95–98] Modern MOSFET transistors have made the possibility of field switching on the timescale of RP dynamics a reality. This technique was first used in a method called SEMF-CIDNP (switched external magnetic field chemically induced dynamic nuclear polarization).[96]

Rapidly switched external magnetic fields have proven highly useful in examining MFEs directly using TROA techniques described above. A magnetic field can be rapidly switched on or off at any point in time after RP formation (with rise times of less than 10 ns possible for fields of up to 40 mT).[95] By measuring the product yield as the timing of the field pulse is varied relative to laser excitation, the time dependence of RP concentration can be determined directly. This allows the problem of overlapping absorptions to be readily overcome and has been used in both the investigation of rapid dephasing processes in micellized RPs [57,98,99] and in the study of f-pair MFEs[95] (see later).

8.8   THE LIFE CYCLE OF RADICAL PAIRS IN HOMOGENEOUS SOLUTION

Unlike constrained RPs, those in homogeneous solution present a unique situation in that radicals that escape from their geminate cage to bulk solution have the possibility of encountering other radicals that have shared the same fate. Such an encounter leads to the formation of a new RP. Such an RP is formed through free diffusion and is referred to as a freely diffusing RP or f-pair. Until recently, only a single publication had experimentally investigated MFEs in f-pairs and observed a field effect on the second-order recombination rate constant.[100] Recent experiments have employed multiple techniques to show unequivocally distinct MFEs from g- and f-pairs and also to investigate the MARY dependence of f-pairs.[95]

8.8.1   Differentiating G-Pairs and F-Pairs

G-pair and f-pair MFEs should exist on distinct timescales when the g-pairs are all generated simultaneously (e.g., through the use of a nanosecond laser flash). For neutral RPs in viscous isotropic solution, this period should be tens of nanoseconds or so. Any MFEs established during the geminate reaction phase should be time invariant thereafter. F-pair MFEs should occur throughout the subsequent reaction period in which radicals randomly encounter one another in solution. In order to observe these two different MFEs in isolation, a SEMF TROA method was employed. First, the lifetime of the geminate period of the reaction was established in a magnetic field switching shift experiment.

Figure 8.9 shows the pulse schemes employed in the experiment and the experimental data obtained. Essentially as the time delay between RP generation and application of the magnetic field increases, the RP concentration falls and the effect of the field is reduced. In this manner, the point at which the geminate phase of the reaction is complete is readily established. The experiment reveals that the geminate period is complete within 100 ns at zero field. This is perhaps longer than might be expected in a cyclohexanol solution, but is likely due to the fact that the RP formation step of hydrogen abstraction from the solvent is relatively slow and so the birth of the RP is smeared out in the early time period. With the geminate phase of the reaction defined, a second experiment then examined the reaction kinetics under two field switching regimes. In the first, the magnetic field is switched on only during the geminate period, and in the second, no field is applied in this early period (indeed for the first 400 ns to ensure the geminate period is completely over) but then the field is rapidly switched on and applied for the duration of the rest of the reaction. Figure 8.10 shows the time dependence of the field effect for these two switching schemes.

For the geminate field, a MFE of about 4% is developed immediately and is unchanged throughout the rest of the reaction period. For the field applied only after the geminate phase is complete, the f-pair MFE is observed to slowly build from nothing to about 4% over the next 60 μs. Thus, it is clear that MFEs from g-pairs are distinct in terms of timescale. There are some other important differences between g- and f-pairs that are confirmed in these experiments.

(1) Regardless of the of the spin multiplicity with which a g-pair is born, f-pairs will always show MFEs of the same sense as a triplet-born g-pair. F-pairs are formed by randomly encountering RPs. If RPs encounter in a singlet state, then they will immediately react and be removed. However, radicals encountering in a triplet state are initially unreactive and do not lead to product formation. However, these unreactive encounters may be turned into reactive ones through the now familiar spin-state evolution of the RP. If, for example, a

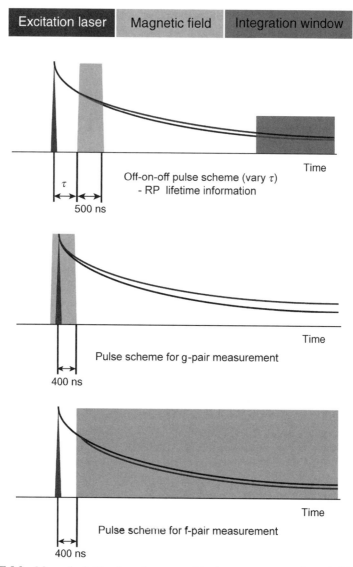

FIGURE 8.9 Magnetic field pulse schemes used in the measurement of g-pair lifetime and the separation of g- and f-pair RP kinetics.

magnetic field with a magnitude in the normal MFE region is applied, then f-pairs encountering in a $T_{\pm 1}$ state are unable to evolve into a singlet state and remain unreactive, whereas f-pairs encountering in a singlet state have no opportunity for spin-state evolution after encounter. Strictly speaking, it should be remembered that random encounters will not be in pure spin states but will always involve RPs with a certain degree of singlet/triplet character. In an applied field, RPs can encounter in pure $T_{\pm 1}$ states or mixed S–T_0 ones.

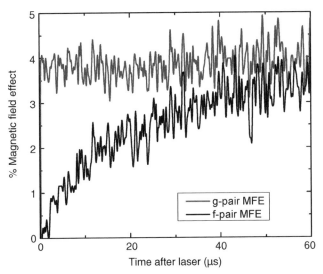

FIGURE 8.10 Time dependence of the g- and f-pair MFEs for the photoexcitation of benzophenone in cyclohexanol solution.

(2) For molecular precursors that produce an unsymmetrical RP, the pair members are predefined in a g-pair. However, f-pairs produced in such a reaction may be composed of any combination of the two types of radical. Figure 8.11 shows the different combinations possible. This means that the MARY dependence of f-pair MFEs must depend on the members that compose the f-pair. As the MARY curve can reveal something about the nature of its members based on the $B_{1/2}$ value and the size of the LFE, the MARY curve can be used to probe the nature of the f-pair throughout the reaction period. This has recently been demonstrated for the first time for the photolysis of 2-hydroxy-4'-(2-hydroxyethoxy)-2-methylpropiophenone in cyclohexanol.[95] The RP has large HFCs

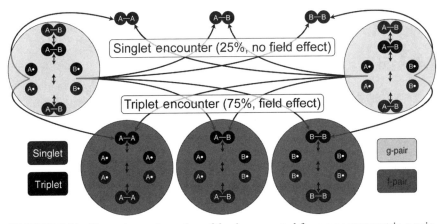

FIGURE 8.11 Formation and reaction of f-pairs generated from a nonsymmetric g-pair.

in one radical and negligible ones in the other giving rise to a substantial geminate LFE. As the reaction progresses, the size of the LFE decreases while the MFE increases. By subtracting the geminate MFE from the later time MFE (which contains contributions from both g- and f-pairs), it is possible to look at the MARY dependence of f-pairs. In this system, the LFE disappears completely and the $B_{1/2}$ value reveals (by comparison of the MARY curve with simulations for the possible pair combinations) that the pair is composed of like (benzoyl) radicals with very small HFCs. Use of rapid field switching in such experiments could improve the sensitivity of the f-pair measurements, and thus this method may have potential for measuring the kinetics of RP members that are spectroscopically silent through the MARY dependence of g- and f-pairs measured via their original geminate partners.

The observation of MFEs in f-pairs is important as this is how most ordinary chemical reactions take place. Most recombination reactions of organic free radicals do not occur through photochemically generated RPs with initially pure spin states, but instead through the random encounter of radicals in solution (think of the classic termination reaction in free radical chain reactions).

8.9   SUMMARY

The study of RPs in solution has had a fruitful history and we now have a very good understanding of the behavior of RPs in this environment and the way that they respond to applied magnetic fields. This knowledge provides an essential basis on which to understand MFEs in more complex systems. At the time of writing, there are a number of areas in which such MFEs are taking a high profile: in the quest to explain the magnetosensitivity of birds and other animals,[38–45] in the question of whether magnetic fields can have desirable or deleterious effects on human health [35–37] on our understanding and design of chemical systems to efficiently harvest light,[101] and in the behavior of conducting and electroactive polymers.[102–109] The spin selectivity of radical reactions has a subtle but profound effect across many aspects of chemistry.

REFERENCES

1. Closs, G. *J. Am. Chem. Soc.* **1969**, *91*, 4552.
2. Kaptein, R.; Oosterhoff, J. *Chem. Phys. Lett.* **1969**, *4*, 195.
3. Kaptein, R.; Oosterhoff, L. *Chem. Phys. Lett.* **1969**, *4*, 214.
4. Sagdeev, R. et al. *Org. Magn. Reson.* **1973**, *5*, 603.
5. Wakasa, M.; Hayashi, H. *Chem. Phys. Lett.* **2001**, *340*, 493.
6. Wakasa, M.; Hayashi, H. *Mol. Phys.* **2002**, *100*, 1099.
7. Brocklehurst, B. et al. *Chem. Phys. Lett.* **1974**, *28*, 361.
8. Brocklehurst, B. *J. Chem. Soc. Faraday Trans. 2* **1976**, *72*, 1869.

9. Schulten, K.; Staerk, H.; Weller, A.; Werner, H.; Nickel, B. *Z. Phys. Chem. Neue. Fol.* **1976**, *101*, 371.

10. Schulten, Z.; Schulten, K. *J. Chem. Phys.* **1977**, *66*, 4616.

11. Steiner, U. E.; Ulrich, T. *Chem. Rev.* **1989**, *89*, 51.

12. Mclauchlan, K. A.; Steiner, U. E. *Mol. Phys.* **1991**, *73*, 241.

13. Woodward, J. *Prog. React. Kinet. Mec.* **2002**, *27*, 165.

14. Nagakura, S.; Hayashi, H.; Azumi, T.; *Dynamic Spin Chemistry: Magnetic Controls and Spin Dynamics of Chemical Reactions*; Kodansha/Wiley: Tokyo/New York, 1998.

15. Atherton, N. M. *Principles of Electron Spin Resonance*; Ellis Horwood/PTR Prentice Hall: New York/London, **1993**; p. ix, 585 pp.

16. Schweiger, A.; Jeschke, G.; *Principles of Pulse Electron Paramagnetic Resonance*; Oxford University Press: Oxford, **2001**; p. xxvi, 578 pp.

17. Eaton, G. R.; Eaton, S. S.; Salikhov, K. M.; *Foundations of Modern EPR*; World Scientific Publishing Company: 1998.

18. Hirota, N.; Weissman, S. I. *J. Am. Chem. Soc.* **1964**, *86*, 2538.

19. Itoh, K.; Hayashi, H.; Nagakura, S. *Mol. Phys.* **1969**, *17*, 561.

20. Ferruti, P. et al. *J. Am. Chem. Soc.* **1970**, *92*, 3704.

21. Adrian, F. J. *Rev. Chem. Intermed.* **1979**, *3*, 3.

22. Murai, H.; Kuwata, K. *Chem. Phys. Lett.* **1989**, *164*, 567.

23. Batchelor, S.; Heikkila, H.; Kay, C.; Mclauchlan, K.; Shkrob, I. *Chem. Phys.* **1992**, *162*, 29.

24. Jeevarajan, A.; Fessenden, R. *J. Phys. Chem.* **1992**, *96*, 1520.

25. O'dea, A.; Curtis, A.; Green, N.; Timmel, C.; Hore, P. *J. Phys. Chem. A* **2005**, *109*, 869.

26. Efimova, O.; Hore, P. J. *Biophys. J.* **2008**, *94*, 1565.

27. Igarashi, M.; Sakaguchi, Y.; Hayashi, H. *Chem. Phys. Lett.* **1995**, *243*, 545.

28. Petrov, N. *J. Phys. Chem. A* **1998**, *102*, 7878.

29. Petrov, N. *High Energy Chem.* **2006**, *40*, 22.

30. Petrov, N.; Borisenko, V.; Starostin, A.; Alfimov, M. *J. Phys. Chem.* **1992**, *96*, 2901.

31. Petrov, N.; Shushin, A.; Frankevich, E. *Chem. Phys. Lett.* **1981**, *82*, 339.

32. Hamasaki, A.; Yago, T.; Wakasa, M. *J. Phys. Chem. B* **2008**, *112*, 14185.

33. Knapp, E. W.; Schulten, K. *J. Chem. Phys.* **1979**, *71*, 1878.

34. Schulten, K.; Wolynes, P. *J. Chem. Phys.* **1978**, *68*, 3292.

35. Ahlbom, A. et al. *Br. J. Cancer* **2000**, *83*, 692.

36. Draper, G.; Vincent, T.; Kroll, M.; Swanson, J. *Br. Med. J.* **2005**, *330*, 1290.

37. Greenland, S.; Sheppard, A.; Kaune, W.; Poole, C.; Kelsh, M. *Epidemiology* **2000**, *11*, 624.

38. Ahmad, M.; Galland, P.; Ritz, T.; Wiltschko, R.; Wiltschko, W. *Planta* **2007**, *225*, 615.

39. Begall, S.; Červený, J.; Neef, J.; Vojtěch, O.; Burda, H. *Proc. Natl. Acad. Sci. USA* **2008**, *105*, 13451.

40. Henbest, K. B. et al. *Proc. Natl. Acad. Sci. USA* **2008**, *105*, 14395.

41. Maeda, K. et al. *Nature* **2008**, *453*, 387.

42. Ritz, T.; Thalau, P.; Phillips, J.; Wiltschko, R.; Wiltschko, W. *Nature* **2004**, *429*, 177.

43. Ritz, T.; Adem, S.; Schulten, K. *Biophys. J.* **2000**, *78*, 707.

44. Ritz, T.; Dommer, D. H.; Phillips, J. B. *Neuron* **2002**, *34*, 503.

45. Schulten, K.; Swenberg, C.; Weller, A. *Z. Phys. Chem. Neue. Fol.* **1978**, *111*, 1.

46. Timmel, C. R.; Henbest, K. *Philos. Trans. R. Soc. A* **2004**, *362*, 2573.

47. Stass, D.; Tadjikov, B.; Molin, Y. *Chem. Phys. Lett.* **1995**, *235*, 511.

48. Stass, D. V.; Lukzen, N.; Tadjikov, B.; Molin, Y. *Chem. Phys. Lett.* **1995**, *233*, 444.

49. Rodgers, C.; Norman, S.; Henbest, K.; Timmel, C.; Hore, P. *J. Am. Chem. Soc.* **2007**, *129*, 6746.

50. Fischer, H. *Chem. Phys. Lett.* **1983**, *100*, 255.

51. Batchelor, S. N.; Mclauchlan, K. A.; Shkrob, I. A. *Chem. Phys. Lett.* **1993**, *214*, 507.

52. Vink, C.; Woodward, J. *J. Am. Chem. Soc.* **2004**, *126*, 16730.

53. Woodward, J.; Vink, C. *Phys. Chem. Chem. Phys.* **2007**, *9*, 6272.

54. Hayashi, H.; Sakaguchi, Y.; Kamachi, M.; Schnabel, W. *J. Phys. Chem.* **1987**, *91*, 3936.

55. Maeda, K.; Suzuki, T.; Arai, T. *RIKEN Rev.* **2002**, *44*, 85.

56. Hayashi, H.; Nagakura, S. *Bull. Chem. Soc. Jpn.* **1984**, *57*, 322.

57. Miura, T.; Murai, H. *J. Phys. Chem. A* **2008**, *112*, 2526.

58. Batchelor, S.; Kay, C.; Mclauchlan, K.; Shkrob, I. *J. Phys. Chem.* **1993**, *97*, 13250.

59. Justinek, M.; Grampp, G.; Landgraf, S. *Phys. Chem. Chem. Phys.* **2002**, *4*, 5550.

60. Grampp, G.; Justinek, M.; Landgraf, S. *Mol. Phys.* **2002**, *100*, 1063.

61. Justinek, M.; Grampp, G.; Landgraf, S.; Hore, P.; Lukzen, N. *J. Am. Chem. Soc.* **2004**, *126*, 5635.

62. Grampp, G.; Hore, P.; Justinek, M.; Landgraf, S.; Lukzen, N. *Res. Chem. Intermed.* **2005**, *31*, 567.

63. Tadjikov, B.; Stass, D.; Molin, Y. *Chem. Phys. Lett.* **1996**, *260*, 529.

64. Toropov, Y.; Sviridenko, F.; Stass, D.; Doktorov, A.; Molin, Y. *Chem. Phys.* **2000**, *253*, 231.

65. Verkhovlyuk, V.; Morozov, V.; Stass, D.; Doktorov, A.; Molin, Y. *Chem. Phys. Lett.* **2003**, *378*, 567.

66. Verkhovlyuk, V.; Stass, D.; Lukzen, N.; Molin, Y. *Chem. Phys. Lett.* **2005**, *413*, 71.

67. Kalneus, E.; Kipriyanov, A.; Purtov, P.; Stass, D.; Molin, Y. *Dokl. Phys. Chem.* **2007**, *415*, 170.

68. Nishızawa, K.; Sakaguchi, Y.; Hayashi, H.; Abe, H.; Kido, G. *Chem. Phys. Lett.* **1997**, *267*, 501.

69. Hayashi, H. *J. Chin. Chem. Soc.* **2002**, *49*, 137.

70. Wakasa, M.; Nishizawa, K.; Abe, H.; Kido, G.; Hayashi, H. *J. Am. Chem. Soc.* **1999**, *121*, 9191.

71. Lawler, R.; Evans, G. *Ind. Chim. Belg.* **1971**, *36*, 1087.

72. Adrian, F. *J. Chem. Phys.* **1970**, *53*, 3374.

73. Adrian, F. *J. Chem. Phys.* **1971**, *54*, 3912.

74. Adrian, F. *J. Chem. Phys.* **1971**, *54*, 3918.

75. Kubo, R. *J. Phys. Soc. Jpn.* **1969**, *S26*, 1.

76. Kitahama, Y.; Kimura, Y.; Hirota, N. *B. Chem. Soc. Jpn.* **2000**, *73*, 851.

77. Tarasov, V. et al. *J. Am. Chem. Soc.* **1995**, *117*, 110.

78. Tsentalovich, Y. et al. *J. Phys. Chem. A* **1997**, *101*, 8809.

79. Gamliel, D.; Levanon, H.; *Stochastic Processes in Magnetic Resonance*; World Scientific: 1995.

80. Steiner, U.; Wu, J. Q. *Chem. Phys.* **1992**, *162*, 53.

81. Stass, D. V.; Woodward, J. R.; Timmel, C. R.; Hore, P. J.; Mclauchlan, K. A. *Chem. Phys. Lett.* **2000**, *329*, 15.

82. Timmel, C.; Woodward, J.; Hore, P.; Mclauchlan, K.; Stass, D. *Meas. Sci. Technol.* **2001**, *12*, 635.

83. Woodward, J. R.; Jackson, R. J.; Timmel, C. R.; Hore, P. J.; Mclauchlan, K. A. *Chem. Phys. Lett.* **1997**, *272*, 376.

84. Woodward, J.; Timmel, C.; Mclauchlan, K.; Hore, P. *Phys. Rev. Lett.* **2001**, *87*, ARTN 077602.

85. Henbest, K.; Kukura, P.; Rodgers, C. T.; Hore, P. J.; Timmel, C. *J. Am. Chem. Soc.* **2004**, *126*, 8102.

86. Woodward, J.; Timmel, C. R.; Hore, P.; Mclauchlan, K. *Mol. Phys.* **2002**, *100*, 1181.

87. Batchelor, S.; Mclauchlan, K.; Shkrob, I. *Chem. Phys. Lett.* **1991**, *181*, 327.

88. Batchelor, S.; Mclauchlan, K.; Shkrob, I. *Mol. Phys.* **1992**, *77*, 75.

89. Batchelor, S.; Mclauchlan, K.; Shkrob, I. *Mol. Phys.* **1992**, *75*, 501.

90. Mclauchlan, K.; Nattrass, S. *Mol. Phys.* **1988**, *65*, 1483.

91. Aich, S.; Basu, S. *J. Chem. Soc. Faraday Trans.* **1995**, *91*, 1593.

92. Choudhury, S.; Basu, S. *Chem. Phys. Lett.* **2005**, *408*, 274.

93. Dey, D.; Bose, A.; Chakraborty, M.; Basu, S. *J. Phys. Chem. A* **2007**, *111*, 878.

94. Sengupta, T.; Basu, S. *Spectrochim. Acta A* **2004**, *60*, 1127.

95. Woodward, J.; Foster, T.; Salaoru, A.; Vink, C. *Phys. Chem. Chem. Phys.* **2008**, *10*, 4020.

96. Bagryanskaya, E.; Gorelik, V.; Sagdeev, R. *Chem. Phys. Lett.* **1997**, *264*, 655.

97. Kouskov, V.; Sloop, D.; Weissman, S.; Lin, T. S. *Chem. Phys. Lett.* **1995**, *232*, 165.

98. Suzuki, T.; Miura, T.; Maeda, K.; Arai, T. *J. Phys. Chem. A* **2005**, *109*, 9911.

99. Miura, T.; Maeda, K.; Arai, T. *J. Phys. Chem. A* **2006**, *110*, 4151.

100. Margulis, L.; Khudyakov, I.; Kuzmin, V. *Chem. Phys. Lett.* **1985**, *119*, 244.

101. Weiss, E.; Tauber, M.; Ratner, M.; Wasielewski, M. *J. Am. Chem. Soc.* **2005**, *127*, 6052.

102. Davis, A.; Bussmann, K. *J. Vac. Sci. Technol. A* **2004**, *22*, 1885.

103. Francis, T.; Mermer, O.; Veeraraghavan, G.; Wohlgenannt, M. *New J. Phys.* **2004**, *6*, ARTN 185.

104. Ito, F.; Ikoma, T.; Akiyama, K.; Kobori, Y.; Tero-Kubota, S. *J. Am. Chem. Soc.* **2003**, *125*, 4722.

105. Ito, F.; Ikoma, T.; Akiyama, K.; Watanabe, A.; Tero-Kubota, S. *J. Phys. Chem. B* **2005**, *109*, 8707.

106. Iwasaki, Y. et al. *Phys. Rev. B* **2006**, *74*, ARTN 195209.

107. Kalinowski, J.; Cocchi, M.; Virgili, D.; Fattori, V.; Di, M. P. *Phys. Rev. B* **2004**, *70*, ARTN 205303.

108. Murai, H.; Ishigaki, A.; Hirooka, K. *Mol. Phys.* **2006**, *104*, 1727.

109. Sakaguchi, Y. et al. *Mol. Phys.* **2006**, *104*, 1719.

9

CHEMICAL TRANSFORMATIONS WITHIN THE PARAMAGNETIC WORLD INVESTIGATED BY PHOTO-CIDNP

MARTIN GOEZ

Institut für Chemie, Martin-Luther-Universität Halle-Wittenberg, Halle/Saale, Germany

9.1 INTRODUCTION

The paramagnetic world, populated by radicals and biradicals, is an important terrain to be traversed in the course of many chemical reactions, in particular photoreactions. Of the numerous techniques used to obtain glimpses into that world, magnetic resonance yields the most detailed information about the structures and structural changes of its inhabitants. EPR spectroscopy, which is dealt with elsewhere in this book, observes paramagnetic species directly. This chapter is concerned with an indirect method, chemically induced dynamic nuclear polarization,[1–3] or CIDNP for short, which detects and characterizes paramagnetic species through the imprint they leave on the nuclear spin states; this signature is produced by the hyperfine interactions between the unpaired electrons and the nuclear spins and is observed in diamagnetic products at later stages of the reaction.

Experimentally, CIDNP denotes the occurrence of anomalous line intensities (enhanced absorption or emission) when NMR spectra are recorded during a chemical reaction. These anomalies are due to spin polarizations of the reaction products, that is, populations of their nuclear spin states that deviate from thermodynamic equilibrium. Only reactions involving radical pairs or biradicals exhibit CIDNP.

Carbon-Centered Free Radicals and Radical Cations, Edited by Malcolm D. E. Forbes
Copyright © 2010 John Wiley & Sons, Inc.

The CIDNP phenomenon was discovered in 1967.[4,5] Two years later, the hitherto unknown radical pair mechanism[6,7] was shown to be responsible for CIDNP, as well as for many other effects a magnetic field has on radical reactions. Within the ensuing 40 years, CIDNP has evolved into a powerful tool for the study of radical reactions.

Why should one resort to an indirect method when a direct method exists or, to put it less provocatively, what are potential advantages of CIDNP spectroscopy over EPR spectroscopy? The answer is that the independence of generation and detection of a physical observable, which is the essence of an indirect method, has three very attractive consequences.

First, these two steps are separated in time, so can occur on quite different timescales. This discrepancy is especially pronounced with CIDNP where the polarizations are created during the lifetime of intermediate spin-correlated species, on a nanosecond timescale or faster, but persist in the diamagnetic products for a time on the order of T_1, which means seconds for protons. Hence, CIDNP captures shorter lived intermediates and responds to faster processes than does EPR[8,9] although detection proper is by NMR spectroscopy, which is normally considered a very slow method.

Second, these two steps can take place (and with CIDNP do take place) in different molecular surroundings. As a result, a CIDNP experiment acts like a double resonance experiment, which greatly facilitates assignments of the observables on a molecular level. An EPR spectrum yields a set of hyperfine coupling constants but does not, by itself, reveal which constant belongs to which nucleus. In contrast, the CIDNP experiment encodes those hyperfine coupling constants as polarization intensities and detects that information in an NMR spectrum of the products, so immediately correlates each hyperfine coupling constant with a particular nucleus.[10,11]

Third, the observables can be regarded as labels that are affixed at an early stage of the reaction. By reading them out at a later stage, the pathways connecting those stages can be traced out. Because the magnetic energies are so tiny, the nuclear polarizations are nearly ideal labels that affect neither the energetics nor the kinetics of all chemical processes after their generation. This feature makes CIDNP especially valuable for the investigation of complex reaction mechanisms.[12,13]

This chapter is concerned with chemical reactions that occur while the system is still in the paramagnetic world. After an explanation of the radical pair mechanism and a brief treatment of experimental details, three case studies are presented that illustrate the application of CIDNP to transformations of radicals into other radicals and to interconversions of biradicals.

9.2 CIDNP THEORY

CIDNP effects are described qualitatively and quantitatively by the radical pair mechanism, which is depicted in Chart 9.1.[6,7]

As the name implies, the key intermediates are radical pairs. The property that causes two radicals to be a pair instead of two independent radicals is a correlation of

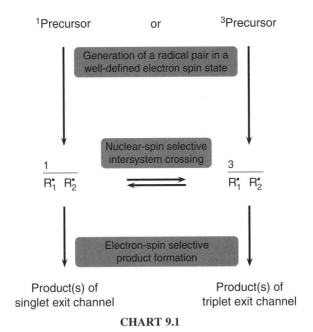

CHART 9.1

their electron spins. When organic molecules with an even number of electrons produce radicals by either of the two possible pathways, homolytical bond cleavage or electron transfer, *two* radicals are inevitably born in one such process, and the principle of spin conservation in chemical reactions demands that their joint spin state be the same as that of their precursor. In contrast to widespread chemical thinking, radical pairs are thus much more common species than are single radicals. Depending on the spin state of its precursor, a pair can only be born in the singlet state $|S\rangle$ or one of the three triplet states, $|T_{+1}\rangle$, $|T_0\rangle$, or $|T_{-1}\rangle$, where the subscript gives the z spin projection. The spin correlation is symbolized by an overline. Only $|S\rangle$ and $|T_0\rangle$ are relevant for the radical pair mechanism in high magnetic fields, so the superscripts "1" or "3" are used as shorthand notations of these two spin states.

Diffusion causes a rapid separation of the two radicals constituting the pair to distances where the interaction between them is negligible. However, they do not invariably lose each other for good; instead, the probability is quite substantial that they will reencounter at some later point of time. It is important to realize that such a reencounter involves the *same* two radicals that were initially in contact, and not just chemically equivalent others. The described sequence of events is called a diffusive excursion.

What are the implications for the spin state? In contact, that is, both at the moment of pair formation and at the moment of a later reencounter, the eigenstates of the pair are $|S\rangle$ and $|T_0\rangle$. For the separated, noninteracting radicals, that is, during the longest part of a diffusive excursion, the proper eigenstates are the two independent doublets $|D_{R_1^{\bullet}}\rangle$ and $|D_{R_2^{\bullet}}\rangle$. Nevertheless, $|S\rangle$ and $|T_0\rangle$ remain valid basis functions also in that situation and allow a description of the joint spin state $|\psi\rangle$ of the radical pair by a linear

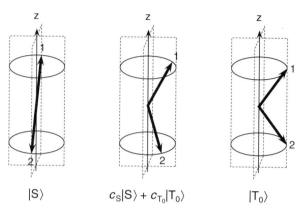

$|S\rangle$ $c_S|S\rangle + c_{T_0}|T_0\rangle$ $|T_0\rangle$

FIGURE 9.1   Vector model for intersystem crossing of radical pairs in high magnetic fields. Left, singlet state; centre, superposition state; right, state $|T_0\rangle$. The labels "1" and "2" denote the electron spins of the two radicals. Further explanation, see text.

combination

$$|\psi\rangle = c_S|S\rangle + c_{T_0}|T_0\rangle \tag{9.1}$$

with time-dependent coefficients c_S and c_{T_0}. During a diffusive excursion, $|\psi\rangle$ undergoes coherent oscillations between singlet and triplet.

This oscillation can be visualized most conveniently with the aid of vector diagrams as shown in Fig. 9.1.[14] The electron spin of each radical is depicted as an arrow. In the singlet state, the two arrows are antiparallel, and in state $|T_0\rangle$, they lie in a plane that also contains the z axis. In the absence of any interaction between the unpaired spins, each arrow precesses independently under the influence of Zeeman and hyperfine interactions. Unless the two radicals are absolutely identical, with respect to both their chemical structure and the spin state of each nucleus, the two precession frequencies differ, and the two arrows gradually get out of phase. The result, a superposition state as described by Equation 9.1, is also displayed in Fig. 9.1. When the diagrams are taken to rotate with the average of the two precession frequencies and the two starting states are drawn with a phase shift of 90°, the projections of the superposition state onto the two orthogonal planes gives the instantaneous singlet and triplet character of the pair. In high magnetic fields, the intersystem crossing frequency ω_{isc} follows from Equation 9.2,

$$\omega_{\mathrm{isc}} = \frac{1}{2\hbar}\left\{ (g_1 - g_2)\beta B_0 + \sum_i a_i m_i - \sum_k a_k m_k \right\} \tag{9.2}$$

where g_1 and g_2 are the g values of the two radicals, β is the Bohr magneton, B_0 is the magnetic field of the NMR spectrometer, a_i and a_k are the hyperfine coupling constants of the nuclei in the first and the second radical, and m_i and m_k are the z

spin components of those nuclei. The difference $g_1 - g_2$ is abbreviated as Δg. As the first radical one has to choose the one containing the nucleus whose polarization is to be observed or to be calculated.

The term Δg provides an intersystem crossing mechanism that is independent of the nuclei, but the nuclear spins come into play through the hyperfine terms of Equation 9.2. Because nuclear spin levels m_j lie symmetrically around zero, the intersystem crossing frequencies also lie symmetrically around the constant contribution of the Δg mechanism. When the radical pairs are born, their spin states are populated according to the Boltzmann distribution, which for attainable magnetic fields means that their populations are practically equal. Hence, in an ensemble of radical pairs, a certain fraction contains nuclear spin states that cause faster intersystem crossing, and an equally large fraction contains nuclear spin states that cause slower intersystem crossing. Approximating the coherent oscillation of Fig. 9.1 by a rate process, the "slower" nuclear spins are seen to be enriched in the radical pairs of the starting multiplicity and the "faster" ones in the pairs of the other multiplicity. In effect, the nuclear spins are, therefore, sorted between the two multiplicities of the pairs.

At the moment of an encounter, the feasibility of a reaction of the two radicals with each other (a so-called geminate reaction) is used to probe the state of the system: A singlet pair can only give a singlet product, and a triplet pair only a triplet product; in the majority of cases, the former is possible while the latter is not. Those pairs not capable of undergoing a geminate reaction can enter on another diffusive excursion during which some of them will separate for good ("escape") to give the so-called free radicals, which finally decay by reactions with other species. This partitioning of the radical pairs between geminate products and ecape products continues until they have all disappeared. Because not only the electron spin state but also the nuclear spin state is conserved in a chemical reaction, the surplus of "slower" nuclear spins shows up in the product(s) of the same multiplicity as that of the precursor and that of the "faster" ones in the products of the other multiplicity. The resulting over- and underpopulations of the nuclear spin states in the two kinds of products are thus of opposite signs but exactly equal magnitudes.

Quantitative calculations of CIDNP intensities can be performed with the stochastic Liouville equation (Eq. 9.3),[15]

$$\frac{\partial}{\partial t}\varrho = -\frac{i}{\hbar}[\mathbf{H}, \varrho] + D\Gamma\varrho + \mathbf{K}\varrho + \mathbf{R}\varrho \qquad (9.3)$$

where ρ is the density matrix of the radical pair. The four terms on the right-hand side of Equation 9.3 describe evolution under the the the spin Hamiltonian $\mathbf{H}$, interdiffusion ($D\Gamma$, diffusion operator), electron-spin selective chemical reactivity (operator $\mathbf{K}$), and electron spin relaxation (operator $\mathbf{R}$). In general, Equation 9.3 can only be solved numerically. However, by assuming that during a diffusive excursion there are abrupt transitions between regions where $\mathbf{H}$ is completely dominated by the exchange interaction and regions where the exchange interaction is zero, an approximate treatment is possible.[16] In the very simplest form, the nuclear-spin dependent probability F^* that a triplet precursor yields a singlet geminate product is given by Equation 9.4,

$$F^* \approx \sqrt{\frac{|\omega_{isc}|d^2}{D}} \qquad (9.4)$$

where ω_{isc} is the intersystem crossing frequency of Equation 9.2, d the sum of the molecular radii, and D the interdiffusion coefficient. The CIDNP spectrum is obtained by calculating the product yields for all nuclear spin configurations with Equation 9.4 and taking differences across all the levels connected by NMR transitions.

As can be shown by inserting Equation 9.2 into Equation 9.4 and expanding to first order, the CIDNP intensity of a particular nucleus is often approximately proportional to its hyperfine coupling constant. The so-called polarization pattern, that is, the relative CIDNP intensities of the different nuclei in a product, thus carries very similar information as the EPR spectrum (and, in addition, also reveals the signs of the coupling constants), so allows the identification and structural characterization of the paramagnetic intermediates.[17]

Very simple rules connect the polarization phases (i.e., signs) with the magnetic parameters on one hand and with the pathway from the precursor to the observed product in Chart 9.1 on the other hand.[18] Although only semiquantitative, these relationships are extremely valuable assets of CIDNP spectroscopy. In the present context, only net effects—meaning that all NMR lines of a nucleus are scaled by the same factor—are relevant. Their phase Γ_n ($\Gamma_n = +1$, absorption; $\Gamma_n = -1$, emission) is given by Equation 9.5,

$$\Gamma_i = \mu \times \varepsilon \times \text{sign}\,(\Delta g) \times \text{sign}\,a_i \qquad (9.5)$$

where $\mu = +1$ is assigned to a triplet precursor, $\mu = -1$ to a singlet precursor, $\varepsilon = +1$ to the singlet exit channel, and $\varepsilon = -1$ to the triplet exit channel; the magnetic parameters Δg and a_i are those entering Equation 9.2.

9.3 EXPERIMENTAL METHODS

Although CIDNP effects were first observed in thermal reactions, photochemical generation of radical pairs gives much better control, so it is not surprising that photo-CIDNP studies account for almost all CIDNP work published today.

The only pieces of hardware needed for photo-CIDNP are a light source and an unmodified NMR spectrometer. Pulsed lasers are most convenient for illumination, as they allow both time-resolved experiments (when the laser flash is followed by an acquisition pulse after a variable time delay)[19] and steady-state ones (when the laser is triggered with a high repetition rate, thus providing quasi-continuous excitation).[20] All the examples of this work draw on the second variant. Nevertheless, they yield kinetic information about much faster processes than would be observable by direct time-resolved CIDNP because they use the spin-correlated lifetime (which is on the order of a nanosecond) as an intrinsic clock.

For these experiments, gated illumination for a period much shorter than the nuclear T_1 was applied, and the polarizations were then sampled by an NMR pulse. This protocol results in a very good signal to noise ratio because the number of absorbed photons is large, and the CIDNP intensities are undisturbed by nuclear spin relaxation in the products. However, these experiments would also record the equilibrium NMR signals of unreacted molecules. This was avoided by special pulse sequences,[21] and all spectra shown below are completely free from such background signals and represent pure polarizations only.

9.4 RADICAL—RADICAL TRANSFORMATIONS DURING DIFFUSIVE EXCURSIONS

When one of the radicals of a pair undergoes a reaction to give another radical while on a diffusive excursion, the spin correlation persists but the intersystem crossing frequency of Equation 9.2 abruptly switches to a different value at that point of time because in general the new radical will have different magnetic parameters. Although only one radical is involved chemically, the effect is thus the exchange of one pair for another, which explains the name "pair substitution" that was given to this phenomenon. The earliest example of such a process, the decarbonylation of an acyl radical, was reported in 1972.[22]

Because CIDNP arises through a coherent oscillation, pair substitution does not simply produce a superposition of the individual CIDNP intensities of the first and the second pair in isolation. Figure 9.2 shows this for two consecutive pairs RP_1 and RP_2 with one proton in the first radical and a triplet precursor. In this figure, the two vectors of Fig. 9.1 have been projected onto the xy plane. Further projections of the two arrows onto the axis they were initially aligned with and on the axis perpendicular to that then yield the triplet and the singlet character of the pair, respectively. Assume Δg to be positive for the first and zero for the second pair, and the hyperfine coupling constant to be zero in the first and positive in the second pair. (Although these conditions may sound contrived, similar examples have been reported.)[23]

It is evident that a proton without a hyperfine coupling cannot acquire polarization because its spin state does not influence the intersystem crossing rate. This is the case for RP_1. Perhaps less obvious is the fact that a g-value difference of zero, as in RP_2, also does not lead to any polarization; the reason is the symmetrical evolution of the electron spin state for the two nuclear spin states.

Hence, each pair on its own is incapable of producing CIDNP. If, however, the spin state after the life of RP_1 serves as the starting condition for evolution under the spin Hamiltonian of RP_2, CIDNP results because for the α spin state of the proton the hyperfine interaction reinforces the Zeeman interaction whereas for the β spin state these two interactions oppose each other. The resulting CIDNP is the same as that of a hypothetical pair with an average Δg and an average hyperfine coupling constant, with the lifetimes of the two pairs as weighting factors for the averaging.

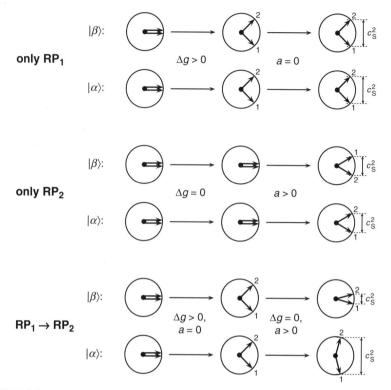

FIGURE 9.2 Vector models (projections) illustrating pair substitution. The labels "1" and "2" denote the electron spin of the first and the second radical of the pairs. The observed proton is contained in the first radical. Its spin state, $|\alpha\rangle$ or $|\beta\rangle$, is displayed at the respective leftmost projection. The radical pairs are born in the triplet state, and the product is formed from the singlet state; c_S^2 gives the singlet character. First radical pair RP_1, positive g-value difference, zero hyperfine coupling constant; second radical pair RP_2, equal g values, positive hyperfine coupling constant. For the situations without pair substitution, the spin evolutions under the influence of the Zeeman and the hyperfine interaction have been separated for clarity. Further explanation, see text.

The example presented in the following serves to illustrate the effects of pair substitution. However, the experiments were originally carried out with quite a different purpose in mind, namely, to prove that for the reaction under study the route giving rise to the polarizations is the only, or at least the predominant, pathway to the observed products. This addresses a controvertial issue when CIDNP is applied to the study of reaction mechanisms: Although it is universally accepted that CIDNP arises only from radical pairs or biradicals, the question is often raised whether the observation of CIDNP allows a definitive conclusion as to the reaction mechanism, the counter argument being that, because by their nature polarizations are amplified signals, the CIDNP experiment might have captured a minor side reaction while the greater part of the reaction proceeds via nonradical intermediates. This objection can be invalidated by preparing the CIDNP-active intermediates by an indirect route that bypasses any other intermediates potentially capable of leading to the observed products; if the CIDNP-active intermediates so obtained yield the same products with the same polarization intensities, other reaction routes are insignificant or nonexistent.[12]

The described problem was encountered in investigations of *cis–trans* isomerizations and cycloadditions of donor olefins D and acceptor olefins A in acetonitrile.[24,25] All polarizations, both of the cycloadducts and of the starting and isomerized olefins, could be traced to radical ion pairs $\overline{D^{\bullet+} A^{\bullet-}}$ formed by photoinduced electron transfer. As, however, exciplexes are frequently discussed as percursors to the products in such systems,[26–28] and CIDNP does not respond to exciplexes because no diffusive separation is possible, the question as to the relative contributions of the radical ion and exciplex pathways arose. To answer it, we employed photoinduced electron transfer sensitization (PET-sensitization).[29]

PET-sensitization means that the desired radical pair $\overline{D^{\bullet+} A^{\bullet-}}$ is produced indirectly by first generating another radical pair $\overline{D^{\bullet+} X^{\bullet-}}$ using an auxiliary sensitizer X and then exchanging $X^{\bullet-}$ for $A^{\bullet-}$ by a thermal electron transfer. X is chosen such that the photophysical parameters and redox potentials bar all other pathways except the PET-sensitized one. Of particular significance for the above mechanistic question is that neither D nor A are excited; hence, an exciplex $(D \cdots A)^*$ cannot be formed. Chart 9.2 juxtaposes the direct and the PET-sensitized formation of $\overline{D^{\bullet+} A^{\bullet-}}$.

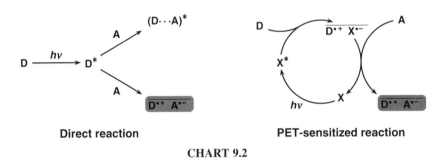

Direct reaction   **PET-sensitized reaction**

CHART 9.2

The rate of the exchange of the radical anions $X^{\bullet-}$ and $A^{\bullet-}$ depends on the concentration of A. If that exchange is slow, it can only involve the long-lived free radicals $X^{\bullet-}$; in that case, CIDNP generation has long come to a close, and PET sensitization has no influence on the polarizations in D and in the products of D, which only stem from the pair $\overline{D^{\bullet+} X^{\bullet-}}$. In contrast, if the exchange is fast on the CIDNP timescale, the intermediacy of the pair $\overline{D^{\bullet+} X^{\bullet-}}$ does not reveal itself in the CIDNP spectrum because that pair is too short lived for polarization generation; in that situation, all polarizations arise from the second pair $\overline{D^{\bullet+} A^{\bullet-}}$ even though this is not formed directly.

These limiting situations can be seen in Fig. 9.3. The system consisted of 9-cyanophenanthrene as the sensitizer X, *trans*-anethole as the donor D, and diethylfumarate as the acceptor A. The excitation wavelength, 357 nm, was chosen such that

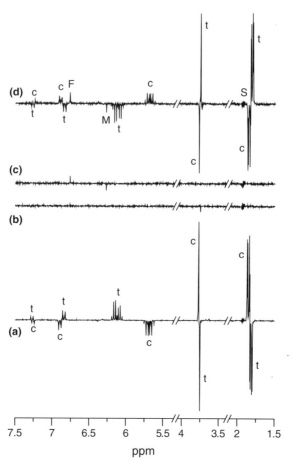

FIGURE 9.3 Photo-CIDNP spectra in the system 9-cyanophenanthrene (X, 3 mM), *trans*-anethole (D, 20 mM), and diethylfumarate (A, 110 mM), excitation at 357 nm. Adjacent signals groups "t" and "c" belong to the same proton in the starting and the isomerized anethole. The labels "F" and "M" denote the olefinic protons in A and its isomerized form, and "S" is an imperfectly suppressed solvent signal. Trace (a), only X and D; trace (b), only D and A; trace (c), only X and A; trace (d), all three components. Further explanation, see text. Reproduced from Ref. 29 with permission, copyright (C) 2006 the PCCP Owner Societies.

D and A do not absorb. The analysis most conveniently focuses on the CIDNP signals of the starting and isomerized donor olefin, which all result from geminate reactions of the radical pairs. Their polarization pattern reflects the hyperfine coupling constants of the *trans*-anethole radical cation. The precursor multiplicity is singlet ($\mu = -1$), the starting olefin is recovered by reverse electron transfer of singlet pairs ($\varepsilon = +1$), and the isomerized olefin is formed via reverse electron transfer of triplet pairs ($\varepsilon = -1$) to give the anethole triplet, which then decays to both isomers. The reason why the CIDNP intensities of the starting and the isomerized anethole are identical despite different chemical yields is well understood.[25]

Trace (a) shows the outcome of an experiment without the acceptor olefin, i.e., for the first of the two limiting cases discussed above. The polarizations necessarily stem from the pair $\overline{D^{\bullet +} X^{\bullet -}}$, which has a positive Δg, that is, $g\,(D^{\bullet +}) > g\,(X^{\bullet -})$.

Trace (b) represents a control experiment on a mixture D and A, without the sensitizer X but all other conditions identical. The absence of any CIDNP signals, together with the observation that the same mixture produces strong polarizations at other excitation wavelengths (e.g., 308 nm), proves that the first step of any photore-action at 357 nm is the excitation of X.

Trace (c) belongs to yet another control experiment: X in the presence of only A (at the same concentration as in traces (b) and (d)) yields small CIDNP signals of the starting and isomerized acceptor olefin, which stem from radical pairs $\overline{X^{\bullet +} A^{\bullet -}}$. However, Stern-Volmer experiments show that this reaction channel is blocked in the presence of D, which is a much better quencher of X^*.

The PET-sensitization experiment uses a very high concentration of A to make the exchange of $X^{\bullet -}$ for $A^{\bullet -}$ fast on the CIDNP timescale, that is, realizes the second of the two above-mentioned limiting cases. It is displayed as trace (d). The polarizations of the donor olefin are seen to be a mirror images of those in trace (a), which is simply due to the fact that the polarizations now arise from the pair $\overline{D^{\bullet +} A^{\bullet -}}$, for which Δg is negative. The phase inversion is thus a spectacular demonstration of the radical exchange in the three-component system.

The PET-sensitized experiment of trace (d) can be compared to the direct photoreaction of D and A at 308 nm, that is, without X. When the amounts of light absorbed are matched in these two cases, the absolute CIDNP intensities are found to be very similar, not only for the olefins but also for the cycloaddition products. Hence, the pathway via radical ion pairs is the predominant pathway to these products.

It is obvious that in between the two extremes of no radical-anion exchange (trace (a)) and very fast exchange (trace (d)) there must be a range where the exchange rate falls within the timescale of CIDNP generation, and for which, therefore, pronounced pair substitution effects are expected. This can be conveniently explored by varying the concentration of the acceptor olefin A. Best suited for a quantitative analysis is the strongly polarized β proton of the anethole (see Fig. 9.3; *trans,* about 6.2 ppm; *cis,* about 5.7 ppm). The normalized integrals over these signals have been displayed in Fig. 9.4 as functions of the concentration of A.

A description of pair substitution by a numerical solution of Equation 9.3, after appropriate modification, is always feasible. However, that frequently employed procedure has one major drawback: it depends on many parameters, some of which are often not known very precisely. The usual remedy is to determine them by a multiparameter nonlinear fit, but the uniqueness of a many-component solution vector obtained in that way is questionable if the curves do not have very characteristic shapes, as in Fig. 9.4. As an alternative approach, one can exploit the fact that the same parameters are contained in the polarization intensities in the limits of no pair substitution and of infinitely fast pair substitution. For the system of Fig. 9.3, recasting the equations in terms of these experimental quantities leads to a closed-form expression[29] that contains most parameters implicitly and only has a single adjustable parameter, namely, the rate constant of pair substitution divided by the intersystem

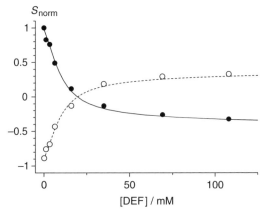

FIGURE 9.4   Signal integrals of the anethole β protons under the conditions of Fig. 9.3d, but for varying concentrations [DEF] of the acceptor olefin A. Filled symbols and solid curve, *trans*-D (signal at about 6.2 ppm); open symbols and dashed curve, *cis*-D (signal at about 6.2 ppm). All integrals were normalized to the value for *trans*-D in the absence of A. The curves are fit functions of a theoretical model. Further explanation, see text. Reproduced from Ref. 29 with permission, copyright (C) 2006 the PCCP Owner Societies.

crossing frequency of the first radical pair (Eq. 9.2). The fit curves of Fig. 9.4 were obtained in that way and allow a reliable determination of that kinetic parameter.

9.5   RADICAL—RADICAL TRANSFORMATIONS AT REENCOUNTERS

The very first use of the polarization pattern to identify the paramagnetic intermediates was in photosensitized (sensitizer, A) reactions of tertiary aliphatic amines DH, for example, triethylamine.[17] Although the gross reaction is a one-step hydrogen abstraction

$$A^* + DH \rightarrow AH^\bullet + D^\bullet \tag{9.6}$$

to give an α-amino alkyl radical $D^\bullet$, for example, $>N-CH^\bullet-CH_3$, CIDNP spectra do not always exhibit the polarization pattern of that radical but in many cases that of an amine radical cation $DH^{\bullet+}$, for example, $>N^{\bullet+}-CH_2-CH_3$, instead.

In $D^\bullet$ the hyperfine coupling constants of both the α and the β protons are large and have opposite signs whereas in $DH^{\bullet+}$ only the α protons have a substantial hyperfine coupling. These distributions are translated into an up/down (or down/up) pattern for the neutral radical, and a pattern of polarized α and unpolarized β protons for the radical cation. Evidently, these two very distinct patterns allow the determination of the intermediate the polarizations stem without knowledge of the other parameters entering Equation 9.5 (i.e., Δg, μ, and ε). The bottom and top traces of Fig. 9.5 show examples of these two cases in the same reaction product, *N, N*-diethylvinylamine V (for the formula see below, Chart 9.3).

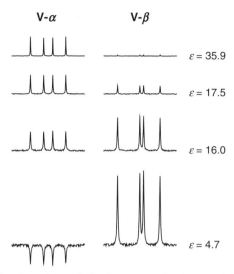

FIGURE 9.5   Polarity-dependent polarization patterns in photosensitized hydrogen abstractions from triethylamine DH (sensitizer A, 9,10-anthraquinone). For the formulas, see Chart 9.3. Shown are the signals of the olefinic α and β protons of the product N, N-diethylvinylamine, V-α (6.05 ppm) and V-β (3.45 ppm), as functions of the relative permittivity ε (given at the right). Top, pure acetonitrile-d_3; bottom, pure chloroform-d_3; other traces, mixtures of these two solvents. All spectra were normalized with respect of the absolute amplitude of V-α. Further explanation, see text.

The occurrence of the radical ion pattern in a polar solvent indicates that the photoreaction involves an electron transfer step

$$A^* + DH \rightarrow A^{\bullet -} + DH^{\bullet +} \tag{9.7}$$

under these circumstances. However, a direct reaction of the radical cation $DH^{\bullet +}$ to give the vinylamine V is chemically inconceivable; the direct precursor to V must be the neutral radical $D^\bullet$. Hence, a deprotonation of $DH^{\bullet +}$ yielding the conjugated base $D^\bullet$ is a key step of the electron transfer-induced formation of V.

Extensive studies of the sensitizer dependence[30] and the solvent dependence[31] of the polarization patterns led to the identification of two parallel pathways of that deprotonation. One is a *proton transfer within the spin-correlated radical pairs,* with the radical anion $A^{\bullet -}$ acting as the base. The other is a *deprotonation of free radicals,* in which case the proton is taken up by surplus starting amine DH. Furthermore, evidence was obtained from these experiments that even in those situations where the polarization pattern suggests a direct hydrogen abstraction according to Equation 9.6 these reactions proceed as two-step processes, electron transfer (Eq. 9.7) followed by deprotonation of the radical cation by either of the described two routes. The whole mechanism is summarized by Chart 9.3 for triethylamine as the substrate. Best suited for an analysis is the product V.

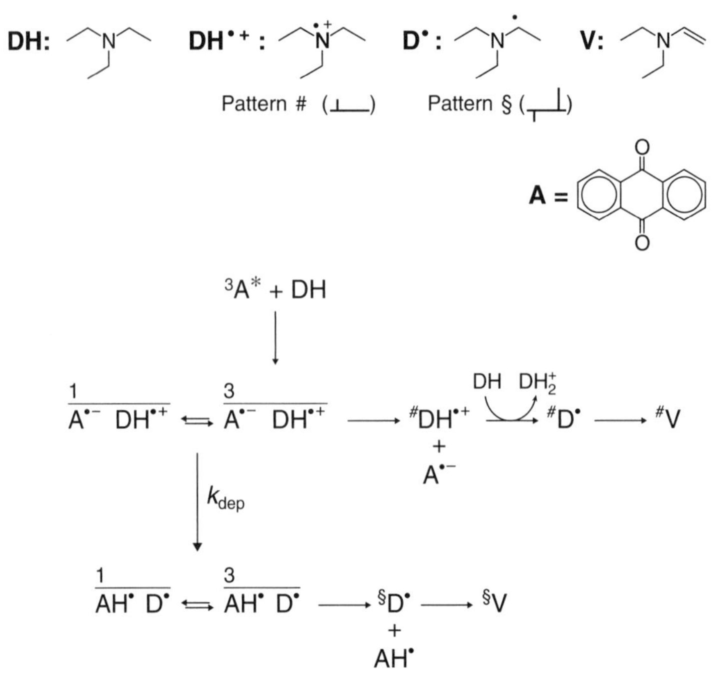

CHART 9.3

As in the preceding section, a pair substitution is central to the scheme, and there are again two limiting cases depending on the rate of that step relative to the spin-correlated life: If in-cage deprotonation is negligibly slow on the CIDNP timescale, the second radical pair is not formed, so V results from deprotonation of the radical cations after escape from the first pair; the escaping radical cations, and hence the product V, exclusively bear the polarization pattern of $DH^{\bullet+}$. If, on the other hand, in-cage deprotonation is extremely fast on the CIDNP timescale, the first radical pair has no chance of developing polarizations before it is transformed into the second radical pair, so CIDNP develops only in the latter pair; therefore, the escaping neutral radicals exclusively bear the polarization pattern of $D^{\bullet}$, which is then carried over to the product V. This example illustrates the power of CIDNP: Although the same product, V, can be formed by two parallel routes, a clear distinction is possible because these routes yield different polarization patterns.

In contrast to the PET sensitizations discussed in the preceding section, the concentrations do not influence the rate constant of the in-cage proton transfer, k_{dep}, because that key step involves the radical pair as an entity, so is a first-order process. However, k_{dep} can be changed by varying the thermodynamic driving force. With different sensitizers,[30] this can be effected only in rather large steps, but a much finer gradation can be realized by the relative permittivity of the reaction medium,[31,32] which only influences the free energy of the radical ion pair but leaves that of the pair of neutral radicals essentially unaffected.

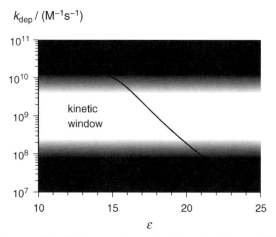

FIGURE 9.6 Photosensitized hydrogen abstraction from triethylamine DH by 9,10-anthra-quinone A (for the formulas, see Chart 9.3). The rate constant of in-cage deprotonation k_{dep}, as obtained from the polarity pattern, is shown as a function of the relativity permittivity ε of the reaction medium (mixtures of acetonitrile and chloroform). The timescale of the CIDNP effect provides a kinetic window, within which such a quantitative treatment is applicable. Further explanation, see text.

An interesting threshold behavior is found. Within a transition range from about -90 kJ/mol to about -130 kJ/mol, the polarization pattern completely changes from that of the radical ion pair to that of the pair of neutral radicals. In-cage deprotonation is thus negligibly slow compared to the "natural" life of the radical ion pair on the more positive side of this range and too fast to be observable on the more negative side. Two spectra within that range have also been included in Fig. 9.5 and serve to illustrate the gradual transmutation of the patterns.

Within the kinetic window provided by the transition regime, the pair substitution process is amenable to a quantitative analysis. Again, the equations can be converted to a form that contains the parameters of the two radical pairs implicitly and expresses them by the polarization patterns in the two limiting cases.[32] The use of the patterns, that is, polarization *ratios,* has the advantage that all factors influencing the absolute size of the polarizations cancel. Figure 9.6 shows the result for the sensitizer 9,10-anthraquinone. It is seen that the kinetic window spans a chemically very important range, namely, from about the diffusion-controlled limit to about two orders of magnitude below that limit.

9.6 INTERCONVERSIONS OF BIRADICALS

As far as CIDNP is concerned, the most important difference between radical pairs and biradicals is that the interdiffusion of the paramagnetic centres is restricted in the latter case. This has two implications.

First, the exchange interaction often does not fall off to zero. On the one hand, this reduces the efficiency of intersystem crossing between $|S\rangle$ and $|T_0\rangle$. On the other hand, it opens up another intersystem crossing pathway, namely, between $|S\rangle$ and $|T_{-1}\rangle$ (or, rarely, $|T_{+1}\rangle$) because the potential energy curves of these states intersect at some point, and the system spends more time in that region if diffusion is not free.

Second, there is usually no escape channel because the two radical termini cannot separate completely. This usually causes a fundamental difference between CIDNP from radical pairs and CIDNP from biradicals: Being nuclear spin sorting as described above, intersystem crossing between $|S\rangle$ and $|T_0\rangle$ crucially relies on both exit channels leading to different products, whereas intersystem crossing between $|S\rangle$ and $|T_{\pm 1}\rangle$ occurs by simultaneous electron–nuclear spin flips, and so creates net nuclear polarizations without the need of different exit channels.

Hence, in most cases, biradicals do not exhibit radical-pair type CIDNP but $S-T_{-1}$-type CIDNP. The two variants are easily distinguishable because the former gives rise to both absorptive and emissive polarizations with, ideally, a grand total of zero (compare Fig. 9.3), while the latter manifests itself by the same phase, usually emission, of all CIDNP signals.

The 1,5-biradical BR_1 formed during the Paterno–Büchi reaction of excited benzo-quinone B with quadricyclane Q (for the formulas, see Chart 9.4) provides one of the extremely rare examples of a short-chain biradical that produces CIDNP of the radical pair type.[33] Extracts of the CIDNP spectra are displayed in Fig. 9.7. The occurrence of both absorption and emission in the same product is clear evidence for this mechanism of polarization generation.

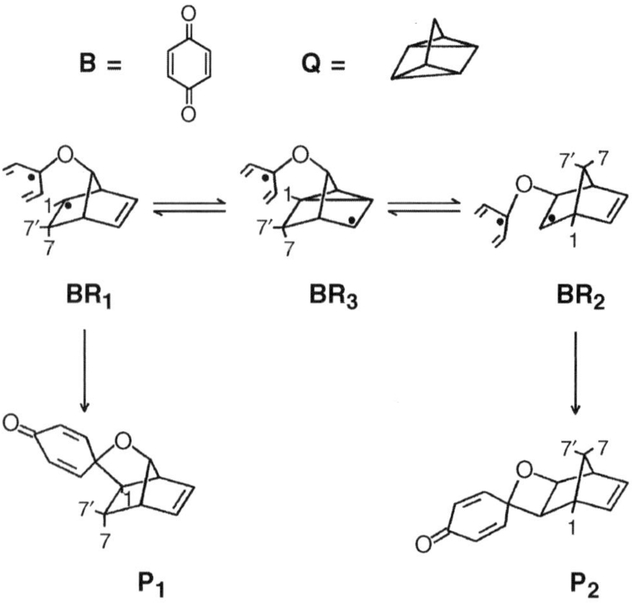

CHART 9.4

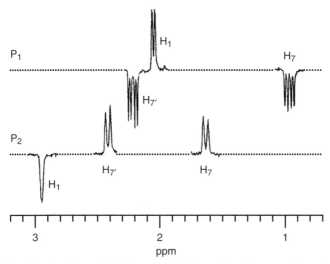

FIGURE 9.7 CIDNP effects in the Paterno–Büchi reaction of benzoquinone B with quadricyclane Q to give the oxolane P_1 and the oxetane P_2. Shown are the most strongly polarized signals, of H_1, H_7, and $H_{7'}$, in the two products. The formulas and the assignment are given in Chart 9.4. For clarity, the spectra of P_1 and P_2 have been separated, and all other signals have been blanked. Further explanation, see text.

The reaction affords two products, an oxolane P_1 and an oxetane P_2, which exhibit a mirror-image relationship of their CIDNP patterns. The three most strongly polarized signals, of H_1, H_7, and $H_{7'}$, with intensity ratios of about -2 to $+3$ to $+3.5$, have been shown in the figure; all the other protons are also polarized, but more weakly. The observed pattern is found to be in excellent agreement with the relative proton hyperfine coupling constants of the neutral benzosemiquinone radical and of the *tert*-butoxybicyclo[2.2.1]heptenyl radical, which were tested as model compounds for the two radical moieties.[33] The biradical BR_1 is thus the source of the polarizations. It is formed in a triplet state, its singlet exit channel produces the oxolane P_1, and its triplet exit channel the oxetane P_2.

Evidently, P_1 is obtained from BR_1 by a straightforward combination of the two radical centres, but an extensive skeleton rearrangement must occur on the route to the product P_2. Because it is natural to assume that no carbon–hydrogen bonds are broken in that process, the polarization of each proton serves as a label of the carbon atom it is attached to. The mechanism displayed in Chart 9.4 sums up the result, and identifies the structural changes as a cyclopropylmethyl–homoallyl rearrangement of the quadricylcane-derived moiety.

Both the biradical with the cyclopropylmethyl structure, BR_2, and that with the other homoallyl structure, BR_3, are too short lived for CIDNP generation, the former because of steric strain and the latter because it is a 1,4-biradical of the Paterno–Büchi type, that is, a structure that is known to undergo fast and efficient intersystem crossing without the participation of the nuclear spins. The latter property explains not only the absence of any polarizations from BR_3 but also the fact that $S–T_0$-type polarizations

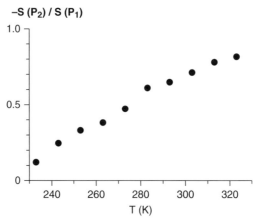

FIGURE 9.8 Ratio $-S(P_2)/S(P_1)$ of the CIDNP signals of H_7 in the two products P_1 and P_2 formed by the Paterno–Büchi reaction of benzoquinone with quadricyclane as function of the experimental temperature T. Further explanation, see text.

from BR_1 are observable at all in that system: because BR_3 acts as a chemical sink, the rearrangement of BR_1 to give BR_3 provides the analog to an escape reaction; hence, those nuclear spin states that decrease the rate of intersystem crossing in BR_1 are enriched in BR_3, and thus in P_2, while those that increase that rate end up preferentially in P_1.

As opposed to the previous examples, the rate of the "pair substitution" $BR_1 \rightleftharpoons BR_2 \rightleftharpoons BR_3$ can be varied by neither the reactant concentrations nor the solvent polarity because it is intramolecular and only involves neutral species. However, the ratio of polarizations of corresponding protons in P_1 and P_2 exhibits a pronounced temperature dependence,[34] which is shown in Fig. 9.8 and can be explained in the following way. Ideally, these opposite polarizations should have exactly equal magnitudes, but their ratio deviates from -1 if nuclear spin relaxation in the paramagnetic intermediates is taken into account. Biradicals with nuclear spin states that slow down intersystem crossing of BR_1 live longer, so their nuclear spins suffer a stronger relaxation loss.

The rate-limiting step of the complicated biradical interconversions is the formation of the three-membered ring, that is, the transformation of BR_1 into BR_2, which follows an Arrhenius law. The known temperature dependence of dipolar relaxation provides a gauge against which the activation energy of the rearrangement can be measured. Again, a polarization ratio is analyzed, so all other influences on the absolute CIDNP intensities—in particular, the temperature dependence of the CIDNP effect itself—cancel. By setting up an appropriate kinetic model, the relative rates of intersystem crossing of BR_3 and BR_1 (about $5:1$), the rate of formation of BR_2 relative to intersystem crossing of BR_1 (about $4:1$), and the activation energy of the transformation $BR_1 \rightarrow BR_2$ (17 kJ/mol) are obtained from these experiments.[34]

9.7 CONCLUSIONS

These case studies illustrate the power of CIDNP spectroscopy. Short-lived paramagnetic intermediates can be identified because their EPR spectrum remains frozen in as a polarization pattern of the nuclear spins in much longer-lived secondary species, and the pathways of their subsequent reactions can be traced out because these polarizations behave as nearly ideal labels. As the examples have shown, transformations of radical pairs into other radical pairs, with or without the participation of a third molecule as a scavenger, and transformations of biradicals can all be investigated by this method, which yields information that is often inaccessible by other techniques.

REFERENCES

1. Steiner, U. E.; Ulrich, T. *Chem. Rev.* **1989**, *89*, 51–147.
2. Goez, M. *Adv. Photochem.* **1997**, *23*, 63–163.
3. Goez, M. *Annu. Rep. NMR Spectrosc.* **2009**, *66*, 77–147.
4. Bargon, J.; Fischer, H.; Johnsen, U. *Z. Naturforsch. A* **1967**, *22*, 1551–1555.
5. Ward, H. R.; Lawler, R. G. *J. Am. Chem. Soc.* **1967**, *89*, 5518–5519.
6. Kaptein, R.; Oosterhoff, L. J. *Chem. Phys. Lett.* **1969**, *4*, 195–197.
7. Closs, G. L. *J. Am. Chem. Soc.* **1969**, *91*, 4552–4554.
8. Eckert, G.; Goez, M.; Maiwald, B.; Mueller, U. *Ber. Bunsen-Ges.* **1996**, *100*, 1191–1198.
9. Goez, M.; Rozwadowski, J.; Marciniak, B. *J. Am. Chem. Soc.* **1996**, *118*, 2882–2891.
10. Roth, H. D. *Z. Phys. Chem.* **1993**, *180*, 135–158.
11. Goez, M.; Rozwadowski, J.; Marciniak, B. *Ang. Chem., Int. Ed.*, **1998**, *37*, 628–630.
12. Eckert, G.; Goez, M. *J. Am. Chem. Soc.* **1994**, *116*, 11999–12009.
13. Eckert, G.; Goez, M. *J. Am. Chem. Soc.* **1999**, *121*, 2274–2280.
14. Ward, H. R. *Acc. Chem. Res.* **1972**, *5*, 18–24.
15. Freed, J. H.; Pedersen, J. B. *Adv. Magn. Reson.* **1976**, *8*, 1–84.
16. Pedersen, J. B. *J. Chem. Phys.* **1977**, *67*, 4097–4102.
17. Roth, H. D.; Manion, M. L. *J. Am. Chem. Soc.* **1975**, *97*, 6886–6888.
18. Kaptein, R. *J. Chem. Soc., Chem. Commun.* **1971**, 732–733.
19. Miller, R. J.; Closs, G. L. *Rev. Sci. Instrum.* **1981**, *52*, 1876–1885.
20. Goez, M. *Chem. Phys. Lett.* **1992**, *188*, 451–456.
21. Goez, M.; Mok, K. H.; Hore, P. J. *J. Magn. Reson.* **2005**, *177*, 236–246.
22. Kaptein, R. *J. Am. Chem. Soc.* **1972**, *94*, 6262–6269.
23. Den Hollander, J. A. *Chem. Phys.* **1975**, *10*, 167–184.
24. Goez, M.; Eckert, G. *J. Am. Chem. Soc.* **1996**, *118*, 140–154.
25. Goez, M.; Eckert, G. *Helv. Chim. Acta* **2006**, *89*, 2183–2199.
26. Lewis, F. D. *Acc. Chem. Res.* **1979**, *12*, 152–158.
27. Caldwell, R. A.; Creed, D. *Acc. Chem. Res.* **1980**, *13*, 45–50.

28. Mattes, S. L.; Farid, S. *Acc. Chem. Res.* **1982**, *15*, 80–86.

29. Goez, M.; Eckert, G. *Phys. Chem. Chem. Phys.* **2006**, *8*, 5294–5303.

30. Goez, M.; Sartorius, I. *J. Am. Chem. Soc.* **1993**, *115*, 11123–11133.

31. Goez, M.; Sartorius, I. *Chem. Ber.* **1994**, *127*, 2273–2276.

32. Goez, M.; Sartorius, I. *J. Phys. Chem. A* **2003**, *107*, 8539–8546.

33. Goez, M.; Frisch, I. *J. Am. Chem. Soc.* **1995**, *117*, 10486–10502.

34. Goez, M.; Frisch, I. *J. Phys. Chem. A* **2002**, *106*, 8079–8084.

10

SPIN RELAXATION IN Ru-CHROMOPHORE-LINKED AZINE/DIQUAT RADICAL PAIRS

Matthew T. Rawls[1], Ilya Kuprov[2], C. Michael Elliott[3], and Ulrich E. Steiner[4]

[1]National Renewable Energy Laboratory, Golden, CO, USA

[2]Chemistry Department, University of Durham, Durham, UK

[3]Department of Chemistry, Colorado State University, Fort Collins, CO, USA

[4]Fachbereich Chemie, Universität Konstanz, Konstanz, Germany

10.1 INTRODUCTION

Photosynthesis is the inspiration for all efforts to harvest solar energy ranging from solar cells to light-induced water splitting. Historically, synthetic molecular systems that undergo light-induced electron transfer reactions have featured prominently in efforts to functionally mimic photosynthesis.[1–9] Like the natural system, many of the intermediates and transient products of these light-induced electron transfer reactions are radicals; thus, they are subject to spin-chemical effects.

While artificial photosynthetic mimics come in many manifestations, our efforts have focused predominantly on the class of molecules represented by the structure in Fig. 10.1. These molecules consist of a visible-light-absorbing chromophore in the form of a trisbipyridineruthenium(II) complex (C) linked by flexible polymethylene chains to one or more electron donors (D) and an electron acceptor (A). The electron acceptor is an N,N'-dialkylated-2,2'-bipyridine (a so-called "diquat"); and the electron donors are N-alkylated phenothiazines. The diquat type acceptor was chosen because

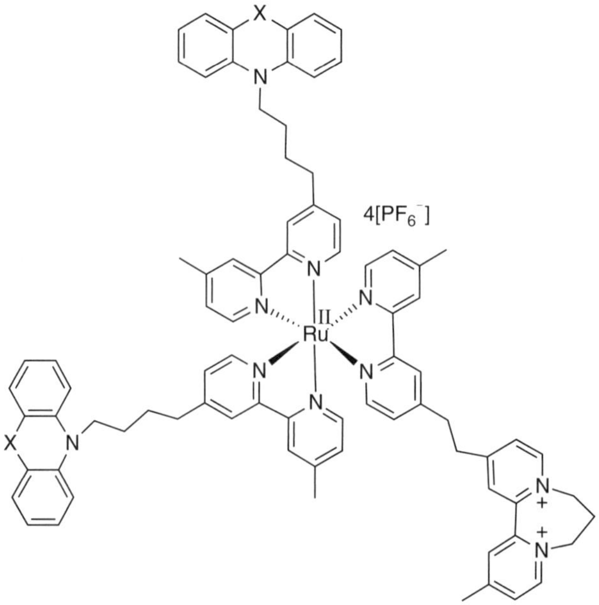

FIGURE 10.1 Donor–Chromophore–acceptor triad. X = a chalcogenide atom: oxygen, sulfur, or selenium.

it is possible to tune its redox potential by almost a half a volt by simply altering the length of the alkyl chain connecting to two quaternary nitrogens. Three different but closely related donors are considered in the following discussions: phenothiazine (PTZ), phenoxazine (POZ), and phenoselenazine (PSZ). Finally, the fact that the structure represented in Fig. 10.1 contains two rather than a single donor moiety is a matter of synthetic expediency and is of no functional consequence. Subsequently, we will refer to structures analogous to the one in Fig. 10.1 as DCA triads.

In the ground state, these DCA triads are singlets. Upon irradiation with visible light, a metal-to-ligand charge transfer (MLCT) transition occurs in which an electron from a metal-based d orbital is transferred to a π^* orbital on one of the bipyridine rings. In less than a picosecond, this state undergoes intersystem crossing to yield a triplet, ^{3}MLCT.[10] It is from this ^{3}MLCT state that a series of intramolecular electron transfer steps ensues leading ultimately to the charge separated state (CSS). The steps in this electron transfer cascade are first the oxidative quenching of the ^{3}MLCT state by the diquat acceptor followed by transfer of an electron from one of the donors to the oxidized chromophore; thus, in the resulting CSS the acceptor is reduced, one of the donors is oxidized and the chromophore is returned to the ground state.[11] This specific class of DCA triads is quite unusual in that the CSS state is formed with essentially unity quantum efficiency irrespective of the identity of the specific azine donor (i.e., PTZ, POZ, or PSZ) or diquat acceptor.[11] Detailed models of the energetics and kinetics of the CSS formation and decay back to the ground state in these specific DCA triad systems have been provided earlier.[11–21]

Of particular relevance to the present discussion is the observation that the CSS, which is a biradical cation, is formed with essentially pure triplet spin correlation.[12] For energetic reasons, this triplet radical pair cannot recombine to form the ^{3}MLCT state and can only form the singlet ground state. Therefore, direct recombination is spin forbidden. Moreover, because the radical pair which constitute the CSS product can separate only to a limited distance, essentially every CSS recombination event is between the same geminate radical pair—in other words, every reduced acceptor is ultimately oxidized by the donor radical cation that was formed from the same initial photochemical event. The spin behavior of the DCA triad CSS can be effectively explained by application of the relaxation mechanism of Hayashi and Nagakura.[22]

Scheme 10.1 shows how this model is applied to the DCA triad CSS under consideration.[12,23] The reduced diquat acceptor and oxidized azine donor are "normal" organic radicals that can physically separate within CSS to at least 20 Å. Therefore, at zero applied magnetic field, the triplet CSS (^{3}CS) and singlet CSS (^{1}CS) are nearly degenerate and isotropic hyperfine coupling provides a path to mix these states allowing for CSS decay. In the zero field case, this mixing and transition occur rapidly relative to the singlet decay (k_s). Upon application of a magnetic field, Zeeman splitting of the ^{3}CS occurs yielding three energy states: a magnetic field independent T_0 state and field dependent T_- and T_+ states. The path to the ground state for the T_0 state remains unchanged while at small fields the splitting of the field dependent states exceeds the hyperfine coupling energy, thus diminishing singlet–triplet mixing from those states and retarding their rate of decay to the ground state.[12]

The model in Scheme 10.1 suggests that application of a magnetic field to the CSS should result in biexponential decay kinetics consisting of a field independent (T_0) portion along with a field dependent portion (T_- and T_+). Within the CSS of these DCA triads the oxidized donor and the reduced acceptor each exhibit intense, isolated

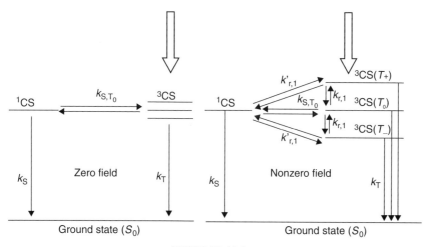

SCHEME 10.1

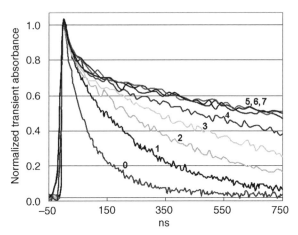

FIGURE 10.2 Field dependent transient absorption for CSS of DCA. (0) 0 mT, (1) 10 mT, (2) 25 mT, (3) 50 mT (4) 100 mT, (5) 300 mT, (6) 600 mT, (7) 1900 mT

features in their respective UV–vis absorption spectra. Thus, it is possible to probe the formation and decay of the CSS by transient absorption spectroscopy at the wavelengths corresponding to either component of the CSS (i.e., the oxidized donor or reduced acceptor). Figure 10.2 is a transient absorption measurement taken for a representative DCA triad at 520 nm that corresponds to the oxidized phenothiazine donor absorption. At time $t = 0$, the excitation laser initiates charge separation leading to CSS formation (subsequently also referred to as the radical pair) and finally to a decay to the ground state. With zero applied magnetic field, the decay of the CSS to the ground state is approximately monoexponential ($\tau = 100$ ns).[12,23] Upon application of an external magnetic field, the decay becomes distinctly biexponential with a field dependent portion and a field independent portion. This result is a clear demonstration of the applicability of the model in Scheme 10.1 to the DCA system. There are numerous examples in the literature of systems that demonstrate an increase in decay lifetime upon application of a magnetic field. However, no system previously reported has shown such a clear and quantitative demonstration of the predictions of the Nagakura relaxation model: monoexponential decay at zero field with biexponential decay at nonzero field with the predicted weighting (1/3 field independent–2/3 field dependent).

Two magnetic field effect (MFE) studies of CSS formation and decay in these DCA triads have been previously reported.[12,23] In the first, a series of DCA triads were synthesized with exclusively phenothiazine donors (X = S in Fig. 10.1). The linkages separating the donor and acceptor moieties from the chromophore were altered over the series along with the acceptor redox potential. Results from this study primarily shed light on how the maximum separation distance in the radical pair influenced the spin chemistry. Consideration from this study led to the generation of a second set of DCA species wherein a single optimal DCA structure was chosen (i.e., the one shown in Fig. 10.1) and the identity of donor species was varied (phenoxazine X = O, phenothiazine X = S, and phenoselenazine X = Se). This set of triads was chosen because they are isostructural, they are qualitatively similar in their CSS formation behavior, the donors have only slightly different (from one another) redox potentials, but vary in the atomic number of the heteroatom and, thus, the expected spin-orbit coupling.

In order to understand the magnetic field dependence of spin relaxation in the CSS, the contributions of various mechanisms need to be considered. These comprise the effects of anisotropic hyperfine coupling, g tensor anisotropy and spin-rotational interaction. All these mechanisms depend on specific magnetic properties of the two radical moieties forming the radical pair in the CSS. In particular, we need the components of the anisotropic hyperfine coupling tensor A_{ii} of the magnetic nuclei and the components g_{ii} of the g tensors of the two radicals. Our approach was to analyze the EPR spectra in liquid solution in as much detail as possible. Since, however, spectra in liquid solution yield only the orientationally averaged isotropic parameters, the measurements were complemented by quantum chemical theoretical calculations yielding all the tensor components, and to use the isotropic information from the spectra to validate the more detailed information from the calculations.

10.2 EPR FOR THE ISOLATED IONS

A detailed discussion of the contribution of various relaxation mechanisms determining the magnetic-field dependence of the decay of the charge separated state in the class of DCA species shown in Fig. 10.1 will follow. However, prior to this discussion it is necessary to establish some properties of the isolated radical species (reduced diquat and the oxidized azine species). The primary information required for these two radical species can be obtained through CW EPR measurements. Unambiguous hyperfine data can be found in the literature for the reduced diquat radical cation.[24] Hyperfine coupling constants for the diquat radical are reported as $(n = 2)$ $(2 \times a_N = 0.4\,\text{mT}, \ 4 \times a_H = 0.35\,\text{mT}, \ 2 \times a_H = 0.31\,\text{mT}, \ 2 \times a_H = 0.25\,\text{mT}, \ 2 \times a_H = 0.06\,\text{mT}, \text{ and } 2 \times a_H = 0.03\,\text{mT})$.

The hyperfine information for the series of azine radical donors requires deeper consideration. Gilbert et al. report detailed hyperfine information for unmethylated azine donors.[25,26] However, the methylated donors provide a better model for application to the DCA species under consideration. Also, the Gilbert studies do not conclusively explain the experimental EPR data, noting a trend of reduced EPR resolution with donor heteroatom size, without conclusive explanation. We therefore have conducted a detailed study of the cw EPR for the oxidized methyl azine radical cation series (MePOZ$^{+\cdot}$, MePTZ$^{+\cdot}$, and MePSeZ$^{+\cdot}$).

Synthesis of the methylated donors can be found in the literature.[27] All EPR measurements were recorded with a Bruker EMX 200U spectrometer with a 4102ST resonator and a variable temperature TE102 cavity. Spectra of solid DPPH were collected immediately following each sample run for g value calibration. Solutions of each of the donor radicals were prepared in an N_2 glovebox. Upon mixing of the sample with the oxidant, the sample tube was sealed, removed from the box and frozen in liquid nitrogen prior to the experiment. AlCl$_3$ oxidant was used in nitromethane solvent for the MePOZ and MePTZ donors. The MePSZ radical was formed in toluene solvent with thallium acetate and trifluoroacetic acid. The radical species remained stable under the given conditions for approximately 20 min at room temperature. Variables impacting signal broadening (concentration of analyte, microwave power, frequency, temperature, frequency modulation) were arrayed to achieve maximum signal resolution. A concentration of 0.06 mM with room temperature data collection proved optimal for all three species.

Figure 10.3 shows the results for the EPR measurements on the three oxidized donor radical cations. The loss of hyperfine resolution across the series (as noted previously by Gilbert et al.) is obvious within this methylated series.[26] The isotropic

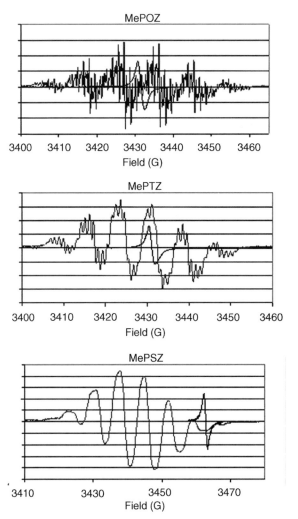

FIGURE 10.3 EPR spectra for the oxidized donor azine series. Each spectrum is overlaid with the relevant DPPH calibration spectrum.

TABLE 10.1   EPR Parameters[a] of Me-PXZ Cation Radicals

	MePOZ	MePTZ[c]	MePSZ
a(1N)	9.18	7.56 (7.49)	6.82
a(3H)	9.07	7.22 (7.24)	7.30
a(2H)	2.99	2.17(2.11)	2.13
a(2H)	1.49	0.96 (0.98)	0.92
a(2H)	0.56	0.75 (0.73)	0.77
a(2H)	0.57	0.31 (0.24)	0.43
a(^{77}Se)[b]			23.5 (25)[d]
G	2.0040	2.0052 (2.0052)	2.0153
Linewidth	0.30	0.30	0.55

[a] Hyperfine couplings and linewidths in Gauss.
[b] Natural abundance of 7.6%.
[c] Values from Ref. 25.
[d] Value for phenoselenazinium cation from Ref. 26.

magnetic parameters and homogeneous linewidth data of the azine radicals are presented in Table 10.1.

The spectral parameters were obtained by fitting the experimental EPR spectra using the WinSim package.[28] In order to determine the hyperfine linewidth, the specific modulation amplitude of the CW detection mode had to be taken into account. To this end, the theoretical spectrum with optimized Lorentzian linewidth was used as a starting point. Then the theoretical spectrum with the same hyperfine parameters but a reduced Lorentzian linewidth was calculated and transformed taking into account the broadening due to the CW modulation technique. The reduced Lorentzian linewidth was varied until the best correlation between experimental and modulation-transformed theoretical spectrum was obtained.

In view of the visible differences in resolution of the EPR spectra, it seems astonishing that the linewidth for the MePTZ radical cation is the same as that for MePOZ. However, as is borne out by the simulations, the reason for this deceiving impression is essentially exclusively due to the differences in hyperfine structure that gives more congested lines in the case of MePTZ. For MePSZ there is a noticeable increase in linewidth, but here again the apparent low resolution of the spectrum is largely due to the hyperfine structure. A secondary contribution is the underlying spectrum of 7.6% of the radicals with the magnetic ^{77}Se isotope.

The increase in linewidth between MePOZ/MePTZ to MePSZ is due to the spin-orbit coupling effect of Se. (In the case of MePTZ, the increased SOC with respect to MePOZ is not manifest in the linewidth, because other relaxation mechanisms dominate).

10.3   CALCULATION METHODS FOR EPR OF THE ISOLATED IONS

Molecular geometries were optimized using Gaussian03 at a DFT B3LYP/6-31G (2d,2p) level of theory using SCI-PCM solvation model (water). Effective core potential basis set (CRENBL ECP) was used for selenium. Hyperfine couplings were computed using DFT B3LYP/EPR-II method in SCI-PCM water with a CRENBL ECP basis set for Se. The calculation of g tensors was carried out in ADF2007 using collinear unrestricted ZORA DFT PBE/TZP method.

10.3.1 Calculation of g Tensor Components

Accurate simulation of radical g tensors has only recently become possible, courtesy of accurate solvent models,[29] ZORA approximation to Dirac equation,[30] and the arrival of an accurate treatment of spin-orbit coupling within the density functional theory.[31] All these methods are implemented in the ADF 2007 package, which was used for the g tensor calculations reported in Table 10.2.

As these results show, there is very good agreement between the theoretical and the experimental values of the isotropic g factors. Thus, the anisotropic components obtained from the quantum chemical calculations may also be considered as reliable. These components are needed in the following two mechanisms contributing to spin relaxation.

The relaxation mechanism depending on orientational modulation of the Zeeman interaction is determined by the anisotropy of the radicals' g tensors. In an axially symmetric approximation, this anisotropy is given by

$$\Delta g = g_{\|} - g_{\perp} \tag{10.1}$$

In Ref. 23, we have assumed that $g_{\|}$ equals the value g_e of the free electron,

$$g_{\|} = g_e \tag{10.2}$$

Hence,

$$g_{\perp} = \frac{(3g_{iso} - g_e)}{2} \tag{10.3}$$

TABLE 10.2   Theoretical Components of Radical g Tensors and Derived Quantities

	Phenoxazine	Phenothiazine	Phenoselenazine
g_{xx}	2.00199	2.00189	1.99982
g_{yy}	2.00411	2.00792	2.01654
g_{zz}	2.00433	2.00607	2.02774
g_{iso}	2.0035	2.0053	2.0147
g_{exp}	2.0040	2.0052	2.0153
Δg	−0.00223	−0.00510	−0.02232
$\frac{3}{2}(g_e - g_{iso})$	−0.00176	−0.00449	−0.01860
$\overline{\delta g^2}$	7.49E-06	4.60E-05	8.56E-04
$\frac{9}{2}(g_e - g_{iso})^2$	6.23E-06	4.03E-05	6.92E-04

and

$$\Delta g = \frac{3}{2}(g_{iso} - g_e) \tag{10.4}$$

As the data in the table show, the assumption of axial symmetry is well founded. Comparing the Δg values based on the approximation (10.3) with the quantum chemically derived values according to Equation (10.1) there is also fair agreement.

The spin-rotational interaction is controlled by $\overline{\delta g^2}$, the sum of square deviations between the individual tensor components of g and the free electron value:

$$\overline{\delta g^2} = \sum (g_{ii} - g_e)^2 \tag{10.5}$$

Using, again, approximations (10.2) and (10.3) we obtain

$$\overline{\delta g^2} = \frac{9}{2}(g_{iso} - g_e)^2 \tag{10.6}$$

As is shown in Table 10.2, the agreement between the results according to the exact Equation (10.5) and the approximation (10.6) is fairly good.

10.3.2 Calculation of Hyperfine Coupling Constants

10.3.2.1 Ab Initio **Hyperfine Coupling Constants: General Notes**   The electron-nucleus dipolar interaction that gives rise to hyperfine coupling naturally falls into two categories: the *anisotropic* part corresponding to pure dipolar interaction between electron and nucleus at large separations and the *isotropic* part resulting from a short-range dipolar interaction and stemming from the fact that because of finite size of the nucleus, the spherical average of point-to-point dipolar interaction is not zero. Because this second part results from short-range interactions, it is often referred to as *contact interaction.*[32]

The expression for the anisotropic part of hyperfine coupling involves an integral over the spatial distribution of the unpaired electron, which is relatively easy to compute accurately even at a relatively low level of theory.[33] The contact term, however, includes a delta-function that chips out the wave function amplitude at the nucleus point. The latter is quite difficult to compute both because standard Gaussian basis sets do not reproduce the wavefunction cusp at the nucleus point and because additional flexibility has to be introduced into the core part of the basis to account for the now essential core valence interaction.[34]

The basis sets and methods adequate for computing contact hyperfine couplings are relatively recent; the best general-purpose methods (DFT B3LYP with EPR-II or EPR-III basis set) were introduced in 1995[35] and 1996.[36] EPR-II is a double-ζ and EPR-III a triple-ζ basis set augmented by polarization and diffuse functions with core s-type Gaussians uncontracted and a number of tight s-type functions added. Furthermore, contraction coefficients were specifically optimized to reproduce experimental isotropic hyperfine couplings in a standard set of molecules. Notably, the basis sets have been optimized for computing hyperfine couplings specifically with B3LYP exchange correlation functional, and an attempt to use a different functional, even a fairly good one like B3PW91 or PBE1PBE, usually leads to poorly predicted contact interactions. The anisotropic components, however, are always satisfactory.

The important aspect of *ab initio* hyperfine coupling calculations is the variation of this parameter during molecular vibrations. In the paper that introduces the EPR-II and EPR-III basis sets, Barone and coworkers show that the equilibrium geometry calculation of hyperfine couplings underestimates the large couplings, which are rectified after taking a vibrational average.[36] The practical difficulties of performing this averaging even for medium-sized molecules are, however, formidable with a need to compute a Hessian to perform even the simplest vibrational average. Even though the code performing this procedure was available to the authors, the necessary calculations were well out of reach of SGI Altix 4700 supercomputer, with the result that the larger of the computed hyperfine couplings in the tables reported below are noticeably smaller than the experimental values. Annoying as these small deviations might be, the estimates obtained at the equilibrium geometry are still very reliable.[37]

10.3.2.2 Theoretical Values of Isotropic and Anisotropic Hyperfine Coupling Constants
The theoretical values for the various magnetic nuclei in the azine radials are given in Table 10.3. The correlation of experimental and theoretical values is plotted in Fig. 10.4. The numbering of the protons is given as follows:

As exhibited by Fig. 10.4, the correlation between theoretical and experimental values of the isotropic hyperfine coupling constants is generally very good. However, the

TABLE 10.3 Experimental and Theoretical Values of Isotropic Hyperfine Coupling Constants[a]

	Phenoxazine		Phenothiazine		Phenoselenazine	
	Experimental[b]	DFT	Expt[b]	DFT	Experimental[b]	DFT
$a(^{14}N)$	9.18	6.89	7.56	6.12	5.75	8.06
$a(CH_3)$	9.07	9.50[c]	7.22	8.11[c]	7.30	7.83[c]
$a(H_{(4)},H_{(g)})$	2.99	−2.89	2.17	−2.34	2.13	−2.20
$a(H_{(2)}H_{(10)})$	1.49	−1.34	0.96	−1.06	0.92	−1.07
$a(H_{(3)},H_{(9)})$	0.57	−0.63	0.75	−0.57	0.77	−0.41
$a(H_{(5)}(H_{(7)})$	0.56	0.63	0.31	0.50	0.43	0.49

[a] Values in Gauss.
[b] Experiment is insensitive to HFC sign.
[c] Average for the three protons in the minimal energy conformation of the methyl group.

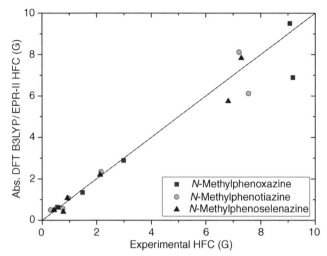

FIGURE 10.4 Experimental versus theoretical hyperfine couplings in phenoxazine, phenothiazine, and phenoselenazine. The three points deviating significantly from the bisector line belong to the ^{14}N nuclei. This deviation is likely caused by the lack of vibrational averaging in the calculation.

nitrogen hyperfine couplings are estimated somewhat too low. As an explanation, one has to consider that in all structures nitrogen is significantly bent out of plane. Thus, some very anharmonic motion is expected to exist along the direction perpendicular to that plane. Vibrational averaging should be taken into account in order to improve the theoretical predictions.

The anisotropic part of the nitrogen hyperfine interaction tensors are given in Table 10.4. As assumed in Ref. 23, these are well described by axial symmetry. The ratio between the axial anisotropy parameter A_N and the isotropic value is fairly constant. However, the value of 2.7–2.8 for this ratio is significantly larger than previously assumed (1.1–1.7[23]). The anisotropy parameters for the ^{1}H-nuclei at the ring positions are in the range of 1–3 Gauss. Thus, neglecting their effect in the

TABLE 10.4 Theoretical Values of the Eigenvalues of the Anisotropic Part of the Nitrogen Hyperfine Interaction Tensors[a]

	Phenoxazine	Phenothiazine	Phenoselenazine
$A_{N,xx}$	−6.40	−5.60	−5.30
$A_{N,yy}$	−6.30	−5.50	−5.20
$A_{N,zz}$	12.70	11.10	10.50
ΔA_N	19.05	16.65	15.75
$A_{N,\mathrm{iso}}$	6.89	6.12	5.75
$\Delta A_N/A_{N,\mathrm{iso}}$	2.8	2.7	2.7

[a] Values are given in Gauss; the z-coordinate is perpendicular to the molecular plane, x is parallel to the long axis and y along the short axis of the molecule.

anisotropic hyperfine relaxation mechanism, which is proportional to A^2, seems well justified. The hyperfine coupling of the methyl group is expected to be near-isotropic due to its rotational motion.

In the diquat radical, the A_N values of the two nitrogens amount to 9.0 G. The anisotropy axes are approximately perpendicular to the rings. Since these rings are, however, significantly rotated against each other due to the strain exhibited by the $(CH_2)_3$ bridge between the nitrogens, the two anisotropy axes are tilted toward each other by an angle of 53°. Thereby, the combined effect of the two nitrogen couplings in the anisotropic hyperfine relaxation mechanism must be significantly reduced relative to a strictly coaxial situation.

10.4 IMPLICATIONS FOR SPIN-RELAXATION IN LINKED RADICAL PAIRS

It is of interest to see how the information on the magnetic properties of the individual radicals relates to the spin-chemically observed spin relaxation in the Ru-chromophore linked radical pairs. As detailed in Ref. 23, the relaxation rate constant k_r defined in Scheme 10.1 can be extracted from the magnetic field dependence of the radical pair recombination kinetics. The pertinent data points for DCA-POZ and DCA-PSZ are shown in Fig. 10.5 together with the theoretical field dependence of k_r and its various contributions. For DCA-PTZ, no significant deviation from the DCA-POZ case is observed.

Theoretically, T_1 relaxation of individual radicals a and b is related to a k_r contribution in the radical pair by[38]

$$k_r = \frac{1}{2T_{1,a}} + \frac{1}{2T_{1,b}} \tag{10.7}$$

Our theoretical analysis shall start with the high field behavior of k_r for the DCA-PSZ case. In this radical pair, k_r must be dominated by the short T_1 time of the PSZ radical. From the linewidth, that exceeds those of the radicals POZ and PTZ by 0.25 G, we derive a specific mechanistic contribution to $1/T_2$ of $3.8 \times 10^6 \, s^{-1}$ in this radical.[39] It seems natural to assign this to the spin-orbit coupling effect of selenium in this radical, which is also borne out by the values of the g tensor components. This should lead to a large contribution of spin-rotational relaxation for which $1/T_1 = 1/T_2$.[40] With this interpretation of the excess linewidth and using Equation 10.7, we conclude that for DCA-PSZ k_r should not drop below approximately $2 \times 10^6 \, s^{-1}$. Taking into account various other magnetic field dependent contributions, we found that the magnetic-field independent part for DCA-PSZ amounts to $3.4 \times 10^6 \, s^{-1}$ (cf. Fig. 10.5). This value is close to what is expected ($1.9 \times 10^6 \, s^{-1}$) for the contribution of spin-rotational relaxation from the PSZ radical. Considering the uncertainty of linewidth determined from the CW EPR spectra the agreement seems fair. Adopting $3.4 \times 10^6 \, s^{-1}$ as the spin-rotational contribution to k_r we can estimate a relation between the effective hydrodynamic radius r of the PSZ moiety and the effective viscosity exhibited by it, by

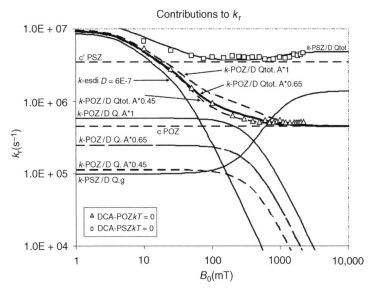

FIGURE 10.5 Contributions to k_r in the theoretical simulation of spin relaxation in the radical pairs from DCA-POZ and DCA-PSZ evaluated under the assumption of $k_T = 0$ (data points). The full simulations are represented by the curves denoted k-POZ/DQtot.A*0.45 and k-PSZ/DQtot, respectively. The contribution from the *esdi* mechanism (k-esdi $D = 6E-7$) corresponds to an effective translational diffusion constant of $D = 6 \times 10^{-7}$ cm^{2}s^{-1}. The curves indicated as k-POZ/DQ.A* represent the contributions of the *ahfi* mechanism, the * indicating the factor by which the theoretical anisotropy parameter A is reduced. The curve indicated by k-PSZ/DQ.g denotes the contribution due to the g tensor anisotropy in the PSZ radical. The constant values c POZ and c′ PSZ represent the field-independent contributions to k_r. For details of the calculation cf. Ref. 23.

using the theoretical relations[23]

$$k_{r,\text{sri}} = \frac{1}{2T_{1,\text{sri}}} = \frac{1}{2T_{2,\text{sri}}} = \frac{1}{18} \frac{\overline{\delta g^2}}{\tau_c} \qquad (10.8)$$

and

$$\tau_c = \frac{4\pi\eta r^3}{3kT} \qquad (10.9)$$

From $k_r = 3.4 \times 10^6$ s^{-1}, we obtain $\tau_c = 1.38 \times 10^{-11}$ s and $\eta r^3 = 13.8$ cP Å^3. Assuming an effective radius between 2.5 and 3.0 Å that seems geometrically reasonable, the effective viscosity would range between 0.50 and 0.89 cP. For 1,2-dichloroethane, the solvent used in the experiments $\eta = 0.464$ cP at 25 °C. Since, however, the radical is linked to the complex by a tetramethylene chain one might well expect some increase of the effective viscosity for rotational diffusion.

To assess the spin-relaxational contribution $k_{r,ahf}$ of anisotropic hyperfine coupling we only take the contribution from the nitrogen nuclei into account. This approach is justified on the basis of the theoretical DFT calculations showing that the contribution of anisotropic coupling to the ^1H spins is at least by one order of magnitude smaller. Details of the calculations of $k_{r,ahf}$ for the radical pairs are given in Ref. 23. In Fig. 10.5, the curve denoted k-POZ/DQ.A*1 shows the theoretical field dependence of $k_{r,ahf}$ using the full anisotropic hyperfine couplings as obtained from the DFT calculations. It is apparent that this contribution is quite negligible at low fields and at high fields. So, different mechanisms must contribute in these field regions. At low fields, the main contribution can be assigned to electron spin dipolar interaction (k_{esdi}) between the two radicals which is stochastically modulated by the relative translational diffusion of the two radicals toward each other. This may be parametrized by the distances of closest approach and farthest separation and an effective diffusion constant. For the distance boundaries, limits of 5.5 and 20 Å, respectively, seem reasonable on geometrical grounds. Adopting an effective diffusion constant D of 6×10^{-7} cm^2/s, the low-field behavior of k_r is well reproduced (for details of the calculations of k_{esdi} Ref. 23). Such an effective value of D is much smaller than to be expected for free diffusion of the radicals in the solvent 1,2 dichloroethane. However, values of such order of magnitude may be expected for pairs of molecules covalently linked by a flexible bridge of the size of the DCA-PXZ systems.[41,42] A molecular dynamic simulation for our systems should be desirable for a more quantitative modeling of the $k_{r,esdi}$ contribution.

As is shown in Fig. 10.5, anisotropic hyperfine interaction can only contribute in the intermediate field region. At high fields, even in the case of the DCA-POZ radical pair where the spin-orbit coupling effect on g is too small to cause a significant contribution of spin-rotational relaxation, a field-independent contribution of $k_r \approx 4.5 \times 10^5$ s^{-1} is present. This applies also to the DCA-PTZ radical pair. As for the source of this field-independent contribution, we can only speculate. A vibrational mechanism with very short correlation time has been invoked.[23] Adding the contribution $k_{r,ahfi}$ of the anisotropic hyperfine mechanism to that of the esdi mechanism and the magnetic-field independent contribution obtained from k_r at high fields, the theoretical prediction when using the values of the anisotropic hyperfine constants from the DFT calculations in the intermediate field range lead to k_r values that are clearly higher than observed (cf. the curve denoted k-POZ/DQ.tot.A*1). In order to become compliant with the experimental field dependence of k_r, we have to reduce the hyperfine anisotropy values by at least a factor of 0.45 (cf. Fig. 10.5). The reason for this discrepancy is not clear at present. Since evaluation of the CW spectra is not sufficient for this purpose, it would be illuminating to access the T_1 times of the free radicals by time-resolved pulsed EPR methods.

Finally, we address the apparent trend of the k_r-values for the DCA-PSZ pair to increase slightly at the high field end. Actually, this trend can be reproduced by the contribution $k_{r,gta}$ of the g tensor anisotropy mechanism, however, using an orientational correlation time τ_c of 8×10^{-12} s that is only about half the value extracted from a fit of the spin-rotational mechanism, as described above. It is possible that distortional motions of the nonplanar azine ring—which has been invoked to account for the deviation of the isotropic ^{14}N hyperfine constants of the azine radicals from the theoretical values predicted by DFT—also contributes to the modulation of the g tensor.

In summary, the field dependence of spin relaxation in the Ru-chromophore linked DCA-PXZ radical pairs can be rationalized well in terms of existing spin relaxation mechanisms and the magnetic parameters of the individual radicals taken from CW EPR spectra and DFT calculations. To complete and refine the picture, it would be most rewarding to measure the field-dependent T_1 times of the individual radicals using pulsed FT EPR and to investigate the dynamic motion of the radical moieties in the Ru-chromophore-DCA-PXZ triad by molecular dynamic calculations.

ACKNOWLEDGMENTS

The help of Dr. Georg Kollmannsberger with the simulation of EPR spectra is gratefully acknowledged. This work was supported (CME, MTR) by the Chemical Sciences, Geosciences and Biosciences Division, Office of Basic Energy Sciences, Office of Science, U.S. Department of Energy (DE-FG02-04ER15591) and (IK) the EPSRC (EP/F065205/1, EP/H003789/1). The authors are grateful to the NSCCS for the generous allocation of CPU time.

REFERENCES

1. Fox, M. A.; Chandon, M., Eds. *Photoinduced Electron Transfer*; Elsevier: Amsterdam, **1988**; Vol. *A-D*.

2. Schanze, K. S.; Walter, K. A. In *Molecular and Supermolecular Photochemistry: Organic and Inorganic Photochemistry*; Ramamurthy, V.; Schanze, K. S., Eds.; Marcel Dekker: New York, **1998**; Vol. *2*, pp 75–127.

3. Kalyanasundaram, K. *Photochemistry of Polypyridine and Porphyrin Complexes*; Academic Press: San Diego, **1992**.

4. Armaroli, N. *Photochem. Photobiol. Sci.* **2003**, *2*, 73–87.

5. Scandola, F.; Chiorboli, C.; Indelli, M. T.; Rampi, M. T. In *Biological and Artificial Supermolecular Systems*; Wiley: Weinheim, **2003**; Vol. *3*, pp 337–403.

6. Gust, D.; Moore, T. A.; Moore, A. L. In *Electron Transfer in Chemistry*, Balzani, V., Ed.; Wiley-VCH, Weinheim, **2001**; Vol. *3*, pp. 272–236.

7. Mattay, J., Ed.; *Topics in Current Chemistry*; Springer-Verlag: Berlin, **1990–1993**; Vol. *156, 158, 159, 163, 168*.

8. Wasielewski, M. R. *Chem. Rev.* **1992**, *92*, 435.

9. Ward, M. D. *Chem. Rev.* **1997**, *26*, 365.

10. Demas, J. N. *Inorganic Chemistry* **1979**, *18*, 3177.

11. Danielson, E.; Elliott, C. M.; Merkert, J. W.; Meyer, T. J. *J. Am. Chem. Soc.* **1987**, *109*, 2519–2520.

12. Klumpp, T.; Linsenmann, M.; Larson, S. L.; Limoges, B. R.; Buerssner, D.; Krissinel, E. B.; Elliott, C. M.; Steiner, U. E. *J. Am. Chem. Soc.* **1999**, *121*, 1076–1087.

13. Larson, S. L.; Cooley, L. F.; Elliott, C. M.; Kelley, D. F. *J. Am. Chem. Soc.* **1992**, *114*, 9504–9509.

14. Larson, S. L.; Elliott, C. M.; Kelley, D. F. *J. Phys. Chem.* **1995**, *99*, 6530–6539.

15. Cooley, L. F.; Larson, S. L.; Elliott, C. M.; Kelley, D. F. *J. Phys. Chem.* **1991**, *95*, 10694–10700.

16. Ryu, C. K.; Wang, R.; Schmehl, R. H.; Ferrere, S.; Ludwikow, M.; Merkert, J. W.; Headford, C. E. L.; Elliott, C. M. *J. Am. Chem. Soc.* **1992**, *114*, 430–438.

17. Schmehl, R. H.; Ryu, C. K.; Elliott, C. M.; Headford, C. L. E.; Ferrere, S. *Adv. Chem. Ser.* **1990**, *226*, 211–223.

18. Cooley, L. F.; Headford, C. E. L.; Elliott, C. M.; Kelley, D. F. *J. Am. Chem. Soc.* **1988**, *110*, 6673–6682.

19. Elliott, C. M.; Freitag, R. A. *J. Chem. Soc., Chem. Rev.* **1985**, 156–157.

20. Elliott, C. M.; Freitag, R. A. *J. Chem. Soc., Chem. Commun.* **1985**, 156.

21. Larson, S. L.; Elliott, C. M.; Kelley, D. F. *Inorg. Chem.* **1996**, *35*, 2070–2076.

22. Hayashi, H.; Nagakura, S. *Bull. Chem. Soc. Jap.* **1984**, *57*, 322–328.

23. Rawls, M. T.; Kollmannsberger, G.; Elliott, C. M.; Steiner, U. E. *J. Phys. Chem. A* **2007**, *111*, 3485–3496.

24. Rieger, A. L.; Rieger, P. H. *J. Phys. Chem.* **1984**, *88*, 5845–5851.

25. Clarke, D.; Gilbert, B. C.; Hanson, P.; Kirk, C. M. *J. Chem. Soc., Perkin Trans. II* **1978**, 1103–1110.

26. Chiu, M. F.; Gilbert, G. C.; Hanson, P. J. *J. Chem. Soc. B* **1970**, 1700–1708.

27. Phenoselenazine in a drybox under N_2 atmosphere was combined with lithium diisopropylamide in THF. Trimethyloxonium tetrafluorobate (1 equivalent) was added slowly and the solution was stirred for 1 h. The solution was then quenched with methanol. Upon silica gel chromatography (20:1 methylene chloride-acetone), Me-PSZ (a white powder) was isolated. The compound was characterized with NMR, TLC, and electrospray mass spectroscopy (M + H 261.0).

28. Duling, D. R. *J. Magn. Reson. B* **1994**, *104*, 105–110.

29. Tomasi, J.; Mennucci, B.; Cammi, R. *Chem. Rev.* **2005**, *105*, 2999–3093.

30. Lenthe, E.; Baerends, E. J.; Snijders, J. G. *J. Chem. Phys.* **1993**, *99*, 4597.

31. Velde, G. T.; Bickelhaupt, F. M.; Baerends, E. J.; Guerra, C. F.; van Gisvergen, J. A.; Snijders, J. G.; Ziegler, T. *J. Comput. Chem.* **2001**, *22*, 931–967.

32. Atkins, P. W.; Friedman, R. S. *Molecular Quantum Mechanics*; 4th ed.; Oxford University Press: Oxford, **2005**.

33. Szabo, A.; Ostlund, N. S. *Modern Quantum Chemistry: Introduction to Advanced Electronic Structure Theory*; Dover: Mineola, NY, **1996**.

34. Helgaker, T.; Jaszunski, M.; Ruud, K. *Chem. Rev.* **1999**, *99*, 293–352.

35. Barone, V. In *Recent Advances in Density Functional Methods*; Chong, D. P., Ed.; World Scientific: Singapore, **1995**, p 287.

36. Rega, N.; Cossi, M.; Barone, V. *J. Chem. Phys.* **1996**, *105*, 11060–11067.

37. Improta, R.; Barone, V. *Chem. Rev.* **2004**, *104*, 1231–1253.

38. Lüders, K.; Salikhov, K. M. *Chem. Phys.* **1987**, *117*, 113–131.

39. In Ref. 23, it was erroneously assigned twice this value.

40. Atkins, P.; Kivelson, D. *J. Chem. Phys.* **1966**, *44*, 169–174.

41. Haas, E.; Katchalski-Katzi, E.; Steinberg, I. Z. *Biopolymers* **1978**, *17*, 11–31.

42. Katchalski-Katzir, E.; Haas, E.; Steinberg, I. Z. *Ann. Rev. NY Acad. Sci.* **1981**, *366*, 44–61.

11

REACTION DYNAMICS OF CARBON-CENTERED RADICALS IN EXTREME ENVIRONMENTS STUDIED BY THE CROSSED MOLECULAR BEAM TECHNIQUE

RALF I. KAISER

Department of Chemistry, University of Hawaii at Manoa, Honolulu, HI, USA

11.1 INTRODUCTION

The chemical dynamics, reactivity, and stability of carbon-centered radicals play an important role in understanding the formation of polycyclic aromatic hydrocarbons (PAHs), their hydrogen-deficient precursor molecules, and carbonaceous nanostructures from the "bottom up" in extreme environments. These range from high-temperature combustion flames[1-6] (up to a few 1000 K) and chemical vapor deposition of diamonds[7-11] to more exotic, extraterrestrial settings such as low-temperature (30–200 K), hydrocarbon-rich atmospheres of planets and their moons such as Jupiter, Saturn, Uranus, Neptune, Pluto, and Titan,[12] as well as cold molecular clouds holding temperatures as low as 10 K.[13]

On Earth, carbonaceous nanoparticles are commonly referred to as soot and are often associated with incomplete combustion processes.[14-16] Soot is primarily composed of nanometer-sized stacks of perturbed graphitic layers that are oriented concentrically in an onion-like fashion.[17] These layers can be characterized as fused benzene rings and are likely formed via agglomeration of polycyclic aromatic

Carbon-Centered Free Radicals and Radical Cations, Edited by Malcolm D. E. Forbes
Copyright © 2010 John Wiley & Sons, Inc.

hydrocarbons.[18] Carbonaceous nanoparticles are emitted to the atmosphere from natural and anthropogenic sources with an average global emission rate of anthropogenic carbon from fossil fuel combustion.[19] Once liberated into the ambient environment, soot particles with sizes of 10–100 nm can be transferred into the lungs by inhalation[20] and are strongly implicated in the degradation of human health,[21] particularly due to their high carcinogenic risk potential. PAHs and carbonaceous nanoparticles are also serious water pollutants of marine ecosystems[22] and bioaccumulate in the fatty tissue of living organisms.[23] Together with leafy vegetables, where PAHs and soot deposit easily, they have been further linked to soil contamination,[24] food poisoning, liver lesions, and tumor growth.[21] Soot particles with diameters up to 500 nm can be transported to high altitudes[25] and influence the atmospheric chemistry.[26] These particles act as condensation nuclei for water ice, accelerate the degradation of ozone, change the Earth's radiation budget,[27] and could lead ultimately to an increased rate of skin cancer on Earth[28,29] and possibly to a reduced harvest of crops.[30] Nienow and Roberts also emphasized the significance of carbonaceous nanoparticles to understand the chemistry of atmospheric pollutants such as sulfur dioxide, nitric acid, and nitrogen oxides.

Whereas on Earth, PAHs and related species such as (benzo[a]pyrene) are considered as highly carcinogenic,[31] mutagenic,[32] and teratogenic, and therefore resemble unwanted by-product in combustion processes, the situation in quite reversed in hydrocarbon-rich atmospheres of planets and their moons, as well as in the interstellar medium.[33] Here, PAH-like species are thought to contribute to the unidentified infrared emission bands (UIBs) observed between 3 and 15 μm.[34] It is also estimated that PAHs and related molecules such as their radicals, ionized PAHs, and heteroaromatic PAH make up to 20% of the total cosmic carbon budget.[35] Also, the role of PAHs in astrobiology should be noted.[33–35]

But despite the key role of carbonaceous nanostructures and their hydrogen-deficient radical precursors in combustion processes and in extraterrestrial settings, the fundamental question "How are these nanoparticles and their precursors actually formed?" has not been conclusively resolved. The majority of mechanistic information on the growth processes has been derived from chemical reaction networks that model the formation of, for instance, PAH-like structures in combustion flames.[36,37] These models suggest that the synthesis of small carbon-bearing molecules together with their radicals is linked to the formation of PAHs and to the production of soot and possibly fullerenes in hydrocarbon flames.[38–43] Various mechanisms have been postulated; those currently in favor are thought to involve a successive buildup of hydrogen-deficient carbon-bearing radicals and molecules via sequential addition steps of small hydrogen-deficient species such as carbon atoms (C), carbon clusters (C_2, C_3), and carbon-bearing doublet radicals including ethynyl (C_2H), cyano radicals (CN), and phenyl radicals (C_6H_5).[44–57] Upon reaction with closed-shell hydrocarbons and their radicals, these elementary reactions do not only form more complex, closed-shell hydrocarbons, but also extremely stable resonantly stabilized free radicals (RSFRs) and aromatic radicals (ARs)[1–6,14–16,58–61] In RSFRs, such as in the propargyl radical (C_3H_3), the unpaired electron is delocalized and spread out over two or more sites in the molecule. This results in a number of resonant electronic structures of comparable importance. Owing to the delocalization, resonantly stabilized free hydrocarbon radicals are—similar to aromatic radicals such as phenyl (C_6H_5)—[62,63] more stable than ordinary radicals.[64] Consequently, RSFRs and ARs can reach high concentrations in flames. These high concentrations make them important reactants to be involved in the formation of soot and PAHs.[38–57]

Here, we review contemporary developments on crossed molecular beam studies that lead to the formation of highly hydrogen-deficient molecules acting as precursors to PAH-like species. In detail, crossed molecular beam reactions of carbon atoms (C), carbon clusters (C_2, C_3), and carbon-bearing doublet radicals including ethynyl (C_2H), cyano radicals (CN), and phenyl radicals (C_6H_5) will be discussed. This chapter is organized as follows. First, a general overview of the crossed molecular beam technique and their unique power is presented in Section 11.2. The experimental setup (crossed beam machine) and multiple supersonic beam sources (ablation, pyrolysis, photolysis) are compiled in Section 11.3. Hereafter, the results of the crossed beam reactions involving carbon-centered radicals are compiled in Section 11.4. Section 11.5 with concluding remarks closes this chapter.

11.2   THE CROSSED MOLECULAR BEAM METHOD

Which experimental approach can best reveal the chemical dynamics of carbon-centered radicals? Recall that since the macroscopic alteration of combustion flames, atmospheres of planets and their moons, as well as of the interstellar medium consists of multiple elementary reactions that are a series of *bimolecular encounters*, a detailed understanding of the mechanisms involved at the most fundamental microscopic level is crucial. These are experiments under *single collision conditions*, in which particles of one supersonic beam are made to "collide" *only* with particles of a second beam. The crossed molecular beam technique represents the most versatile approach in the elucidation of the energetics and dynamics of elementary reactions.[13,65–67] In contrast to bulk experiments, where reactants are mixed, the crossed beam approach has the capability of forming atoms and radicals in separate supersonic beams. In principle, both reactant beams can be prepared in well-defined quantum states before they cross at a specified energy under single collision conditions; these data help to derive, for instance, the reaction mechanism, entrance barriers, information on the reaction intermediates, and product distributions. These features provide an unprecedented opportunity to observe the consequences of a single collision event, excluding secondary collisions and wall effects. The products of bimolecular reactions can be detected via spectroscopic detection schemes such as laser-induced fluorescence (LIF)[68] or Rydberg tagging,[69] ion imaging probes,[70–74] or a quadrupole mass spectrometric detector (QMS) with universal electron impact ionization or photoionization. Crossed beam experiments can therefore help to untangle the chemical dynamics and infer the intermediates and the nascent reaction products *under single collision conditions*. It should be mentioned that recent kinetics experiments pioneered an isomeric-specific detection of reaction products utilizing time-resolved multiplexed photoionization mass spectrometry via synchrotron radiation.[75] Under those experimental conditions, the reaction intermediates may undergo up to a few thousand collisions with the bath molecules so that three-body encounters cannot be eliminated, and true single collision conditions are not provided. On the other hand, in "real" combustion processes, stabilizations due to collisions are important, and they can be only probed in collisional environments. Therefore, crossed beam and kinetics studies must be regarded as highly complementary.

The use of crossed molecular beams has led to an unprecedented advancement in our understanding of fundamental principles underlying chemical reactivity in light elementary reactions such as three-[76–82] and tetraatomic systems.[83–85] These simple systems are prototypical reactions in bridging our theoretical understanding of reactive scattering, via dynamics calculations on chemically accurate potential energy surfaces, with experimental observations.[86] These dynamics calculations are needed to turn the *ab initio* results into quantities that can be compared with experiments. Although interest in these light elementary reactions still continues, with the development of powerful theoretical models, attention is turning to more complex systems of significant practical interest such as in catalysis,[87–89] atmospheric chemistry,[90–92] interstellar[93–95] and planetary chemistry,[96–100] organometallic chemistry,[101–103] and combustion processes.[104–108] Due to the experimental difficulties in generating two unstable reactants simultaneously, only a very few atom–radical reactions have been conducted so far under single collision conditions. These are reactions of ground-state carbon atoms with propargyl by the PI[109] and of oxygen atoms with *t*-butyl (t-C_4H_9),[110] propargyl,[111,112] and allyl (C_3H_5).[113,114] However, the detection of the products of the oxygen atom reactions was limited to those with well-established fingerprints using LIF of the hydroxyl (OH) radical product formed in the hydrogen abstraction channel and of atomic hydrogen using hydrogen atom Doppler profile analysis. The oxygen–allyl and oxygen–methyl system was also investigated by Casavecchia and coworkers.[115]

11.3 EXPERIMENTAL SETUP

11.3.1 The Crossed Beam Machine

The crossed molecular beam method with mass spectrometric detection presents the most versatile technique to study elementary reactions with reaction products of *unknown* spectroscopic properties, thus permitting the elucidation of the chemical dynamics and—in the case of polyatomic reactions—the primary products.[116] The apparatus consists of two source chambers at a crossing angle of 90°, a stainless steel scattering chamber, and an ultrahigh-vacuum tight, rotatable, differentially pumped quadrupole mass spectrometric (QMS) detector that can be pumped down to a vacuum in the high 10^{-13} Torr range (Fig. 11.1).

In the primary source, a *pulsed* beam of open unstable (open shell) species is generated by laser ablation (C, C_2, C_3),[117] laser ablation coupled with *in situ* reaction

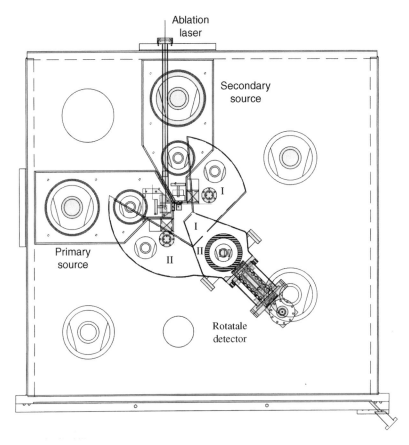

FIGURE 11.1 Top view of the crossed molecular beams machine. Shown are the main chamber, the primary (laser ablation configuration) and secondary source chambers, and the rotatable differentially pumped mass spectrometer detector.

(CN, C_2D),[98,118] photolysis (C_2H, C_2H_3, C_3H_3), [109,119,120] or flash pyrolysis (C_3H_5, C_6H_5).[121,122] The pulsed primary beam is passed through a skimmer into the main chamber; a chopper wheel located after the skimmer and prior to the collision center selects a slice of species with well-defined velocity that reach the interaction region. This section of the beam then intersects a *pulsed* reactant beam released by a second pulsed valve under well-defined collision energies. It is important to stress that the incorporation of *pulsed beams* allows that reactions with often expensive (partially) deuterated chemicals be carried out to extract additional information on the reaction dynamics, such as the position of the hydrogen and/or deuterium loss if multiple reaction pathways are involved. In addition, pulsed sources allow that the pumping speed and hence costs can be reduced drastically.

To detect the product(s), our machine incorporates a triply differentially pumped, *universal* quadrupole mass spectrometric detector coupled to an electron impact ionizer. Here, any reactively scattered species from the collision center after a single collision event has taken place can be ionized in the electron impact ionizer, and—in principle—it is possible to determine the mass (and the gross formula) of all the products of a bimolecular reaction by varying the mass to charge ratio, m/z, in the mass filter. Since the detector is rotatable within the plane defined by both beams, this detector makes it possible to map out the angular (LAB) and velocity distributions of the scattered products. Measuring the time-of-flight (TOF) of the products, that is, selecting a constant mass to charge value in the controller and measuring the flight time of the ionized species, from the interaction region over a finite flight distance at different laboratory angles allows extracting the product translational energy and angular distributions in the center-of-mass (CM) reference frame. This provides insight into the nature of the chemical reaction (direct versus indirect), intermediates involved, the reaction product(s), their branching ratios, and in some cases the preferential rotational axis of the fragmenting complex(es) and the disposal of excess energy into the products' internal degrees of freedom as a function of scattering angle and collision energy. However, despite the triply differential pumping setup of the detector chambers, molecules desorbing from wall surfaces lying on a straight line to the electron impact ionizer cannot be avoided. Their mean free path is of the order of 10^3 m compared to maximum dimensions of the detector chamber of about 1 m. To reduce this background, a copper plate attached to a two-stage closed-cycle helium refrigerator is placed right before the collision center and cooled down to 4 K. In this way, the ionizer views a cooled surface that traps all species with the exception of hydrogen and helium.

What information can we obtain from these measurements? The observables contain some basic information. Every species can be ionized at the typical electron energy used in the ionizer, and, therefore, it is possible to determine the mass and the gross formula of all the possible species produced from the reactions by simply selecting different m/z in the quadrupole mass spectrometer. Even though some problems such as dissociative ionization and background noise limit the method, the advantages with respect to spectroscopic techniques are obvious, since the applicability of the latter needs as a prerequisite the knowledge of the optical properties of the products. Another important aspect is that, by measuring the product velocity distributions, one can immediately derive the amount of the total energy available to the products and, therefore, the enthalpy of reaction of the reactive collision. This is of great help when different structural isomers with different enthalpies of formation can be produced. For a more detailed physical interpretation of the reaction mechanism, it is necessary to transform the laboratory (LAB) data into the center-of-mass system using a forward-convolution routine.[123] This approach initially assumed an angular distribution $T(\theta)$ and a translational energy distribution $P(E_T)$ in the center-of-mass reference frame. TOF spectra and the laboratory angular distribution were then calculated from these center-of-mass functions. The essential output of this process is the generation of a product flux contour map, $I(\theta, u) = P(u) \times T(\theta)$. This function reports the flux of the reactively scattered products (I) as a function of the center-of-mass scattering angle (θ) and product velocity (u) and is called the reactive *differential cross section*. This map can be seen as the *image* of the chemical reaction and contains *all* the information on the scattering process.

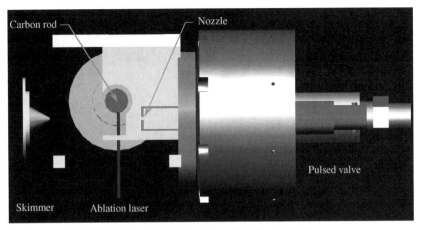

FIGURE 11.2 Schematic view of the laser ablation source.

11.3.2 Supersonic Beam Sources

11.3.2.1 Ablation Source A *pulsed* supersonic beams of ground-state carbon atoms, $C(^3P)$, can be generated via laser ablation of graphite utilizing a home-built ablation source generating reactant concentrations of up to $3 \times 10^{13} \, cm^{-3}$ in the interaction region of the scattering chamber (Fig. 11.2).[124]

Here, a laser beam originating from a Nd:YAG laser is tightly focused onto a rotating rod. The ablated species are seeded into a carrier gas (mostly helium, neon, or argon). Note that no ablation source can produce a pure beam of carbon atoms or of pure dicarbon or tricarbon molecules. However, a careful adjustment of the experimental conditions (laser power, laser focus, delay times, backing pressure) can often maximize one reactant and minimize other coreactants to less than a few percent. It should be stressed that the experimental conditions can also be optimized so that no carbon clusters higher than tricarbon are produced. A chopper wheel located behind the ablation zone and after the skimmer selects beam velocities between 800 and 2950 m/s. Even though the primary beam still contains small amount of, for instance, dicarbon and tricarbon, which may also react with the hydrocarbon reactant in the secondary beam, energy conservation and angular momentum conservation as well as the different center-of-mass angles and the product masses can distinguish the carbon atom channel from those of dicarbon and tricarbon.[125] For completeness, we would like to mention that a simple replacement of the graphite rod by boron or silicon also enables us to establish a supersonic atomic boron and silicon source, respectively.[101,126] Finally, it is important to note that the seeding gases (helium, neon, argon) can be also replaced by molecular nitrogen or deuterium. In those cases, both nitrogen and deuterium act not only as a seeding gas, but also as a reagent. In the ablation center, the ablated species can react with nitrogen or deuterium, thus producing strong supersonic beams of cyano radicals (CN)[118] and d_1-ethynyl radicals,[98] respectively. This *in situ* production of cyano radicals offers numerous advantages compared to photolytic (from ICN or C_2N_2) or pyrolytic (NOCN) cyano

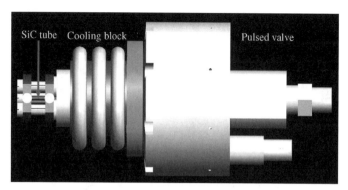

FIGURE 11.3   Schematic view of the pyrolytic source.

radical sources since the photolytic/pyrolytic precursors are either highly toxic or thermally unstable.

11.3.2.2   Pyrolytic Source   Pyrolytic sources can be utilized if radical precursors are available that have very weak bonds so that a thermally induced homolytic bond cleavage process generates two radical species (Fig. 11.3).

In this way, allyl iodide (C_3H_5I) and nitrosobenzene (C_6H_5NO) can be pyrolyzed quantitatively to generate allyl (C_3H_5)[121] and C_6H_5[122] radicals, respectively. For instance, a pulsed supersonic beam of phenyl radicals can be generated via flash pyrolysis of a nitrosobenzene precursor at seeding fractions of less than 0.1% employing a modified Chen source;[127] this unit was coupled to a piezoelectric pulsed valve. Helium buffer gas was introduced into a stainless steel reservoir; the latter kept the nitrosobenzene sample at a temperature of 283 K. The mixture was expanded through a resistively heated silicon carbide tube; the temperature of the tube was estimated to be about 1200–1500 K. At these experimental conditions, the decomposition of the nitrosobenzene molecule to form nitrogen monoxide and the phenyl radical was *quantitative*. Typically, supersonic beams with phenyl radical segments holding velocities of 2800–3500 m/s can be achieved. Again, the advantage of this beam source is the *quantitative* conversion of the precursor molecule into radicals; on the other hand, the heated silicon carbide tube only allows generating "fast" radical beams.

11.3.2.3   Photolytic Source   In case of photochemically active molecules, supersonic beams of radicals can be generated via photodissociation of helium-seeded precursors (Fig. 11.4).

Typically, these are bromide or iodide precursors such as bromoacetylene (C_2HBr),[119] vinyl bromide (C_2H_3Br),[120] and propargyl bromide (C_3H_3Br)[109] to yield supersonic beams of helium-seeded ethynyl, vinyl, and propargyl radicals, respectively. Briefly, the radical precursor is entrained in helium at a ratio of typically 1%. The gas mixture is released by a pulsed valve at 700 Torr backing pressure. A teflon extension with a slit located parallel to the expansion direction of the pulsed beam was

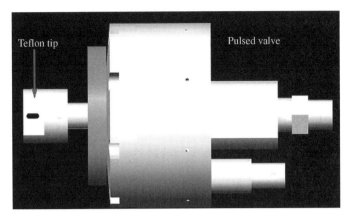

FIGURE 11.4   Schematic view of the photolytic source.

interfaced to the end of the nozzle; this allowed the photolysis laser (for instance, 193 nm in case of bromoacetylene) to be focused to 2 mm by 5 mm downstream of the nozzle. Note that higher concentrations of the precursor must be avoided to eliminate radical–radical recombination reactions in the teflon extension nozzle.

11.4   CROSSED BEAM STUDIES

11.4.1   Reactions of Phenyl Radicals

In this section, we compile the results of our crossed molecular beam experiments of phenyl radicals with acetylene (C_2H_2), ethylene (C_2H_4), methylacetylene (CH_3CCH), allene (H_2CCCH_2), propylene (CH_3CHCH_2), and benzene (C_6H_6).[116] Specific details and the methodology of the crossed beam approach are given for the reactions of phenyl radicals with acetylene, ethylene, and benzene as simplest representatives of unsaturated molecules containing triplet, double, and "aromatic" bonds, respectively. Then, the conclusions are transferred to the methylacetylene, allene, and propylene reactants to elucidate generalized concepts on the chemical dynamics, reactivity, and reaction mechanisms of the reactions of phenyl radicals with unsaturated hydrocarbons in extreme environments.

We observed reactive scattering signal at $m/z = 102$ ($C_8H_6{}^+$) (acetylene),[128] $m/z = 104$ ($C_8H_8{}^+$) (ethylene),[122] and $m/z = 159$ ($C_{12}H_5D_5{}^+$) (d_6-benzene)[129] (Fig. 11.5).

For all systems, ion counts were also detected at lower mass to charge ratios. It is important to stress that at all angles, the TOF spectra recorded at lower m/z values are after scaling identical to those taken at $m/z = 102$, 104, and 159 in the acetylene, ethylene, and d_6-benzene systems, respectively. These findings indicate that the phenyl radical reacts with the unsaturated hydrocarbon molecule only under elimination of a hydrogen atom (acetylene/ethylene) or a deuterium atom (d_6-benzene) to form hydrocarbon molecules of the gross formulae C_8H_6 (acetylene), C_8H_8 (ethylene), and $C_{12}H_5D_5$ (d_6-benzene). Since TOF spectra at lower m/z

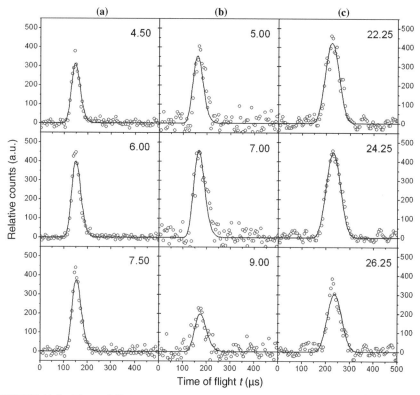

FIGURE 11.5 Time-of-flight (TOF) spectra of ion counts at mass-to-charge rations (m/z) of 102 ($C_8H_6^+$) (a), 104 ($C_8H_8^+$) (b), and 159 ($C_{12}H_5D_5^+$) (c) recorded in the reactions of phenyl radicals with acetylene (a; 99.0 kJ mol^{-1}), ethylene (b; 83.6 kJ mol^{-1}), and D6-benzene (c; 185.0 kJ mol^{-1}) at distinct laboratory angles. The open circles represent the experimental data, and the solid black lines the best fits utilizing the center-of-mass functions as depicted in Figures 7 and 8. The collision energies are given in parenthesis.

values are superimposable to those at the parent ion, signal at these lower mass to charge ratios originate from dissociative ionization of the primary products in the electron impact ionizer of the detector. However, these data alone do not allow us to elucidate the nature of the product isomer formed. The laboratory angular distributions, derived by integrating the TOF spectra at distinct laboratory angles, are depicted in Fig. 11.6.

It is important to note that, for example, in case of the acetylene and ethylene reactions, the hydrogen atom can be lost from either the phenyl radical or from the closed-shell molecule. To pin down the position of the atomic hydrogen loss, we conducted the crossed beam reactions of phenyl radicals with perdeuterated acetylene (C_2D_2) and ethylene (C_2D_4). Considering the d_2-acetylene reactant, the emission of a hydrogen atom from the phenyl group should result in signal at m/z = 104 ($C_8H_4D_2^+$); on the other hand, the release of atomic deuterium is expected to be monitored at

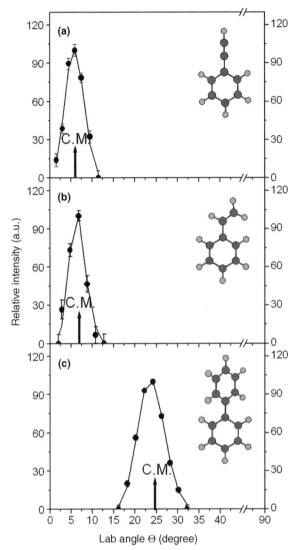

FIGURE 11.6 Laboratory angular distributions (LAB) of 102 ($C_8H_6^+$) (a), 104 ($C_8H_8^+$) (b), and 159 ($C_{12}H_5D_5^+$) (c), recorded in the reactions of phenyl radicals with acetylene (a; 99.0 kJ mol^{-1}), ethylene (b; 83.6 kJ mol^{-1}), and D6-benzene (c; 185.0 kJ mol^{-1}). The collision energies are given in parenthesis. The arrows indicate the corresponding center-of-mass angles. The open circles represent the experimental data, the solid lines the best fits with the center of mass functions shown in Figure 7.

$m/z = 103$ ($C_8H_5D^+$). An analysis of the data taken demonstrated clearly that only deuterium atom losses are observed from d_2-acetylene and d_4-ethylene, indicating that under single collision conditions, the phenyl group stays intact during the reactions; furthermore, the phenyl radical versus deuterium atom replacement pathways are the

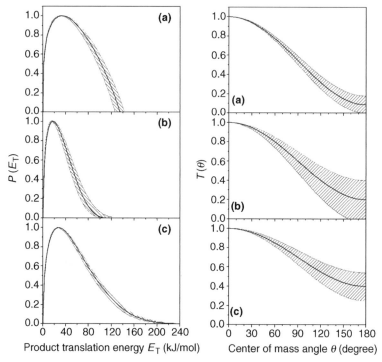

FIGURE 11.7 Corresponding center-of-mass translational energy distributions (left) and angular distributions (right) extracted from the laboratory data as presented in Figures 5 and 6. The solid lines are the best fits and the hatched areas the acceptable fits within the error limits of the experiments.

dominating exit channels for the reactions with perdeuterated acetylene, ethylene, and benzene. However, the laboratory data alone provide only evidence that products of the generic formula C_8H_6 (acetylene), C_8H_8 (ethylene), and $C_{12}H_5D_5$ (d_6-benzene) are formed under single collision conditions. So far, we have no information on the actual product isomer(s) formed.

To gain these important data and to gather crucial mechanistic information, it is important to inspect the derived center-of-mass functions and flux contour maps (Figs 11.7 and 11.8, respectively).

The translational energy distributions, $P(E_T)$ (Fig. 11.7), are illustrated by characteristic maxima of the translational energy, E_{Tmax}, of 125–140 (acetylene), 100–120 (ethylene), and 210–230 kJ/mol (d_6-benzene). Energy conservation dictates that each high-energy cutoff (the maximum translational energy released assuming no energy channels into the internal degrees of freedom) presents the sum of the collision energy plus the reaction exoergicity. Consequently, a subtraction of the collision energies from these high-energy cutoffs yields the reaction energies of the phenyl versus hydrogen/deuterium atom exchange pathways. The reactions are found to be exoergic by 45 ± 11 kJ/mol (acetylene), 25 ± 12 kJ/mol (ethylene), and 35 ± 15 kJ/mol (d_6-benzene). By comparing these experimental data with those obtained from electronic

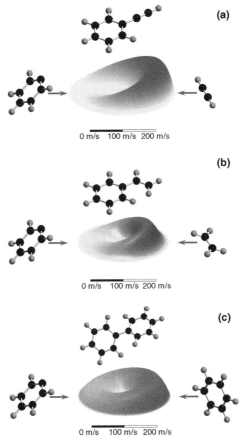

FIGURE 11.8 Center-of-mass velocity contour flux map for the reaction of phenyl radicals (left; 0°) with acetylene (a), ethylene (b), and D6-benzene (c) to form phenylacetylene (a), styrene (b), and D5-diphenyl (c) (right; 180°). The colors connect data points with an identical flux and range from red (highest flux) to yellow (lowest flux). The units of axis are given in ms^{-1} (see legend).

structure calculations and/or from those extracted from the NIST database,[130] we can identify which structural isomer is formed. A detailed comparison indicates the reaction products are phenylacetylene (C_6H_5CCH), styrene ($C_6H_5C_2H_3$), and d_5-diphenyl ($C_6H_5C_6D_5$) in the reactions of phenyl radicals with acetylene, ethylene, and d_6-benzene, respectively. The second set of information that can be extracted from the translational energy distributions originates from the distribution maxima. It is evident that all distributions peak away from zero translational energy at a range of 15–40 kJ/mol. This likely indicates the *exit barriers* when the hydrogen/deuterium atoms are emitted. Considering the reversed addition of hydrogen/deuterium to the closed-shell hydrocarbon molecules phenylacetylene (C_6H_5CCH), styrene ($C_6H_5C_2H_3$), and d_5-diphenyl ($C_6H_5C_6D_5$), *entrance barriers* of this order of magnitude are expected to be present.[131]

As stated in section 11.2, the center-of-mass angular distributions, $T(\theta)$ values, and the flux contour plots present a second set of data (Figs 11.7 and 11.8, respectively). For all systems, fits of the laboratory data were derived with center-of-mass angular distributions showing intensity over the *complete* angular range from $0°$ to $180°$. Each time an angular distribution has intensity over the whole range, this result is evidence that the reactions of the phenyl radicals with acetylene, ethylene, and d_6-benzene follow indirect scattering dynamics via reaction intermediates, here $C_2H_2C_6H_5$, $C_2H_4C_6H_5$, and $C_6D_6C_6H_5$. Recall that the experiments were carried out under single collision conditions; the absence of wall effects and third-body stabilization of these intermediates dictate that *all* intermediates fragment due to the high internal energy content (rovibrational excitation of the intermediates). The energy of the chemical bond formed upon the phenyl radical reaction with the unsaturated hydrocarbon molecule cannot be diverted by a collision with a third body! This helps us to identify the true, primary reaction product that results from fragmentation of the reaction intermediates. It is important to point out that the $T(\theta)$ values are not symmetric with respect to $90°$, but depict an enhanced flux in the forward direction (with respect to the phenyl radical beam). These findings strongly indicate that the lifetime of each reaction intermediate is shorter than their rotation periods (osculating complex behavior).[132]

We would like to summarize these experimental finding and compare these data with recent electronic structure calculations. The relevant parts of the potential energy surfaces (PESs) are compiled in Fig. 11.9 for all phenyl radical reactions studied in our laboratory. Here, the experimental findings correlate very well with the *ab initio* calculations. As suggested by the experimental results, the phenyl radical with its unpaired electron located in a A_1 symmetric σ-like orbital adds via small entrance barriers to the π cloud of acetylene, ethylene, and benzene to form doublet, carbon-centered radical intermediates. It is important to recall that crossed molecular beam experiments can predict the existence of these intermediates based on the shape of the center-of-mass angular distributions; however, the experiments could not pin down the energetics of these collision complexes. Here, the electronic structure calculations provide these missing data and indicate that the collision complexes are stabilized by about 100–70 kJ/mol with respect to the separated reactants. These radical intermediates are rovibrationally excited and fragment via atomic hydrogen pathways to form closed-shell hydrocarbons. The experimentally derived reaction energies agree nicely with those obtained from electronic structure calculations. In addition, the computed exit barriers are confirmed by our experiments and the off-zero peaking of the center-of-mass translational energy distributions.

It should be indicated that the methylacetylene[133] and propylene[134] are more complex reactants than the nonsubstituted counterparts and depict nonequivalent hydrogen atoms at the acetylenic and methyl group (methylacetylene) and at the vinyl and methyl group (propylene). Therefore, even the detection of the atomic hydrogen loss makes it difficult to elucidate if the hydrogen atoms are lost from the methyl group, the acetylenic/vinyl units, or from both positions. In these cases, it is very useful to conduct experiments with partially deuterated reactants d_3-

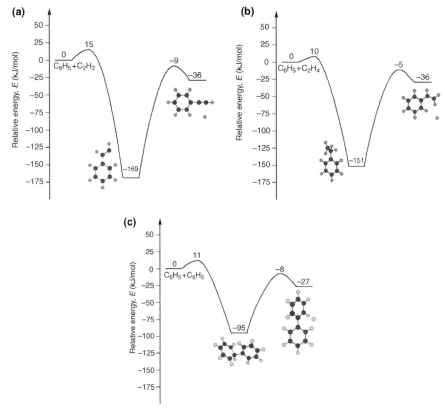

FIGURE 11.9   Schematic potential energy surfaces for the reactions of phenyl radicals with acetylene (a), ethylene (b), and D6-benzene (c) to form phenylacetylene (a), styrene (b), and diphenyl (c).

methylacetylene (CD_3CCH) as well as d_3-propylenes ($CD_3C_2H_3$; $CH_3C_2D_3$). Here, the experiments showed that in the case of the d_3-methylacetylene reaction, only a hydrogen atom loss (from the acetylenic group) was observed; however, in the case of both d_3-propylene reactants, the crossed beam studied depicted explicitly hydrogen atom losses from the methyl group ($CH_3C_2D_3$) and from the vinyl group ($CD_3C_2H_3$) (Fig. 11.9).

Here, the phenyl radical once again attacks the unsaturated bond. However, the steric effect and larger cone of acceptance (the methyl group screens the β carbon atom and makes it less accessible to addition) direct the addition process of the radical center of the phenyl radical to the α carbon atoms of methylacetylene and propylene (the carbon atom holding the acetylenic hydrogen atom). Consequently, crossed beam reactions with complex hydrocarbon molecules can be conducted and valuable information on the reaction pathways can be derived if (partially) deuterated reactions are utilized.

11.4.2 Reactions of CN and C₂H Radicals

The reactions of the isoelectronic cyano and ethynyl radicals with acetylene,[100,135] ethylene,[136] benzene,[97] methylacetylene,[98,137] and allene[138] depict striking similarities, but also important differences to the reactions of the phenyl radical. First, both the cyano and ethynyl radials interact with their radical centers with the π electronic density of the hydrocarbon reactant yielding to an addition of the radical reactant to triple, double, and "aromatic" bonds. Compared to the phenyl radical reactions, which depict distinct entrance barriers to addition, the reactions of cyano and ethynyl radicals have no entrance barrier.[12,13] This is extremely important since a barrierless addition is one of the prerequisites that a chemical reaction occurs in low-temperature environments such as in the interstellar medium and in planetary atmospheres. These addition processes lead to doublet radicals. Note that in case of the cyano radicals, electronic structure calculations also predicted that the cyano radical can add barrierlessly with the nitrogen atom to the hydrocarbon reactant. The resulting isocyano radical intermediates can isomerize via cyclic intermediates yielding ultimately the corresponding nitrile radical intermediates. Similar to the double radicals in the phenyl reactions, the resulting radical intermediates can fragment via atomic hydrogen emission through tight exit transition states. These processes form closed-shell nitriles (cyano radical reactions) as well as highly unsaturated hydrocarbon molecules (ethynyl radical reactants) (Fig. 11.10).

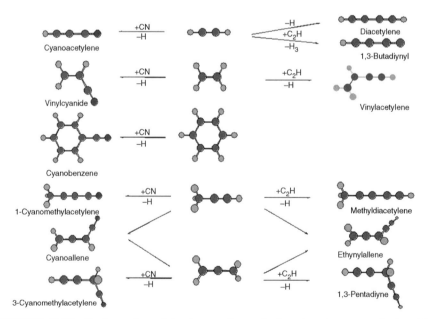

FIGURE 11.10 Products formed in the reactions of ethynyl and cyano radicals with unsaturated hydrocarbon molecules under single collision conditions. The reaction of ethynyl radicals with benzene is currently being studied in our laboratory.

It should be stressed that in case of the ethynyl–acetylene reaction, a molecular hydrogen loss channel synthesizing the 1,3-butadienyl radical is open as well. Since the reactions of cyano and ethynyl radicals have no entrance barrier, are exoergic, and all transition states involved are lower than the energy of the separated reactants, these reaction classes are extremely important to form nitriles and complex unsaturated hydrocarbons in low-temperature environments. On the other hand, the corresponding phenyl radical reactions are—due to the presence of an entrance barrier—closed in those environments. However, the elevated temperature in combustion systems helps to overcome these barriers, thus making phenyl radical reactions important pathways to form aromatic molecules in combustion flames.

11.4.3 Reactions of Carbon Atoms, Dicarbon Molecules, and Tricarbon Molecules

The reactions of ground-state carbon atoms[13] and dicarbon and tricarbon molecules[139] have been reviewed previously. Therefore, only a very brief summary is presented here. The interested reader is referred to the original research papers and review articles for details. Whereas the reactions of the doublet radicals phenyl, cyano, and ethynyl with unsaturated hydrocarbons are dominated mainly by a radical addition, atomic hydrogen elimination mechanism and bimolecular reactions of carbon, dicarbon, and tricarbon are more diverse. In a similar manner to the doublet radicals, carbon, dicarbon, and tricarbon also add to the carbon–carbon double or triple bonds of the unsaturated hydrocarbon reactant. Reactions of carbon atoms and dicarbon molecules have no barrier to addition; however, tricarbon reactions have to pass substantial entrance barriers between 40–110 kJ/mol in the initial addition. Recall that the doublet radicals only add to one carbon atom of the reactant molecule (screening effects direct an addition to the sterically more accessible and hence least substituted carbon atom of the hydrocarbon reactant). However, carbon, dicarbon, and tricarbon can also add to two carbon atoms simultaneously; this can lead to initially cyclic reaction intermediates. These initial addition processes are often followed by complex isomerization steps; in most of the cases, this results in acyclic reaction intermediates that decompose via atomic hydrogen loss. Note that in strong contrast to the phenyl, ethynyl, and cyano radical reactions, in case of carbon, dicarbon, and tricarbon reactions the intermediates may also decompose without exit barrier (e.g., singlet intermediates fragment via simple bond rupture processes). Ultimately, these bimolecular reactions can lead to important "families" of carbon-centered radicals (Figs. 11.11–11.13). These include linear hydrogen-terminated carbon clusters of the formula C_nH ($n = 3$–6), resonantly stabilized free radicals including the C_nH_3 ($n = 3$–6) and C_nH_5 ($n = 4$–5) groups (among them propargyl), 1- and 3-alkyl-substituted propargyl radicals ($R = CH_3$, C_2H, C_2H_3), and 1- and 5-methyl-substituted i-C_4H_3 radicals. These considerations make it clear that the reactivity of carbon atoms and of dicarbon and tricarbon molecules are remarkably diverse and differ substantially from the reaction dynamics of doublet radicals such as ethynyl, cyano, and phenyl with unsaturated hydrocarbons.

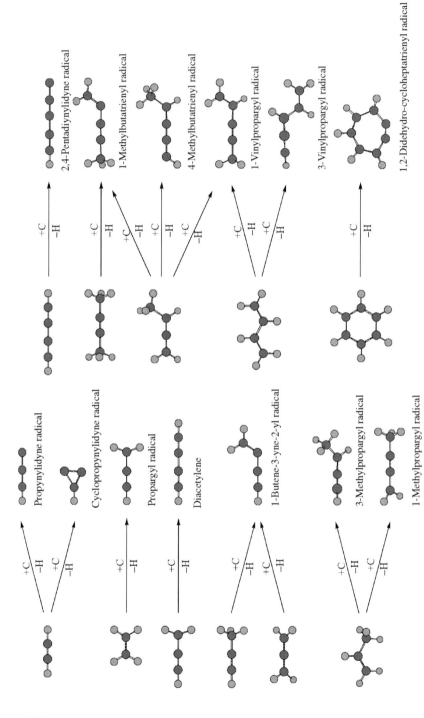

FIGURE 11.11 Products formed in the reactions of ground state carbon atoms with unsaturated hydrocarbon molecules under single collision conditions via hydrogen atom elimination. Note that for the reaction of carbon atoms with acetylene, the tricarbon plus molecular hydrogen elimination channel was observed, too.

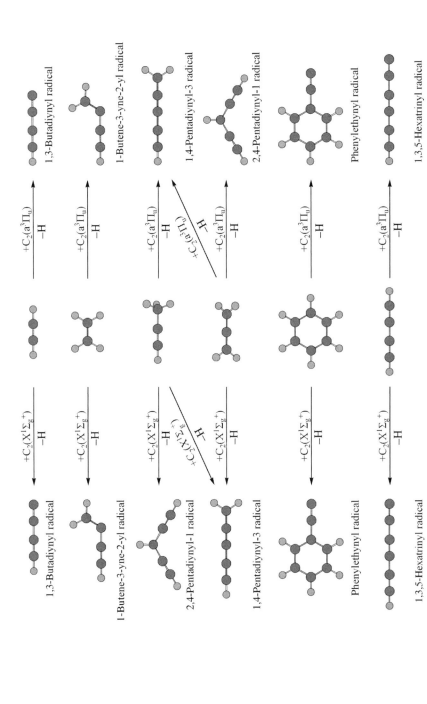

FIGURE 11.12 Products formed in the reactions of ground (left) and excited state dicarbon molecules (right) with unsaturated hydrocarbon molecules under single collision conditions.

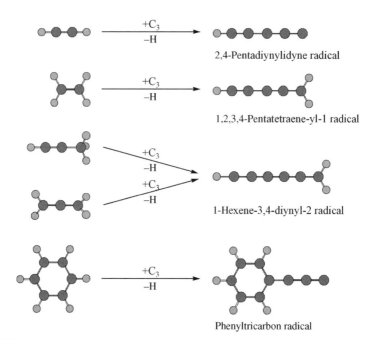

FIGURE 11.13 Products formed in the reactions of ground state tricarbon molecules with unsaturated hydrocarbon molecules under single collision conditions.

11.5 CONCLUSIONS

We have reviewed the reaction dynamics of hydrogen-deficient species such as carbon atoms (C), carbon clusters (C_2, C_3), and carbon-bearing doublet radicals including ethynyl (C_2H), cyano radicals (CN), and phenyl radicals (C_6H_5) with unsaturated hydrocarbons. The crossed beam reactions have provided important information on the reaction dynamics involved (entrance barriers tricarbon and phenyl radical reactions; exit barriers; indirect nature of the reactions proceeding via addition; position of hydrogen versus deuterium losses), on the nature of the reaction intermediates, as well as on the energetics of the reactions and on the identity of the reaction products formed under single collision conditions. We have identified various classes of reaction products, all of the important reaction intermediates in the chemical evolution of combustion flames, hydrocarbon-rich atmospheres of planets and their moons, and in the interstellar medium. These classes are organic nitriles carrying the cyano group, phenyl-substituted closed-shell hydrocarbon molecules acting as precursors to complex PAHs, and hydrogen-deficient, carbon-centered radicals like linear, hydrogen-terminated carbon clusters C_nH ($n = 3$–6), resonantly stabilized free radicals including the groups C_nH_3 ($n = 3$–6) and C_nH_5 ($n = 4$–5), 1- and 3-alkyl-substituted propargyl radicals (R = CH_3, C_2H, C_2H_3), and 1- and 5-methyl-substituted i-C_4H_3 radicals. We also inferred the existence of multiple reaction intermediates on singlet, doublet, and triplet potential energy surfaces. Although in the interstellar medium and in hydrocarbon-rich atmospheres of planets and their moons these intermediates cannot be isolated, these intermediates might be stabilized in high-density environments like combustion flames if the time between collisions of the energized intermediate and the bath molecules is shorter than the lifetime of the reaction intermediates. We hope that these crossed beam experiments trigger future investigations of hitherto poorly studied radical reactions under single collision conditions.

ACKNOWLEDGMENTS

The work reported here has been supported by the U.S. Department of Energy, Basic Energy Sciences (DE-FG02-03ER15411), and by the U.S. National Science Foundation (CHE-0234461; CHE-06278540). The author would also like to thank Coherent, Inc. for support. The author also acknowledges the hard work of postdoctoral fellows Xibin Gu, Fangtong Zhang, and Ying Guo, as well as of Edwin Kawamura (Department of Chemistry, University of Hawaii).

REFERENCES

1. Faraday Discussion 119, Combustion Chemistry: Elementary Reactions to Macroscopic Processes, 2001.

2. Frenklach, M. *Phys. Chem. Chem. Phys.* 2002, *4*, 2028.

3. Baukal, C. E. *Oxygen-Enhanced Combustion*; CRS Press: New York, 1998.

4. Nienow, A. M.; Roberts, J. T. *Annu. Rev. Phys. Chem.* **2006**, *57*, 4.

5. Budyka, M. F.; Zyubina, T. S.; Ryabenko, A. G.; Muradyan, V. E.; Esipov, S. E.; Cherepanova, N. I. *Chem. Phys. Lett.* **2002**, *354*, 93.

6. Lin, Y.-T.; Mishra, R. K.; Lee, S.-L. *J. Phys. Chem. B* **1999**, *103*, 3151.

7. John, P.; Rabeau, J. R.; Wilson, J. I. B. *Diam. Relat. Mater.* **2002**, *11*, 608.

8. Tabata, H.; Fujii, M.; Hayashi, S. *Chem. Phys. Lett.* **2004**, *395*, 138.

9. Shiomi, T.; Nagai, H.; Kato, K.; Hiramatsu, M.; Nawata, M. *Diam. Relat. Mater.* **2001**, *10*, 388.

10. Loewe, A. G.; Hartlieb, A. T.; Brand, J.; Atakan, B.; Kohse-Hoeinghaus, K. *Combust. Flame* **1999**, *118*, 37.

11. Goyette, A. N.; Matsuda, Y.; Anderson, L. W.; Lawler, J. E. *J. Vac. Sci. Technol. A* **1998**, *16*, 337.

12. Kaiser, R. I.; Balucani, N. *Acc. Chem. Res.* **2001**, *34*, 699.

13. Kaiser, R. I. *Chem. Rev.* **2002**, *102*, 1309.

14. Kazakov, A.; Frenklach, M. *Combust. Flame* **1998**, *112*, 270.

15. Ergut, A.; Levendis, Y. A.; Richter, H.; Carlson, J.; In *Fourth Joint Meeting of the U.S. Sections of the Combustion Institute*: Western States, Central States, Eastern States, Philadelphia, PA, 2005.

16. Austin, C. C.; Wang, D.; Ecobichon, D. J.; Dussault, G. *J. Toxicol. Environ. Health A* **2001**, *63*, 437.

17. Stanmore, B. R.; Brilhac, J. F.; Gilot, P.; Carbon **2001**, *39*, 2247.

18. Baird, C. *Environmental Chemistry*; W.H. Freeman Company: New York, 1999.

19. Aubin, D. G.; Abbatt, J. P. *J. Phys. Chem. A* **2003**, *107*, 11030.

20. Sabin, L. D.; Kozawa, K.; Behrentz, E.; Winer, A. M.; Fitz, D. R.; Pankratz, D. V.; Colome, S. D.; Fruin, S. A. *Atmos. Environ.* **2005**, *39*, 5243.

21. Finlayson-Pitts, B. J.; Pitts, J. N. *Science* **1997**, *276*, 1045.

22. Hylland, K. *J. Toxicol. Environ. Health A* **2006**, *69*, 109.

23. Shantakumar, S. et al., *J. Exp. Anal. Environ. Epidemiol.* **2005**, *15*, 482.

24. Oleszczuk, P.; Baran, S. *J. Toxicol. Environ. Health A* **2005**, *40*, 2085.

25. Hinds, W. C. *Aerosol Technology: Properties, Behavior, and Measurement of Airborne Particles*; Wiley: New York, **1999**.

26. Wayne, R. P. *Chemistry of the Atmosphere*; Oxford University Press: Oxford, **2000**.

27. Smith, D. M.; Chughtai, A. R. *Colloids Surf. A* **1995**, *105*, 47.

28. Pedersen, D. U.; Durant, J. K.; Taghizadeh, K.; Hemond, H. F.; Lafleur, A. L.; Cass, G. R. *Environ. Sci. Technol.* **2005**, *39*, 9547.

29. Musafia-Jeknic, T.; Mahadevan, B.; Anon, P.; Pereira, C.; Baird, W. M. *Toxicol. Sci.* **2005**, *88*, 358.

30. Smith, D. M.; Chughtai, A. R. *J. Atmos. Chem.* **1997**, *26*, 77.

31. Denissenko, M. F.; Pao, A.; Tang, M.-S.; Pfeifer, G. P.; Science **1996**, *274*, 430.

32. Durant, J. L.; Busby, W. F.; Lafleur, A. L.; Penman, B. W.; Crespi, C. L. *Mutat. Res. Genet. Toxicol.* **1996**, *371*, 123.

33. Cox, P.; Kessler, M. F. The Universe as Seen by ISO, ESA, SP-427; Noordwijk, **1999**, Vol. 2.

34. Allamandola, L. J.; Hudgins, D. M.; Sandford, S. A. *Astrophys. J.* **1999**, *511*, L115.

35. Dwek, E.; Arendt, R. G.; Fixsen, D. J.; Sodroski, T. J.; Odegard, N.; Weiland, J. L.; Reach, W. T.; Hauser, M. G.; Kelsall, T.; Moseley, S. H.; Silverberg, R. F.; Shafer, R. A.; Ballester, J.; Bazell, D.; Isaacman, R. *Astrophys. J.* **1997**, *475*, 565.

36. Tsang, W.; Wing, I. Fourth Joint Meeting of the U.S. Sections of the Combustion Institute: Western States, Central States, Eastern States, Philadelphia, PA, March 20–23, 2005.

37. Atakan, B.; Lamprecht, A.; Kohse-Hoeinghaus, K. *Combust. Flame* **2003**, *133*, 431.

38. Higgins, K. J.; Jung, H.; Kittelson, D. B.; Roberts, J. T.; Zachariah, M. R. *J. Phys. Chem. A* **2002**, *106*, 96.

39. Mitchell, P.; Frenklach, M. *Phys. Rev. E* **2003**, *67*, 061407.

40. Fialkov, A. B.; Dennebaum, J.; Homann, K. H. *Combust. Flame* **2001**, *125*, 763.

41. Kislov, V. V.; Mebel, A. M.; Lin, S. H. *J. Phys. Chem. A* **2002**, *106*, 6171.

42. Kou, J.; Mori, T.; Kubozono, Y.; Mitsuke, K. *Phys. Chem. Chem. Phys.* **2005**, *7*, 119.

43. Kitajima, A.; Hatanaka, T.; Takeshi, T.; Masao, M.; Torikai, H.; Miyadera, T. *Combust. Flame* **2005**, *142*, 72.

44. Necula, A.; Scott, L. T. *J. Am. Chem. Soc.* **2000**, *122*, 1548.

45. Taylor, R.; Langley, G. J.; Kroto, H. W.; Walton, D. R. M. *Nature* **1993**, *366*, 728.

46. Battin-Leclerc, F. *Phys. Chem. Chem. Phys.* **2002**, *4*, 2072.

47. Richter, H.; Howard, J. B. *Phys. Chem. Chem. Phys.* **2002**, *4*, 2038.

48. Siegmann, K.; Sattler, K. *J. Chem. Phys.* **2000**, *112*, 698.

49. Babushok, V. I.; Miziolek, A. W. *Combust. Flame* **2004**, *136*, 141.

50. Unterreiner, B. V.; Sierka, M.; Ahlrichs, R. *Phys. Chem. Chem. Phys.* **2004**, *6*, 4377.

51. Maricq, M. M. *Combust. Flame* **2004**, *137*, 340.

52. Shebaro, L.; Bhalotra, S. R.; Herschbach, D. *J. Phys. Chem. A* **1997**, *101*, 6775.

53. Dolgonos, G. A.; Peslherbe, G. H. *Int. J. Mass Spectrom.* **2005**, *241*, 261.

54. Apicella, B.; Barbella, R.; Ciajolo, A.; Tregrossi, A. *Combust. Sci. Technol.* **2002**, *174*, 309.

55. Oektem, B.; Tolocka, M. P.; Zhao, B.; Wang, H.; Johnston, M. V. *Combust. Flame* **2005**, *142*, 364.

56. Goroff, N. S. *Acc. Chem. Res.* **1996**, *29*, 77.

57. Homann, K. H. *Angew. Chem., Int. Ed.* **1998**, *37*, 2434.

58. Cataldo, F. *Polyynes: Synthesis, Properties and Applications*; Taylor & Francis: New York, 2006.

59. Richter, H.; Howard, J. B. *Prog. Energy Combust. Sci.* **2000**, *26*, 565.

60. Appel, J.; Bockhorn, H.; Frenklach, M. *Combust. Flame* **2000**, *121*, 122.

61. Kazakov, A.; Frenklach, M. *Combust. Flame* **1998**, *114*, 484.

62. Hausmann, M. Homann, K. H. 22nd International Annual Conference of ICT 1991 (Combust. React. Kinet.), pp. 22/1–22/12.

63. Law, M. E.; Westmoreland, P. R.; Cool, T. A.; Wang, J.; Hansen, N.; Taatjes, C. A.; Kasper, T. *Proc. Combust. Inst.* **2007**, *31*, 565.

64. Hahn, D. K.; Klippenstein, S. J.; Miller, J. A. *Faraday Discuss.* **2001**, *119*, 79.

65. Liu, K. *J. Chem. Phys.* **2006**, *125*, 132307.

66. Lee, Y. T. In *Atomic and Molecular Beam Methods*; Scoles, G., Ed.; Oxford University Press: Oxford, 1988.

67. Casavecchia, P.; Capozza, G.; Segoloni, E. *Modern Trends in Chemical Reaction Dynamics, Part 2, Advanced Series in Physical Chemistry*; 2004; Vol. 14, p 329.

68. Nam, M. J.; Youn, S. E.; Choi, J. H. *J. Chem. Phys.* **2006**, *124*, 104307.

69. Witinski, M. F.; Ortiz-Suarez, M.; Davis, H. F. *J. Chem. Phys.* **2006**, *124*, 094307.

70. Houston, P. L. *J. Phys. Chem.* **1996**, *100*, 12757.

71. Ahmed, M.; Peterka, D. S.; Suits, A. G. *Chem. Phys. Lett.* **1999**, *301*, 372.

72. Liu, X. H.; Gross, R. L.; Suits, A. G. *J. Chem. Phys.* **2002**, *116*, 5341.

73. Townsend, D.; Li, W.; Lee, S. K.; Gross, R. L.; Suits, A. G. *J. Phys. Chem. A* **2005**, *109*, 8661.

74. Li, W.; Huang, C.; Patel, M.; Wilson, D.; Suits, A. G. *J. Chem. Phys.* **2006**, *124*, 011102/1.

75. Meloni, G.; Selby, T. M.; Goulay, F.; Leone, S. R.; Osborn, D. L.; Taatjes, C. A. *J. Am. Chem. Soc.* **2007**, *129*, 14019; Goulay, F.; Osborn, D. L.; Taatjes, C. A.; Zou, P.; Meloni, G.; Leone, S. R. *Phys. Chem. Chem. Phys.* **2007**, *9*, 4291; Osborn, D. L. *Adv. Chem. Phys.* **2008**, *138*, 213; Selby, T. M.; Meloni, G.; Goulay, F.; Leone, S. R.; Fahr, A.; Taatjes, C. A.; Osborn, D. L. *J. Phys. Chem. A* **2008**, *112*, 9366; Taatjes, C. A.; Hansen, N.; Osborn, D. L.; Kohse-Hoeinghaus, K.; Cool, T. A.; Westmoreland, P. R. *Phys. Chem. Chem. Phys.*

2008, *10*, 20; Taatjes, C. A.; Osborn, D. L.; Cool, T. A.; Nakajima, K. *Chem. Phys. Lett.* **2004**, *394*, 19.

76. Alagia, M.; Balucani, N.; Cartechini, L.; Casavecchia, P. *Science* **1996**, *273*, 1519; Skouteris, D.; Werner, H. J.; Aoiz, F. J.; Banares, L.; Balucani, N.; Casavecchia, P. *J. Chem. Phys.* **2001**, *114*, 10622; Balucani, N.; Skouteris, D.; Capozza, G.; Segoloni, E.; Casavecchia, P.; Alexander, M. H.; Capecchi, G.; Werner, H. J. *Phys. Chem. Chem. Phys.* **2004**, *6*, 5007.

77. Lee, S. H.; Dong, F.; Liu, K. *J. Chem. Phys.* **2006**, *125*, 133106/1.

78. Bean, B. D.; Ayers, J. D.; Fernandez-Alonso, F.; Zare, R. N. *J. Chem. Phys.* **2002**, *116*, 6634.

79. Balucani, N.; Capozza, G.; Segoloni, E.; Russo, A.; Bobbenkamp, R.; Casavecchia, P.; Gonzalez-Lezana, T.; Rackham, E. J.; Banares, L.; Aoiz, F. J. *J. Chem. Phys.* **2005**, *122*, 234309.

80. Pederson, L. A.; Schatz, G. C.; Ho, T. S.; Hollebeek, T. *J. Chem. Phys.* **1999**, *110*, 9091; Balucani, N.; Alagia, M.; Cartechini, L.; Casavecchia, P.; Volpi, G. G.; Pederson, L. A.; Schatz, G. C. *J. Chem. Phys.* **2001**, *105*, 2414; Balucani, N.; Cartechini, L.; Capozza, G.; Segoloni, E.; Casavecchia, P.; Volpi, G. G.; Javier, A.; Banares, L.; Honvault, P.; Launay, J. M. *Phys. Rev. Lett.* **2002**, *89*, 013201; Balucani, N.; Casavecchia, P.; Banares, L.; Aoiz, F. J.; Gonzalez-Lezana, T.; Honvault, P.; Launay, J. M. *Jean-Michel. J. Phys. Chem. A* **2006**, *110*, 817.

81. Balucani, N.; Casavecchia, P.; Aoiz, F. J.; Banares, L.; Castillo, J. F.; Herrero, V. J. *Mol. Phys.* **2005**, *103*, 1703; Garton, D. J.; Brunsvold, A. L.; Minton, T. K.; Troya, D.; Maiti, B.; Schatz, G. C. *J. Phys. Chem. A* **2006**, *110*, 1327; Alagia, M.; Balucani, N.; Cartechini, L.; Casavecchia, P.; van Kleef, E. H.; Volpi, G. G.; Kuntz, P. J.; Sloan, J. J. *J. Chem. Phys.* **1998**, *108*, 6698; Gray, S. K.; Petrongolo, C.; Drukker, K.; Schatz, G. C. *Phys. Chem. Chem. Phys.* **1999**, *6*, 1141; Aoiz, F. J.; Banares, L.; Castillo, J. F.; Herrero, V. J.; Martinez-Haya, B.; Honvault, P.; Launay, J. M.; Liu, X.; Lin, J. J.; Harich, S. A.; Wang, C. C.; Yang, X. *J. Chem. Phys.* **2002**, *116*, 10692.

82. Lee, S. H.; Liu, K. *Appl. Phys. B: Lasers Opt.* **2000**, *71*, 627.

83. Alagia, M.; Balucani, N.; Casavecchia, P.; Stranges, D.; Volpi, G. G. *J. Chem. Phys.* **1993**, *98*, 8341.

84. Strazisar, B. R.; Lin, C.; Davis, H. F. *Science* **2000**, *290*, 5493.

85. Takayanagi, T.; Schatz, G. C. *J. Chem. Phys.* **1997**, *106*, 3227.

86. Bowman, J. M.; Schatz, G. C. *Annu. Rev. Phys. Chem.* **1995**, *46*, 169.

87. Hinrichs, R. Z.; Willis, P. A.; Stauffer, H. U.; Schroden, J. J.; Davis, H. F. *J. Chem. Phys.* **2000**, *112*, 4634; Stauffer, H. U.; Hinrichs, R. Z.; Schroden, J. J.; Davis, H. F. *J. Phys. Chem. A* **2000**, *104*, 1107; Schroden, J. J.; Wang, C. C.; Davis, H. F. *J. Phys. Chem. A* **2003**, *107*, 9295; Hinrichs, R. Z.; Schroden, J. J.; Davis, H. F. *J. Am. Chem. Soc.* **2003**, *125*, 860; Schroden, J. J.; Teo, M.; Davis, H. F. *J. Chem. Phys.* **2002**, *117*, 9258; Hinrichs, R. Z.; Schroden, J. J.; Davis, H. F. *J. Phys. Chem. A* **2008**, *112*, 3010.

88. Schroden, J. J.; Davis, H. F. *Modern Trends in Chemical Reaction Dynamics, Advanced Series in Physical Chemistry*, 2004, p. 215.

89. Stauffer, H. U.; Hinrichs, R. Z.; Schroden, J. J.; Davis, H. F. *J. Chem. Phys.* **1999**, *111*, 10758.

90. Perri, M. J.; Van Wyngarden, A. L.; Lin, J. J.; Lee, Y. T.; Boering, K. A. *J. Phys. Chem. A* **2004**, *108*, 7995.

91. Van Wyngarden, A. L.; Mar, K. A.; Boering, K. A.; Lin, J. J.; Lee, Y. T.; Lin, S. Y.; Guo, H.; Lendvay, G. *J. Am. Chem. Soc.* **2007**, *129*, 2866.

92. Lu, Y. J.; Xie, T.; Fang, J. W.; Shao, H. C.; Lin, J. J. *J. Chem. Phys.* **2008**, *128*, 184302.

93. Kaiser, R. I.; Sun, W.; Suits, A. G. *J. Chem. Phys.* **1997**, *106*, 5288; Kaiser, R. I.; Ochsenfeld, C.; Head-Gordon, M.; Lee, Y. T. *Science* **1998**, *279*, 1181.

94. Kaiser, R. I.; Hahndorf, I.; Huang, L. C. L.; Lee, Y. T.; Bettinger, H. F.; Schleyer, P. v. R.; Schaefer, H. F. III; Schreiner, P. R. *J. Chem. Phys.* **1999**, *110*, 6091;Bettinger, H. F.; Schleyer, P. v. R.; Schreiner, P. R.; Schaefer, H. F. III, Kaiser, R. I.; Lee, Y. T. *J. Chem. Phys.* **2000**, *113*, 4250; Hahndorf, I.; Lee, Y. T.; Kaiser, R. I.; Vereecken, L.; Peeters, J.; Bettinger, H. F.; Schleyer, P. v. R.; Schaefer, H. F. III *J. Chem. Phys.* **2002**, *116*, 3248.

95. Kaiser, R. I.; Lee, Y. T.; Suits, A. G. *J. Chem. Phys.* **1996**, *105*, 8705; Kaiser, R. I.; Stranges, D.; Bevsek, H. M.; Lee, Y. T.; Suits, A. G. *J. Chem. Phys.* **1997**, *106*, 4945; Hahndorf, I.; Lee, H. Y.; Mebel, A. M.; Lin, S. H.; Lee, Y. T.; Kaiser, R. I. *J. Chem. Phys.* **2000**, *113*, 9622; Huang, L. C. L.; Lee, H. Y.; Mebel, A. M.; Lin, S. H.; Lee, Y. T.; Kaiser, R. I. *J. Chem. Phys.* **2000**, *113*, 9637; Lee, T. N.; Lee, H. Y.; Mebel, A. M.; Kaiser, R. I. *J. Phys. Chem. A* **2001**, *105*, 1847; Nguyen, T. L.; Mebel, A. M.; Kaiser, R. I. *J. Phys. Chem. A* **2001**, *105*, 3284; Kaiser, R. I.; Nguyen, T. L.; Le, T. N.; Mebel, A. M. *Astrophys. J.* **2001**, *561*, 858; Geppert, W. D.; Naulin, C.; Costes, M.; Capozza, G.; Cartechini, L.; Casavecchia, P.; Volpi, G. G. *J. Chem. Phys.* **2003**, *119*, 10607.

96. Balucani, N.; Cartechini, L.; Alagia, M.; Casavecchia, P.; Volpi, G. G. *J. Phys. Chem. A* **2000**, *104*, 5655.

97. Balucani, N.; Asvany, O.; Chang, A. H. H.; Lin, S. H.; Lee, Y. T.; Kaiser, R. I.; Bettinger, H. F.; Schleyer, P. v. R.; Schaefer, H. F. III, *J. Chem. Phys.* **1999**, *111*, 7457.

98. Kaiser, R. I.; Chiong, C. C.; Asvany, O.; Lee, Y. T.; Stahl, F.; Schleyer, P. v. R.; Schaefer, H. F. *J. Chem. Phys.* **2001**, *114*, 3488.

99. Stahl, F.; Schleyer, P. v. R.; Kaiser, R. I.; Lee, Y. T.; Schaefer, H. F. III, *J. Chem. Phys.* **2001**, *114*, 3476; Stahl, F.; Schleyer, P. v. R.; Schaefer, H. F.; Kaiser, R. I. *Planet. Space Sci.* **2002**, *50*, 685.

100. Kaiser, R. I.; Stahl, F.; Schleyer, P. v. R.; Schaefer, H. F. *Phys. Chem. Chem. Phys.* **2002**, *4*, 2950.

101. Balucani, N.; Asvany, O.; Lee, Y. T.; Kaiser, R. I.; Galland, N.; Hannachi, Y. *J. Am. Chem. Soc.* **2000**, *122*, 11234.

102. Balucani, N.; Asvany, O.; Lee, Y. T.; Kaiser, R. I.; Galland, N.; Rayez, M. T.; Hannachi, Y. *J. Comput. Chem.* **2001**, *22*, 1359.

103. Geppert, W. D.; Goulay, F.; Naulin, C.; Costes, M.; Canosa, A.; Le Picard, S.; Sebastien, D.; Rowe, B. R. *Phys. Chem. Chem. Phys.* **2004**, *6*, 566.

104. Zhang, B.; Shiu, W.; Lin, J. J.; Liu, K. *J. Chem. Phys.* **2005**, *122*, 131102.

105. Liu, X.; Suits, A. G. *Modern Trends in Chemical Reaction Dynamics, Advanced Series in Physical Chemistry*; pp. **2004**, 105–143.

106. Ran, Q.; Yang, C. H.; Lee, Y. T.; Lu, I. C.; Shen, G.; Wang, L.; Yang, X. *J. Chem. Phys.* **2005**, *122*, 044307.

107. Capozza, G.; Segoloni, E.; Leonori, F.; Volpi, G. G.; Casavecchia, P. *J. Chem. Phys.* **2004**, *120*, 4557.

108. Troya, D.; Schatz, G. C.; Garton, D. J.; Brunsvold, A. L.; Minton, T. K. *J. Chem. Phys.* **2004**, *120*, 731.

109. Kaiser, R. I.; Sun, W.; Suits, A. G.; Lee, Y. T. *J. Chem. Phys.* **1997**, *107*, 8713.

110. Choi, J. H. *Int. Rev. Phys. Chem.* **2006**, *25*, 613.

111. Lee, H.; Joo, S. K.; Kwon, L. K.; Choi, J. H. *J. Chem. Phys.* **2003**, *119*, 93374.

112. Kwon, L. K.; Nam, M. J.; Youn, S. E.; Joo, S. K.; Lee, H.; Choi, J. H. *J. Chem. Phys.* **2006**, *124*, 204320.

113. Park, J. H.; Lee, H.; Kwon, H. C.; Kim, H. K.; Choi, Y. S.; Choi, J. H. *J. Chem. Phys.* **2002**, *117*, 2017.

114. Joo, S. K.; Kwon, L. K.; Lee, H.; Choi, J. H. *J. Chem. Phys.* **2004**, *120*, 7976.

115. Leonori, F.; Balucani, N.; Capozza, G.; Segoloni, E.; Stranges, D.; Casavecchia, P. *Phys. Chem. Chem. Phys.* **2007**, *9*, 1307.

116. Gu, X.; Kaiser, R.I. *Acc. Chem. Res.* **2009**, *42*, 290–302.

117. Kaiser, R. I.; Suits, A. G. *Rev. Sci. Instrum.* **1995**, *66*, 5405.

118. Kaiser, R. I.; Ting, J. W.; Huang, L. C. L.; Balucani, N.; Asvany, O.; Lee, Y. T.; Chan, H.; Stranges, D.; Gee, D. *Rev. Sci. Instrum.* **1999**, *70*, 4185.

119. Zhang, F.; Kim, S.; Kaiser, R.I. *Phys. Chem. Chem. Phys.* **2009**, *11*, 4707–4714.

120. Kaiser, R. I.; Ochsenfeld, C.; Stranges, D.; Head-Gordon, M.; Lee, Y. T. *Faraday Discuss.* **1998**, *109*, 183.

121. Guo, Y.; Mebel, A. M.; Zhang, F.; Gu, X.; Kaiser, R. I. *J. Phys. Chem. A* **2007**, *111*, 4914.

122. Zhang, F.; Gu, X.; Guo, Y.; Kaiser, R. I. *J. Org. Chem.* **2007**, *72*, 7597.

123. Weis, M. S. Ph.D. Thesis, University of California, Berkeley, 1986.

124. Gu, X.; Guo, Y.; Kawamura, E.; Kaiser, R. I. *J. Vac. Sci. Technol. A* **2006**, *24*, 505.

125. Gu, X.; Guo, Y.; Kaiser, R. I. *Int. J. Mass Spectrom.* **2007**, *261*, 100.

126. Sillars, D.; Kaiser, R. I.; Galland, N.; Hannachi, Y. *J. Phys. Chem. A* **2003**, *107*, 5149; Kaiser, R. I.; Balucani, N.; Galland, N.; Caralp, F.; Rayez, M. T.; Hannachi, Y. *Phys. Chem. Chem. Phys.* **2004**, *6*, 2205; Bettinger, H. F.; Kaiser, R. I. *J. Phys. Chem. A* **2004**, *108*, 4576; Zhang, F.; Guo, Y.; Gu, X.; Kaiser, R. I. *Chem. Phys. Lett.* **2007**, *440*, 56; Zhang, F.; Chang, A. H. H.; Gu, X.; Kaiser, R. I. *J. Phys. Chem. A* **2007**, *111*, 13305; Zhang, F.; Gu, X.; Kaiser, R. I.; Bettinger, H. F. *Chem. Phys. Lett.* **2008**, *450*, 178; Zhang, F.; Kao, C. H.; Chang, A. H. H.; Gu, X.; Guo, Y.; Kaiser, R. I. *Chem. Phys. Chem.* **2008**, *9*, 95; Zhang, F.; Gu, X.; Kaiser, R. I.; Balucani, N.; Chang, A. H. H. *J. Phys. Chem. A* **2008**, *112*, 3837.

127. Kohn, D. W.; Clauberg, H.; Chen, P. *Rev. Sci. Instrum.* **1992**, *63*, 4003.

128. Gu, X.; Zhang, F.; Guo, Y.; Kaiser, R. I. *Angew. Chem., Int. Ed.* **2007**, *46*, 6866.

129. Zhang, F.; Gu, X.; Kaiser, R. I. *J. Chem. Phys.* **2008**, *128*, 084315/1.

130. http://webbook.nist.gov/chemistry/.

131. Levine, R. D. *Molecular Reaction Dynamics*; Cambridge University Press: Cambridge, 2005.

132. Miller, W. B.; Safron, S. A.; Herschbach, D. R. *Discuss. Faraday Soc.* **1967**, *44*, 108.

133. Gu, X.; Zhang, F.; Guo, Y.; Kaiser, R. I. *J. Phys. Chem. A* **2007**, *111*, 11450.

134. Zhang, F.; Gu, X.; Guo, Y.; Kaiser, R. I. *J. Phys. Chem. A* **2008**, *112*, 3284.

135. Huang, L. C. L.; Lee, Y. T.; Kaiser, R.I. *J. Chem. Phys.* **1999**, *110*, 7119; Huang, L. C. L.; Asvany, O.; Chang, A. H. H.; Balucani, N.; Lin, S. H.; Lee, Y. T.; Kaiser, R. I.; Osamura, Y. *J. Chem. Phys.* **2000**, *113*, 8656.

136. Balucani, N.; Asvany, O.; Chang, A. H. H.; Lin, S. H.; Lee, Y. T.; Kaiser, R. I.; Osamura, Y. *J. Chem. Phys.* **2000**, *113*, 8643.

137. Huang, L. C. L.; Balucani, N.; Lee, Y. T.; Kaiser, R. I.; Osamura, Y. *J. Chem. Phys.* **1999**, *111*, 2857.

138. Balucani, N.; Asvany, O.; Kaiser, R. I.; Osamura, Y. *J. Phys. Chem. A* **2002**, *106*, 4301.

139. Kaiser, R. I.; Lee, T. L.; Nguyen, T. L.; Mebel, A. M.; Balucani, N.; Lee, Y. T.; Stahl, F.; Schleyer, P. v. R.; Schaefer, H. F. III, *Faraday Discuss.* **2001**, *119*, 51; Gu, X.; Guo, Y.; Zhang, F.; Mebel, A. M.; Kaiser, R. I. *Faraday Discuss.* **2006**, *133*, 245; Balucani, N.; Mebel, A. M.; Lee, Y. T.; Kaiser, R. I. *J. Phys. Chem. A* **2001**, *43*, 9813; Kaiser, R. I.; Yamada, M.; Osamura, Y. *J. Phys. Chem. A* **2002**, *106*, 4825; Kaiser, R. I.; Balucani, N.; Charkin, D. O.; Mebel, A. M. *Chem. Phys. Lett.* **2003**, *382*, 112; Guo, Y.; Gu, X.; Balucani, N.; Kaiser, R. I. *J. Phys. Chem. A* **2006**, *110*, 6245; Gu, X.; Guo, Y.; Mebel, A. M.; Kaiser, R. I. *J. Phys. Chem. A* **2006**, *110*, 11265; Guo, Y.; Gu, X.; Zhang, F.; Mebel, A. M.; Kaiser, R. I. *J. Phys. Chem. A* **2006**, *110*, 10699; Guo, Y.; Kislov, V. V.; Gu, X.; Zhang, F.; Mebel, A. M.; Kaiser, R. I. *Astrophys. J.* **2006**, *653*, 1577; Gu, X.; Guo, Y.; Zhang, F.; Mebel, A. M.; Kaiser, R. I. *Chem. Phys. Lett.* **2007**, *436*, 7; Gu, X.; Guo, Y.; Zhang, F.; Mebel, A. M.; Kaiser, R. I. *Chem. Phys.* **2007**, *335*, 95; Gu, X.; Guo, Y.; Zhang, F.; Mebel, A. M.; Kaiser, R. I. *Chem. Phys. Lett.* **2007**, *444*, 220; Guo, Y.; Gu, X.; Zhang, F.; Mebel, A. M.; Kaiser, R. I. *Phys. Chem. Chem. Phys.* **2007**, *9*, 1972; Gu, X.; Guo, Y.; Mebel, A. M.; Kaiser, R. I. *Chem. Phys. Lett.* **2007**, *449*, 44.

12

LASER FLASH PHOTOLYSIS OF PHOTOINITIATORS: ESR, OPTICAL, AND IR SPECTROSCOPY DETECTION OF TRANSIENTS

Igor V. Khudyakov[1] and Nicholas J. Turro[2]

[1]*Bomar Specialties, Torrington, CT, USA*

[2]*Department of Chemistry, Columbia University, New York, NY, USA*

12.1 INTRODUCTION

Photopolymerization has been a rapidly growing area of science and technology during the past several decades. Materials produced by photopolymerization are widely used as coatings, imaging compositions, adhesives, rapid prototyping, electronics, and optics.[1a] The most common is so-called "UV cure," which is the polymerization of formulations, including vinyl monomers/oligomers, under the irradiation of UV light.[1a] Irradiation occurs in the presence of photoinitiator(s) (PI) that produce free radicals. In this chapter, we will deal with PIs for free radical polymerization (FRP). Moreover, we will mainly discuss PIs of Type I,[1b] that is, compounds that undergo photodissociation to form free radicals under UV–vis irradiation:

$$AB \xrightarrow{h\nu} A^{\cdot} + B^{\cdot}$$

The obvious requirement for a PI is that it should absorb UV light emitted by available light source and produce free radicals with a high yield throughout the cure. "UV cure

Carbon-Centered Free Radicals and Radical Cations, Edited by Malcolm D. E. Forbes
Copyright © 2010 John Wiley & Sons, Inc.

of coatings" means the formation of a polymeric film from a (viscous) liquid in the course of photopolymerization. There are many other demands for industrial PI, for example, low migration, low toxicity, lack of hazardous by-products of the reactions of PIs, lack of odor, etc.

In this chapter, we will focus on photogeneration of free radicals of PIs and some elementary reactions of these radicals. This chapter does not aim to provide an exhaustive coverage of the subject; but tried to outline the current status of this area, accomplishments and unresolved problems. For more specific information on one or another reaction or a process, the reader is referred to cited literature.

12.2 PHOTODISSOCIATION OF INITIATORS

12.2.1 Quantum Yields of Free Radicals in Nonviscous Solutions

The quantum yield of the formation of free radicals Φ_{diss} in the photodissociation step is an important characteristic of a PI. A large family of PIs manufactured by Ciba Additives under the names Irgacure® and Darocur® have a benzoyl fragment in their molecular structure:

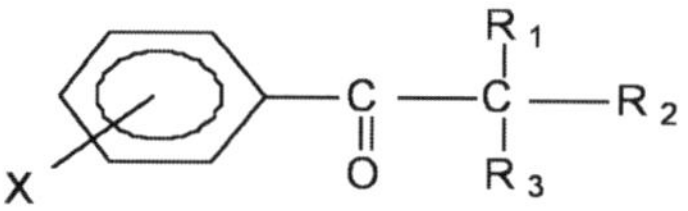

These molecules have a high quantum yield for the formation of their triplet state and usually rapidly dissociate into free radicals in the triplet state undergoing α-cleavage[2,3a,b]:

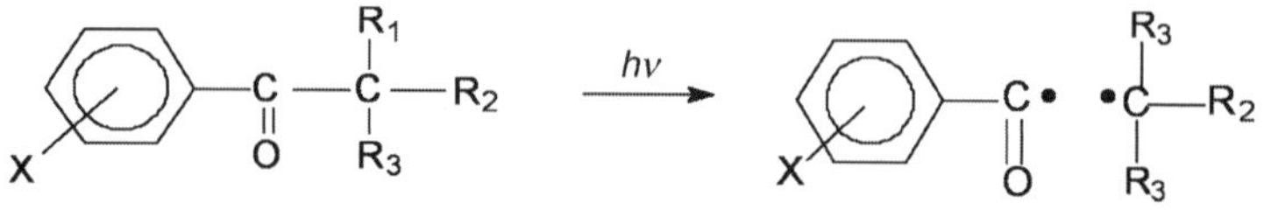

This is termed the Norrish I type process for ketones.[4]

Photophysical characteristics of PIs (Scheme 12.1), especially the quantum yields of their dissociation Φ_{diss}, are very important. Most of the photophysical data were measured by nanoseconds or picoseconds laser flash photolysis (LFP) or phosphorescence at low temperature. The properties of representative PIs are given in Table 12.1.

There can be a substantial difference in the reactivity of n,π^* and π,π^* excited triplet states in α-cleavage. These differences are experimentally characterized by τ_T (triplet lifetime) or k_{diss} (the dissociation rate constant) for α-cleavage.[2] In general, aromatic ketones with n,π^* lowest triplet states undergo much faster α-cleavage and have shorter triplet lifetimes than ketones with π,π^* states.[2] The latter may be dubbed inefficient PIs.[2]

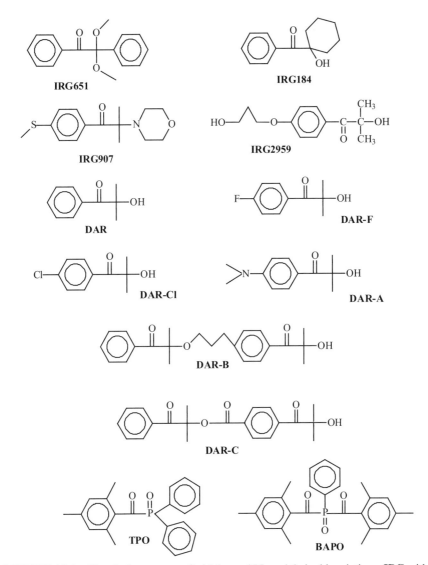

SCHEME 12.1 Chemical structures of widely used PIs and their abbreviations. **IRG** with a number stands for Ciba's designation of a PI: Irgacure 2959, etc.; **DAR** stands for Ciba's Darocur 1173; **TPO** stands for Lucerin TPO of BASF or Darocur TPO of Ciba; **BAPO** stands for bis-phosphine oxide or Irgacure 819 of Ciba.

Picosecond flash photolysis of **DAR** (see Scheme 12.1) allowed the direct measurement of τ_T as 0.4 ns (cyclohexane) and 0.2 ns (acetonitrile.)[3b] The lowest triplet n,π^* state of **DAR** undergoes crossing with the $^3\sigma,\sigma^*$ state of the forming radicals.[3b]

Many PIs (Table 12.1) demonstrate an unusual performance: the rate constant of S_1–T intersystem crossing is lower than k_{diss}.[6] There is no experimental evidence so far that PIs (Table 12.1) dissociate in the S_1 state.[6]

TABLE 12.1 Photophysical Properties of Photoinitiators in Acetonitrile and Other Nonviscous Solvents[a]

PI	τ_T (ns)	E_T (kJ/mol)	Triplet Nature	Φ_{diss}	Reference
IRG651	0.25	–	–	0.95	3a
IRG651	<0.1	278	–	0.52	5
DAR	0.2–0.5	–	n,π^*	~0.8	3a,b
DAR	0.37	299	n,π^*	0.38	2
DAR-F	0.86	301	n,π^*	0.67	2
DAR-Cl	1.0	295	n,π^*	0.60	2
DAR-A	12,000	264	n,π^*	0.03	2
DAR-B	0.31	–	n,π^*	0.33	2
DAR-B	3000	–	n,π^*	0.05	2
IRG907	10	–	–	~0.3	3
IRG2959	12	295	$n,\pi^*/\pi,\pi^*$	0.29	2
TPO	~0.09	262–263	n,π^*	~0.5	6

[a] For determination error of parameters, solvents, and assumptions made under estimation of parameters, see the original publication. τ_T and Φ_{diss} were measured at room temperature.

12.2.2 Cage Effect Under Photodissociation

The *cage effect* means that in the liquid phase, as opposed to the gas phase, molecules undergo not a single collision but a series of collisions or contacts.[7,8] As a result of the cage effect, the fragments produced under photolysis (thermolysis, radiolysis) of a molecule do not promptly separate, and they exist for a while as a dynamic geminate (G) pair. Molecules (radicals and atoms) can also encounter each other in the course of random walking in a solution (F pairs—pairs of radicals met in the solvent bulk as a result of random wandering (an encounter)) and they also undergo a series of contacts. The value of the cage effect (Φ) under pairwise distribution of generated radicals in the liquid phase (G pairs—geminate RP) is the fraction of radical pairs (RPs) that undergo reaction within a pair with formation of a diamagnetic product. The efficiency of thermoinitiators in polymer chemistry is often designated as $f = 1 - \Phi$, where f is a cage escape value. Thus, the efficiency of a Type I PI is $\Phi_{diss} \times f$. Obviously, the ideal efficiency of a PI is 1.0, which means that every absorbed light quantum of the UV light leads to two reactive free radicals that escape the solvent cage. PIs, which dissociate in a triplet state with a formation of a *triplet* G pair, are preferable to those PIs that give singlet pairs. Radicals of a triplet pair cannot immediately recombine or disproportionate due to spin prohibition, and they have to exit a cage ($\Phi \sim 0, f \sim 1$.) A large family of PIs used in industry undergoes a Norrish Type I process with formation of a triplet RP (see the previous section).[9]

Triplet RPs in nonviscous solutions exit the cage with $f \sim 1$. An increase in viscosity leads to an increase in a RP lifetime and slows down molecular diffusivity: these features allow S–T transitions to occur in the RP, and geminate recombination of free radicals is expected to occur, increasing the cage effect Φ.[11] Experimental measurements demonstrate that the cage effect Φ increases with an increase in solvent viscosity.[11–13] An increase of media viscosity, which usually takes place upon polymerization, also results in a decrease in the rate of polymerization as the process approaches completion.[14]

Cage effect dynamics or kinetics of geminate recombination was observed for the first time under photodissociation of a C−C dimer of aromatic radicals in a viscous media.[12] A suggestion has been made that at least in a number of studied cases the mutual diffusion coefficient of radicals in the pair is approximately 10 times lower than the sum of macroscopic diffusion coefficients of the individual species. In other words, a geminate recombination proceeds considerably longer than expected.[12,13,15]

Not only does the temperature/viscosity of media affect Φ; another possible way that Φ increases in a system with the same reagents was observed in Ref. 16. Photopolymerization of acrylamide initiated by **IRG2959** (Scheme 12.1) was studied in the aqueous solution in the presence of poly(methacrylic acid) (**PMA**).[16] It was found that **PMA** forms a cluster around the **IRG2959** at pH < 6.9 and that the cluster holds the RP in proximity. As a result, Φ increases, and the rate of photopolymerization decreases.[16]

There is an important question on the initial distance and initial mutual orientation of photogenerated free radicals. The initial distance between atoms formed from diatomic molecules should increase with a decrease in wavelength (increasing energy of the absorbed photon) of photolyzing light and a decrease in solvent viscosity. However, experimental evidence for such a conclusion to multiatomic radicals is not evident due to fast translational and vibrational relaxation. Still, there are data testifying to the fact that with a decrease in viscosity, radicals separate at *larger* initial distances or at least turn one radical against another in a way that reactive atoms are positioned at *larger* distance.[11]

A study of chemically induced dynamic electron polarization, CIDEP (see Section 12.3.3) on F and G pairs of radicals formed under photolysis of a common termo- and photoinitiator 2,2′-azobis(2-methylpropionitrile) (**AIBN**) led to a tentative conclusion that initial spatial separation of 2-cyano-2-propyl radicals *does not depend upon viscosity*.[17] However, it is plausible that the diamagnetic dinitrogen molecule formed under photolysis of **AIBN** (and is invisible by ESR) separates further from a contact RP under photolysis in solvents of lower viscosity. The problem of initial spatial separation and mutual orientation of radicals under photolysis still waits experimental elucidation.

There is an interesting predicted effect of polymer environment on low MW molecules (radicals).[18] Let us assume that radicals strongly interact with polymer chains by attraction or repulsion. Then, the free energy of a polymer system with low MW species decreases when low MW species are located in the proximity of each other.[18] That means that the value of Φ of transient radicals should increase in such polymer solutions. To the best of our knowledge, this effect has not yet been observed experimentally.

12.2.3 The Magnetic Field Effect on Photodissociation

Application of low, moderate, and strong magnetic fields (MFs) affects the escape of free radicals from a (viscous) solvent cage or from a microheterogeneous compartment such as a micelle. The theory of magnetic field effects (MFEs) is well described in a number of review articles.[19]

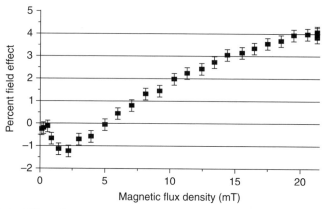

FIGURE 12.1   Effect of an external magnetic field (B) on the yield of free radicals obtained under photodissociation of **IRG2959** (Scheme 12.1) in cyclohexanol.[20]

Figure 12.1 presents data on a low field effect (LFE) and on a moderate MFE on the relative initial concentration of substituted benzoyl radicals obtained under LFP of **IRG2959**. An application of MF of a few millitesla leads to a decrease in the yield of radicals (see Fig. 12.1). This is the result of the action of the LFE, which enhances S–T mixing in the RP through removal of HF spin-state degeneracies.[20,21] At higher magnetic flux densities ($B > 10$ mT), a well-studied HFC mechanism[19] leads to an increase in the radical escape from a triplet RP (Fig. 12.1). LFE leads to a decrease in Φ for both singlet- and triplet-born pairs. Photodissociation in a microheterogeneous solution or in a viscous solvent leads to an increase in the RP lifetime and to the action of the HFC mechanism of MFE.[19] For example, a HFC mechanism was observed for **TPO** (Scheme 12.1) in micelles. At $B \sim 500$ mT the radicals exit, and the yield of radicals escaping from the micelles increases.[22]

At high magnetic fields of $B \geq 1$ T, another mechanism becomes efficient, namely, a Δg mechanism.[19a,c] Action of a Δg mechanism leads to an increase in Φ and a decrease in f.[19a,c]

The RP must live long enough to allow singlet–triplet evolution under MF: that is one of the main demands for observation of each of the three MFEs briefly mentioned above.[13,19–21]

Application of a moderate MF accelerates photopolymerization initiated by PI leading to triplet RPs.[14] The main effect is an increase in f and an increase in the rate of initiation. One should expect a second weak effect leading to deceleration of a chain termination by bimolecular radical reaction. A MFE on an F pair was observed for the first time in Refs 23,24.

12.3   TR ESR DETECTION OF TRANSIENTS

12.3.1   CIDEP Under Photodissociation of Initiators

Laser flash photolysis of PIs in the cavity of an ESR spectrometer is often accompanied by chemically induced dynamic electron polarization, CIDEP, that is, by the formation of radicals with non-Boltzmann population of electron Zeeman levels.[19a,b,g,25] CIDEP manifests itself in enhanced absorption or emission of all or of certain components in the ESR spectra of photogenerated free radicals. Radicals, which manifest CIDEP, are termed polarized. The main mechanisms leading to CIDEP in photoinduced reactions are well established and have been investigated both theoretically and experimentally.[19a,b,25] The most common mechanisms, which are well described in the literature, are triplet mechanism (TM), radical pair mechanism (RPM), and spin correlated RP mechanism (SCRP).[19a,b,25]

Analysis of a CIDEP pattern with time-resolved ESR (TR ESR) spectra provides a solid conclusion to be made on the spin multiplicity of molecular precursors of polarized free radicals (a singlet or a triplet excited molecule) and the tracking of fast reactions of polarized radicals leading to secondary radicals. Thus, TR ESR is a convenient method in mechanistic photochemistry and free radical chemistry. Continuous wave TR ESR (CW TR ESR) devices are widely used for detection of photogenerated radicals. They usually consist of a pulsed ns laser with detection of transients by their ESR spectra with a X band ESR spectrometer in the direct detection mode (no filed modulation).[19a,b,25] Time-resolved Fourier transform ESR (FT ESR) has some advantages and drawbacks with respect to CW TR ESR.[26,27] Rather sophisticated FT ESR devices have become available, and FT ESR studies become more common.

Many research groups have observed TR ESR spectra under photolysis of **DAR**, **IRG651**, **TPO**, and **BAPO** (Scheme 12.1) and of other Type I PIs. Analysis of ESR spectra of the primary radicals formed and their spin adducts allows determination of radical structure. Computer simulation with user-friendly software was used to elucidate the radical structure.

The reactions following photolysis of **IRG651** and **TPO** lead to polarized free radicals of PIs:

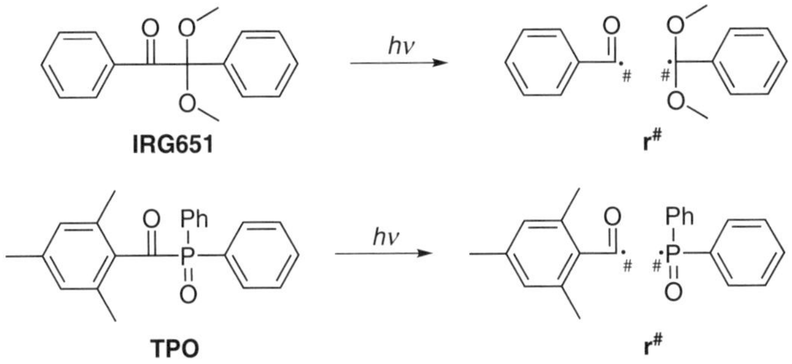

SCHEME 12.2 Photolysis of PIs occurs via a triplet state and leads to polarized reactive free radicals.

The superscript [#] is used in Scheme 12.2 and throughout this chapter to represent spin polarization, a term applied to situations for which a paramagnetic species possesses a population of spin states that is different from the Boltzmann distribution at the temperature of the experiment. Polarization disappears during the radical paramagnetic relaxation time, usually in the microsecond timescale. Here and below, we will

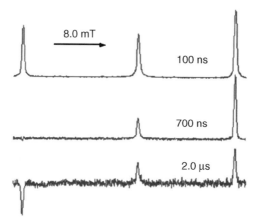

FIGURE 12.2   TR ESR spectra of **TPO** in ethyl acetate were taken at different observation times after a laser pulse. The two outmost components correspond to a large hyperfine coupling (HFC) constant on the P atom. The signal in the center of the spectrum corresponds to 2,4,6-trimethylbenzoyl radical (an envelope of small HFC.)[28,29]

use the symbol $r^{\#}$ (r) to designate polarized (and nonpolarized) free radicals produced by the photolysis of PIs, respectively.

Radicals **r** initiate polymerization. Figure 12.2 shows the TR ESR spectra obtained under photolysis of **TPO** and shows that both primary radicals are polarized.

The chemical structures of other compounds described in this chapter, namely, photosensitizers thioxanthene-9-one (**TX**) and 2-isopropyl thioxanthene-9-one (**ITX**), monomers isobornyl acrylate (**IBOA**), n-butyl methacrylate (**NBA**), methyl acrylate (**MA**), vinyl acrylate (**VA**) and methyl methacrylate (**MMA**) are presented in Scheme 12.3.

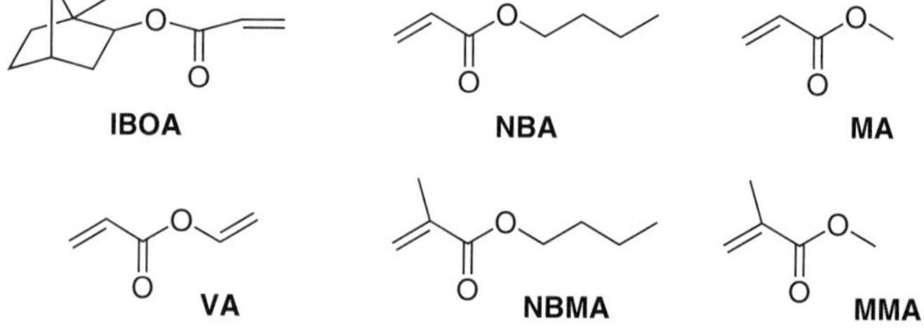

SCHEME 12.3   Chemical structures and designations of sensitizers and (meth)acrylates described in this chapter.

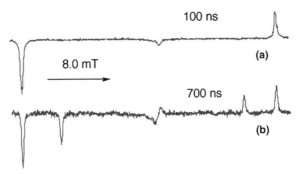

FIGURE 12.3 TR ESR spectra in ethyl acetate of (a) **TPO** in the presence of **TX** and (b) a mixture of **TPO** and **BAPO** in the presence of **ITX**. Laser light was absorbed predominantly by sensitizers.[28]

The energies of the triplet states of sensitizers **TX** (**ITX**) and of PIs **TPO** (**BAPO**) are close to each other ($\sim$260 kJ/mol), allowing for slightly exothermic or thermoneutral T–T energy transfer from sensitizer to PI.[28] Direct photolysis of phosphine oxides results in a well-documented initial strong *absorptive* (A) pattern of ESR spectra (see Fig. 12.2). Sensitization by **TX** or **ITX** of the photolysis of phosphine oxides leads evidently to the same radicals, but an initial polarization pattern is quite different, namely, *emission/absorption* (E/A) pattern (see Fig. 12.3).

Thus, the observation of TR ESR spectra patterns allows the determination of the reaction pathway leading to the same radicals. In the cases studied, it is direct versus sensitized photolysis.

ESR spectrometers at the X band are the most common but are not unique. Figure 12.4 presents experimental TR ESR spectra of **IRG651** (Scheme 12.1) obtained with ESR spectrometers at different frequencies.

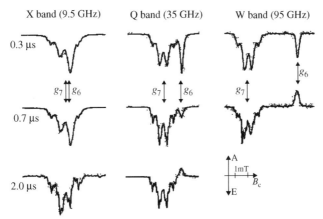

FIGURE 12.4 X, Q, and W band TR ESR spectra taken at different times after laser flash of **IRG651**. Here, g_6 and g_7 are g factors of benzoyl and 1,1-dimethoxybenzyl radicals, respectively.[30]

The data of Figure 12.4 demonstrate that an increase in frequency (and of magnetic flux density B) leads to a larger contribution of Δg radical pair mechanism: the benzoyl radical manifests absorption in the W band spectrum.[30] A high MF leads to an increase in Boltzmann polarization and even allows detection of transient radicals without CIDEP effects. TR ESR spectra of **TPO** and **BAPO** were studied with these three different microwave frequencies (Fig. 12.4) and also with a S band spectrometer (2.8 GHz).[31,55] The results obtained demonstrated a quantitative agreement between theory and experiment of the TM action. Parameters of the excited triplet state of **BAPO** (Scheme 12.1) were estimated from the dependence of the TM polarization versus microwave frequency. In particular, zero field splitting $D_{ZFS} \sim 0.18\,\text{cm}^{-1}$.[31]

Figure 12.5 presents TR ESR and FT ESR spectra obtained under photolysis of **DAR** (Scheme 12.1). One can observe a broadened signal of benzoyl radical in the FT ESR (or a signal of much lower apparent intensity). The intensity of the signals in CW TR ESR is determined by polarization, longitudinal (spin lattice) relaxation time T_1 and by the rate of chemical disappearance of $\mathbf{r}^{\#}$. The intensity of signals in FT ESR is determined by polarization, and phase memory time T_M, which includes T_1, transverse (spin–spin) relaxation time T_2, and a rate of chemical disappearance of $\mathbf{r}^{\#}$. Broad ESR components have short T_M, and they are difficult to observe. Broadening of components in spin adducts is ascribed to a hindered rotation around a C_α–C_β bond or *cis–trans* isomerization (Scheme 12.4).[26,32]

Spin correlated radical pairs, SCRPs, have been observed in micellar solutions and their origin was elucidated by Forbes et al.[25a] These SCRP have been widely studied under *photoreduction* of benzophenone and other electron/hydrogen acceptors mostly

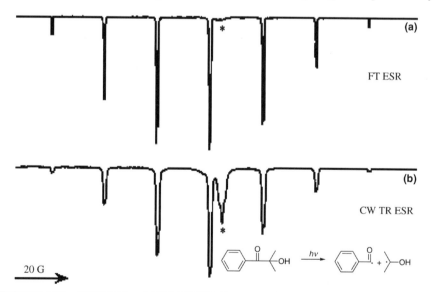

FIGURE 12.5   FT ESR (a) and (CW) TR ESR (b) spectra taken at times of several hundreds of nanoseconds following laser excitation of **DAR** in propan-2-ol solution. The asterisk marks the benzoyl radical.[26]

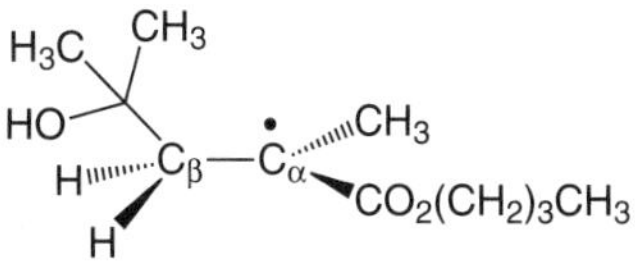

SCHEME 12.4 *Cis–trans* isomerization of acrylate adduct radicals.[26]

in micellar solutions.[25] SCPR in micelles were observed under *photodissociation* 4-*tert*-phenyl-1-hydroxy-1-propyl butyl ketone.[33]

It was suggested that substituted benzyl **r** of **IRG651** (Scheme 12.1) slowly dissociates ($k_{diss} \sim 250 \text{ s}^{-1}$) with the formation of methyl benzoate and methyl radical[34a]:

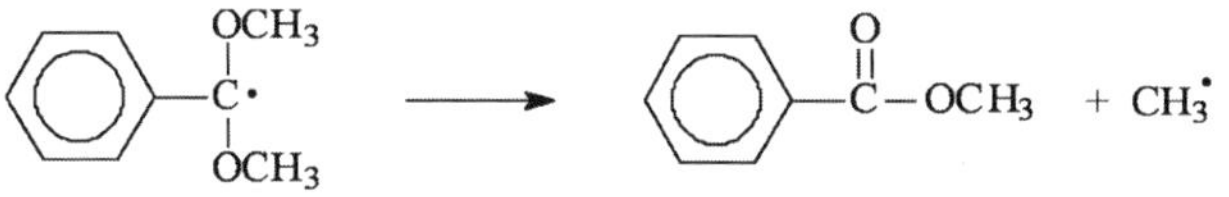

At the same time, this 1,1-dimethoxybeznyl radical **r** undergoes a facile *photo*fragmentation.[34a] Thus, a mechanism of photodecomposition of **IRG651** becomes rather complex and it strongly depends upon light intensity.[34a]

It was found that photoexcited **IRG651** reduces dye Methylene Blue in acrylate media.[34b] It was speculated that a reducing agent is the methyl radical formed under decomposition of **IRG651**.[34b]

Radicals **r** of **BAPO** (Scheme 12.1) are believed to dissociate during their lifetime if they are not promptly intercepted by an acrylate (Scheme 12.5).[10]

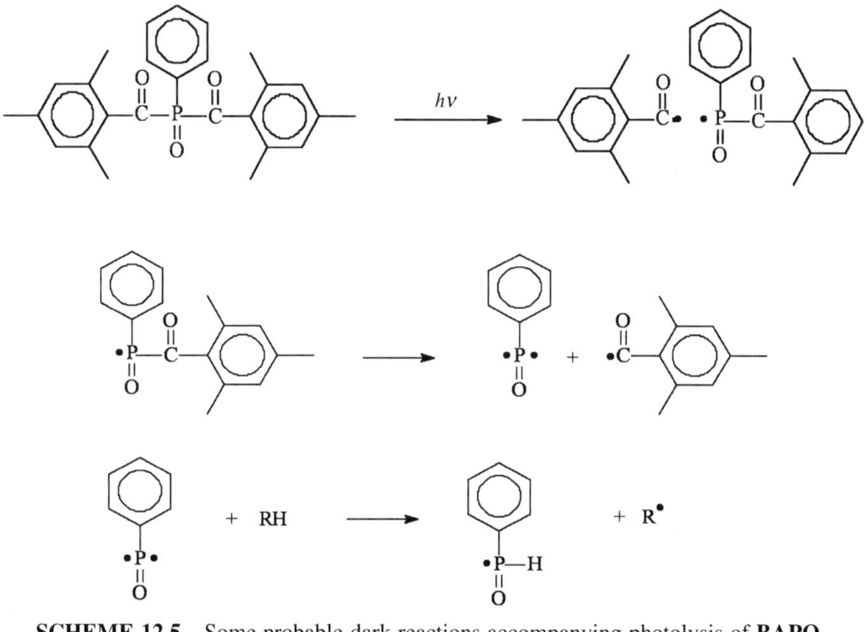

SCHEME 12.5 Some probable dark reactions accompanying photolysis of **BAPO**.

A trivalent phosphorous compound phosphene can abstract hydrogen from a C−H bond with a formation of two radicals. Thus, **BAPO** can produce up to four radicals upon absorption of one photon. (One absorbed *einstein* can lead up to four moles of reactive radicals.) This unusual feature makes **BAPO** a very efficient PI.[10]

IRG2959 (Scheme 12.1) was used as a probe of molecular motion in cotton fibers.[35] TR ESR spectra of **IRG2959** consist of spectra of RPs in a liquid-like and in crystalline environments. Simulation of spectra led to the conclusion that radicals participate in 3D and in 2D motion. Observation of a contribution of SCRP suggested that approximately 50% of radicals are trapped in cages of cotton fibers during the time of the experiment (0.5 µs).[35]

The TR ESR spectrum of **IRG651** in a viscous solvent ethylene glycol was subjected to a detailed analysis in Ref. 36. It was assumed that in ethylene glycol the following holds true: $T_1 \gg T_2$. This assumption allows the simplification of calculations and to obtain for 1,1-dimethoxybenzyl radical $T_2 = 0.8\,\mu\text{s}$.[36]

In conclusion, it is worthwhile mentioning that a number of common PIs without an aromatic carbonyl group in their structure dissociate via an excited singlet state and demonstrate not an E/A but an A/E (absorptive/emissive) CIDEP pattern. An A/E CIDEP pattern was observed under photolysis of **AIBN**[17] and of **HABI** dimers[10,37] (see Fig. 12.6).

12.3.2   Addition of Free Radicals to the Double Bonds of Monomers

Addition of a free radical of an initiator **r** to a monomer (oligomer) **M** is considered as the initiation step of a polymerization. In TR ESR experiments $\mathbf{r}^{\#}$ can be polarized, as well as secondary radicals:

$$\mathbf{r}^{\#} + \mathbf{M} \rightarrow \mathbf{r}\text{-}\mathbf{M}^{\bullet\,\#}$$

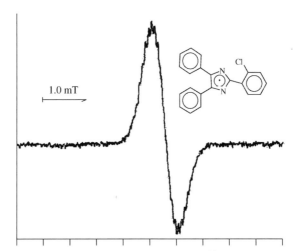

FIGURE 12.6   TR ESR spectrum of **o-Cl-HABI**[10] dimer in cyclohexanol.[37]

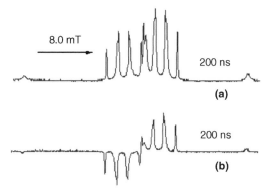

FIGURE 12.7 TR ESR spectra in ethyl acetate of (a) **TPO** in the presence of **IBOA** and (b) **BAPO** in the presence of **TX** and **IBOA**. In the latter case, laser light was absorbed mainly by triplet sensitizers. See Schemes 12.1 and 12.3 for chemical structures of compounds.[28]

If the fast reactions of polarized radicals are fast enough, polarization is preserved and is transferred into secondary radicals or spin adducts. As examples of the addition study, we will consider the reaction of $\mathbf{r}^{\#}$ produced from photolysis of **TPO** with **IBOA** and with **MMA** (see Schemes 12.1 and 12.3 for structures of compounds.)

TR ESR spectra of PIs in the presence of **IBOA** are presented in Fig. 12.7. The observed spectra (Fig. 12.7) consist mainly of polarized signals of spin adducts of a phosphinoyl radical and an acrylate with the following structure: Scheme 12.6 demonstrates the so-called "tail" adduct, where **r** adds to a CH_2 group ("tail"). Tail adducts are the most common in addition to substituted olefins.[38] In this work, only the tail addition was observed.[28]

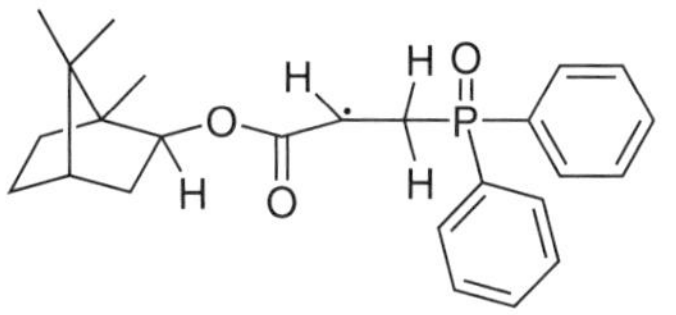

SCHEME 12.6 The structure of the main product of the addition of a phosphinoyl radical to **IBOA**.[28]

(Substituted) benzoyl radicals are known to be much less reactive than phosphinoyl radicals in addition to acrylates, and benzoyl radicals persist contributing to the center of the CW experimental spectra of spin adducts (Fig. 12.7); see Table 12.2 and additional comments below.

The spectra presented in Fig. 12.7 and 12.8 demonstrate that a polarization pattern is preserved during a fast addition of radicals to a monomer. Absorptive (A) polarization observed under direct photolysis of PIs (Fig. 12.2) results in the A pattern of a spectrum of the spin adduct (Fig. 12.7a), and the E/A pattern observed under sensitized photolysis (Fig. 12.3) results in an E/A pattern of a spectrum of the

TABLE 12.2   Addition Rate Constants k_{add} of Radicals of Photoinitiators to Acrylates[a]

Radical	Acrylate[b]	Solvent	k_{add} (M^{-1} s^{-1})	Method[c]	Reference
(structure: 2-hydroxyprop-2-yl radical)	NBA	Acetonitrile	1.3×10^7	LFP	52
(structure: 2-hydroxyprop-2-yl radical)	MA	Propan-2-ol	3.5×10^7	LFP	56
(structure: 2-hydroxyprop-2-yl radical)	MA	Toluene	367 (315 K)	ESR	57
(structure: 2-hydroxyprop-2-yl radical)	VA	Propan-2-ol	7500	ESR	58a
(structure: 2-hydroxycyclohexyl radical)	NBA	Acetonitrile	1.1×10^7	LFP	52
(structure: HO-CH$_2$CH$_2$-O-C$_6$H$_4$-C(O)· radical)	NBA	Acetonitrile	3.5×10^5	LFP	58c
(structure: benzoyl radical)	NBA	Acetonitrile	2.7×10^5	LFP	58c
(structure: 4-methylthio-benzoyl radical)	NBA	Acetonitrile	5.5×10^5	LFP	58c
(structure: morpholino isopropyl radical)	NBA	Acetonitrile	6.1×10^6	LFP	52
(structure: diphenylphosphinoyl radical)	NBA	Acetonitrile	2.9×10^7	LFP	59a
(structure: diphenylphosphinoyl radical)	NBA	Toluene	1.79×10^7 / 1.98×10^7	ESR / LFP	42
(structure: diphenylphosphinoyl radical)	NBA	Ethyl acetate	2.2×10^7	LFP	58a
(structure: 2,4,6-trimethylbenzoyl radical)	NBA	Acetonitrile	1.8×10^5	LFP	58c
(structure: dibenzoyl / phosphine-type radical)	NBA	Toluene	1.2×10^6 / 1.45×10^6	ESR / LFP	42

TABLE 12.2 (*Continued*)

Radical	Acrylate[b]	Solvent	k_{add} (M^{-1} s^{-1})	Method[c]	Reference
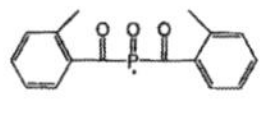 *(structure)*	**NBA**	Toluene	2.2×10^6 2.6×10^6	ESR LFP	42
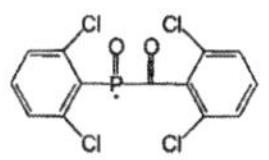 *(structure)*	**NBA**	Toluene	8.8×10^6	ESR	42
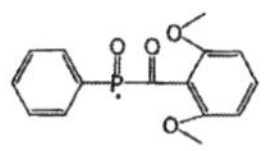 *(structure)*	**NBA**	Acetonitrile	2.4×10^7	LFP	59a
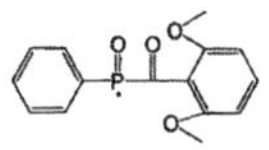 *(structure)*	**NBA**	Toluene	1.25×10^7 1.46×10^7	ESR LFP	42
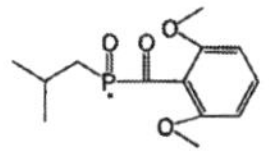 *(structure)*	**NBA**	Toluene	2.3×10^6	ESR	42
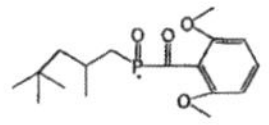 *(structure)*	**NBA**	Acetonitrile	1.5×10^7	LFP	59a
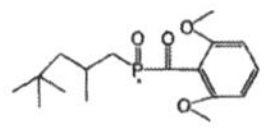 *(structure)*	**NBA**	Toluene	1.1×10^7 1.4×10^7	ESR LFP	42
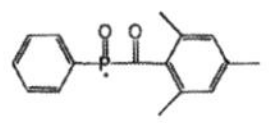 *(structure)*	**NBA**	Acetonitrile	1.1×10^7	LFP	59a
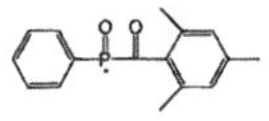 *(structure)*	**NBA**	Toluene	7.7×10^6 8.7×10^6	ESR LFP	42
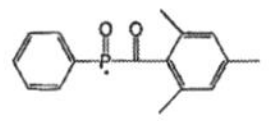 *(structure)*	**NBA**	Acetonitrile	1.1×10^7	LFP	59a
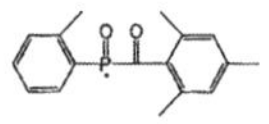 *(structure)*	**NBA**	Toluene	8.0×10^5 9.2×10^5	ESR LFP	42
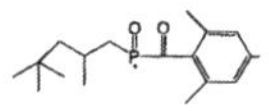 *(structure)*	**NBA**	Acetonitrile	4.0×10^6	LFP	59a
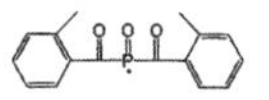 *(structure)*	**NBA**	Toluene	3.73×10^6	ESR LFP	42
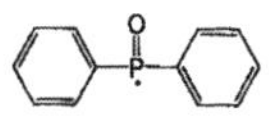 *(structure)*	**VA**	Ethyl acetate	3.3×10^7	LFP	58a

(*continued*)

TABLE 12.2   (*Continued*)

Radical	Acrylate[b]	Solvent	k_{add} (M^{-1} s^{-1})	Method[c]	Reference
	MA	*n*-Hexane	3.3×10^7	LFP	60
	MMA	scCO$_2$; 140 bar	6.1×10^7 (315 K)	ESR	41
	MMA	Acetonitrile	8.1×10^7	ESR	41
	MMA	Toluene	4.0×10^7	ESR	41
	MMA	*n*-Hexane	1.1×10^8	LFP	60
	MMA	*n*-Hexane	2.3×10^7	LFP	60
	MA	*n*-Hexane	1.1×10^7	LFP	60

[a] Experiments done at ambient (room) temperature unless stated otherwise. Usual determination error of rate constants is ~15%. See original references for more detail.
[b] See Scheme 12.3 for structures of monomers.
[c] Methods of determination of k_{add} used: ESR; Laser flash photolysis (LFP) with IR or optical detection.

spin adduct (Fig. 12.7b). Thus, TR ESR allows the establishment of the origin and structure of free radicals, and their reactions as well.

An FT ESR and TR ESR study of the addition of radicals of **DAR** to **NBA**, **NBMA**, and **MMA** (see Schemes 12.1 and 12.3 for abbreviations) allowed the determination of HFC constants of the spin adducts (Fig. 12.8). An E(missive) CIDEP pattern of an ESR spectrum of photolyzed **DAR** (Fig. 12.5) is transferred to the adducts (Fig. 12.8).[26]

Scheme 12.7 presents another case of the addition of $r^{\#}$, this time to the monomer **MMA**, which is very important in industrial polymerizations:[39]

The data shown in Scheme 12.7 confirm that addition of (substituted) benzoyl radicals is slower than that of counter radicals (Scheme 12.7b), and therefore the benzoyl adducts have weak signals in the TR ESR spectra. A similar conclusion was obtained by FT TR ESR study of the photolysis of **IRG2959** in the presence of **NBA** (Schemes 12.1 and 12.3).[40] ESR signals of two products of photolysis of **IRG2959**, namely, 2-hydroxy-2-propyl radical and substituted benzoyl radical, demonstrate quite different time dependences in the presence of acrylate. The signal of 2-hydroxy-2-propyl radical disappears faster with an increase in acrylate concentration, whereas benzoyl radical is practically unaffected by the presence of acrylate in concentrations up to 0.1 M.[40] However, one can obtain only estimations of kinetic rate constants k_{add}

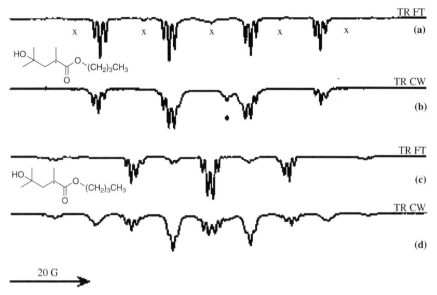

FIGURE 12.8   FT ESR ((a) and (c)) and CW TR ESR ((b) and (d)) obtained under photolysis of **DAR** in the presence of *n*-butyl acrylate **NBA** ((a) and (b)) and *n*-butyl methacrylate **NBMA** ((c) and (d)) in propan-2-ol solution. x: unreacted 2-hydroxy-2-propyl radical; *: benzoyl radical.[26]

in TR ESR experiments by the disappearance of a radical spectrum. Examining a TR ESR component width at a constant delay time allows the determination of k_{add} (rate constant of addition of **r** to a substrate).[41,42]

We conclude that the dominant process under photoinitiation of polymerization of (meth)acrylates by **TPO** (Scheme 12.1) is a tail addition of phosphinoyl radical to a double bond (Scheme 12.8).

Vinyl acrylate (**VA**) is an intriguing compound capable of self-initiation under UV irradiation.[43,44] Photopolymerization of **VA** proceeds faster when PI is added, with other conditions being the same. The question is which double bond is attacked by **r** of PI (Scheme 12.9).

Different reactivity of the two distinct double bonds toward free radicals will lead either to polyacrylate or to poly(1-acryloyloxyethylene) or to a mixed polymer if reactivity of both double bonds is comparable (Scheme 12.10).

One can formally expect the formation of a total of eight adducts under the photolysis of PI in the presence of **VA**: each **r** can form four secondary radicals: two "head to tail" and two "head to head."[29]

Photolysis of **TPO** (Scheme 12.1) in the presence of **VA** leads to the main adduct presented according to Scheme 12.8 above. However, a weak signal, which is ascribed to the addition to a vinyl ether group, is also observed in the TR ESR spectra.[29] The ratio of the intensities of two ESR signals is approximately 20:1. This ratio is in good agreement with the ratio of the rate constant of the addition k_{add} of phosphinoyl radicals to the acrylate group and to the vinyl ether group (in a model compound vinyl

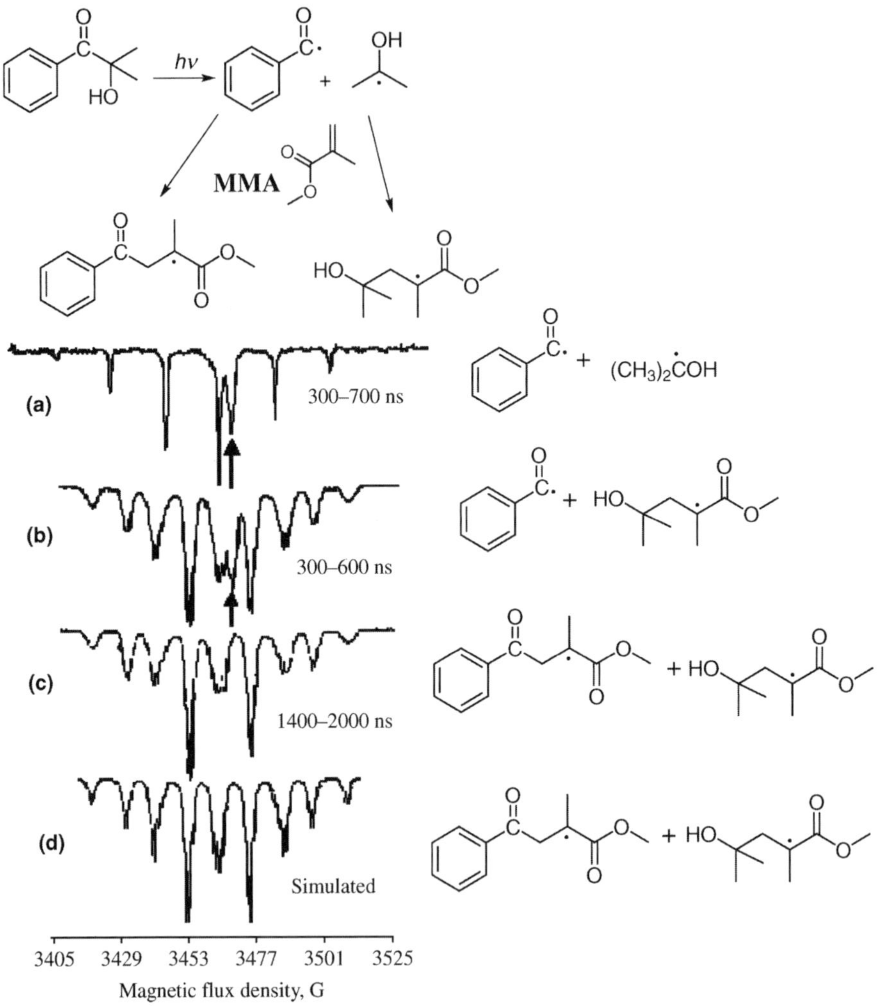

SCHEME 12.7 TR ESR study of addition of **r** of **DAR** to **MMA** (Schemes 12.13) in ethyl acetate. Arrows at the spectra point to a signal of substituted benzoyl radical. The top spectrum was obtained under photolysis of **DAR** only. The second and the third spectra from the top were taken in the presence of **MMA**. The spectra correspond to adducts of both $r^{\#}$ s of **DAR** (predominantly of ketyl radical) to **MMA**. All radicals are polarized; the sign $^{\#}$ is not shown in the scheme. Vertical arrow points to a signal of a benzoyl radical.[39]

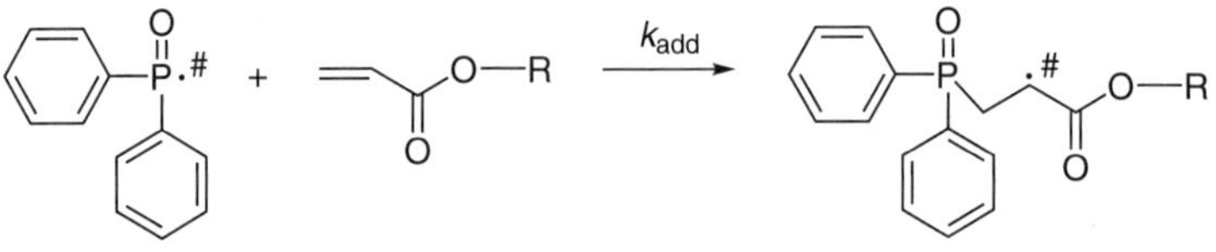

SCHEME 12.8 Reaction of a phosphinoyl radical with an acrylate.

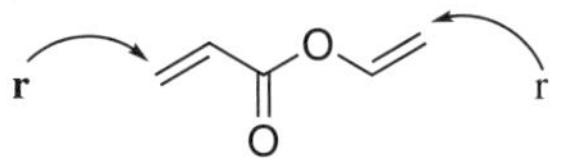

SCHEME 12.9   The two possible additions of radicals to vinyl acrylate.

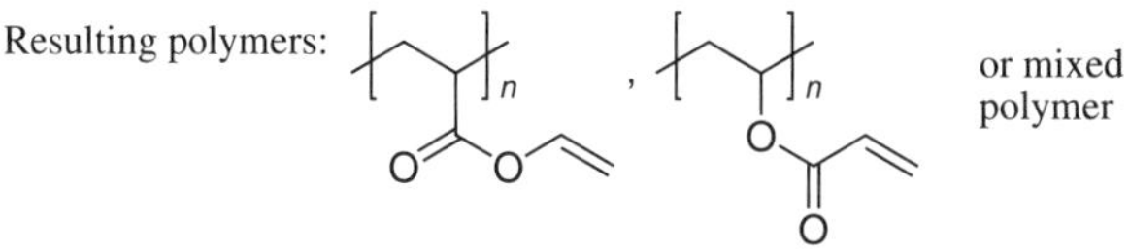

SCHEME 12.10   Structures of polymers that can be formed under free radical polymerization of vinyl acrylate.

pivalate) as $17:1$.[29] Experimental data demonstrate that polyacrylate is the main product of the free radical polymerization of vinyl acrylate.[45]

No adducts of substituted benzoyl radical are observed during the time of monitoring of phosphinoyl adducts. Thus, in experiments with **VA** one main adduct and the second adduct (in the much lower concentration) are observed instead of eight adducts is seen.

TR ESR allows not only structural identification of short-lived radicals but also provides insights into their conformation. Analysis of the ESR spectra of adducts of phosphinoyl radicals to acrylates reveals that these secondary radicals have two conformations (Scheme 12.11):

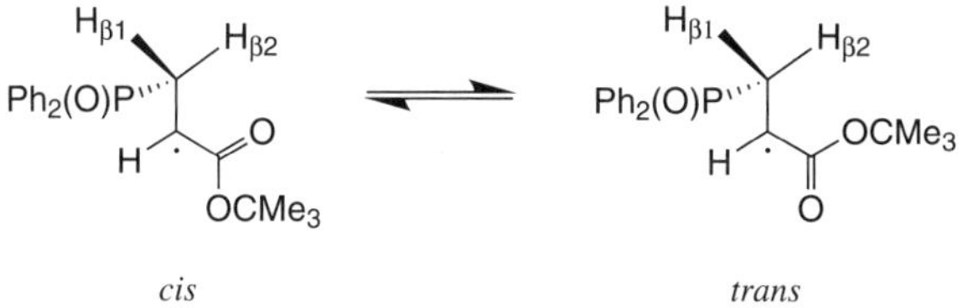

SCHEME 12.11   *Cis–trans* isomerization of acrylate adduct radicals.[29] For a similar scheme, see Scheme 12.4.

We have discussed above (Scheme 12.4) that hindered rotation along a single bond manifests itself as a broadening of TR ESR spectra components.

The same conclusions for the primary and secondary radicals can be achieved by steady-state (SS) irradiation in the ESR cavity termed SS ESR. SS ESR has several advantages and disadvantages as compared to TR ESR. First, SS ESR allows for detection of all free radicals, not necessarily only those in non-Boltzmann population (polarized). Second, SS ESR has a high sensitivity toward paramagnetic species in low concentration because it enables the application of field modulation. SS ESR spectra are presented in the well-known form as the first derivative of the signal. On the other hand, it is not easy to detect highly reactive free radicals by SS ESR due to their low SS concentration.

FT ESR spectra of acrylate radicals in the presence of acetone/propan-2-ol were studied.[46] A suggestion was made that acrylate radicals are vinyl radicals;[46] such a suggestion probably requires further study. Estimations of rate constants were obtained by monitoring the temporal behavior of several components in FT ESR spectra.[46]

12.3.3 Electron Spin Polarization Transfer from Radicals of Photoinitiators to Stable Nitroxyl Polyradicals

Termination of free radical polymerization is a reaction between two macroradicals ($R_n^{\bullet}$):

$$R_n^{\bullet} + R_m^{\bullet} \rightarrow \text{diamagnetic product(s)}$$

or between $R_n^{\bullet}$ and any other reagent **X** leading to the disappearance of reactive radicals:

$$R_n^{\bullet} + X \rightarrow \text{no reactive free radicals}$$

In particular, **X** can be a spin trap[32] or dioxygen. (Peroxyl radicals are inactive in propagation of the polymerization.)

Using TR ESR, a model reaction between the free radicals of PIs and stable nitroxyl radicals of 2,2,6,6-tetramethylpiperidine-*N*-oxyl (**TEMPO**) family was studied. We will abbreviate the **TEMPO** fragment further as **N**. Nitroxyl biradicals (**N-O-N**), had radical termini in proximity to each other (see Scheme 12.12).

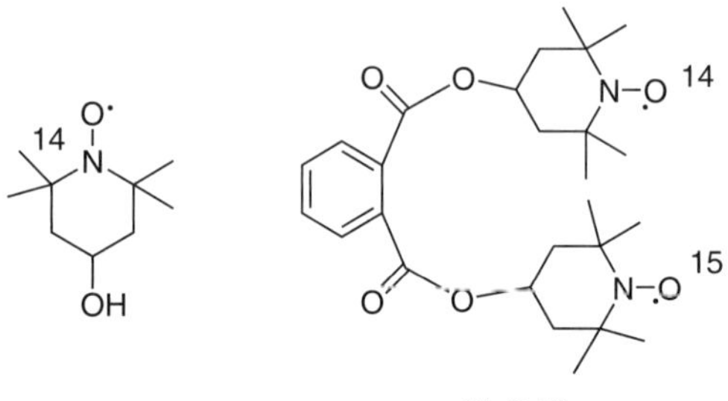

SCHEME 12.12 Chemical structure of 2,2,6,6-tetramethyl-4-hydroxypiperidine-*N*-oxyl (**TEMPOL, N**) and biradical (**N-O-N**). Both symmetric (**N-O-N**) with ^{14}N isotopes and asymmetric (**N-O-N**) with ^{14}N and ^{15}N isotopes such as those presented in the scheme.[47]

Both stable nitroxyls and nitroxyl biradicals are known as inhibitors of free radical polymerization: they intercept reactive free radicals. Scheme 12.13 presents processes occurring upon an encounter of $r^{\#}$ and **N**:

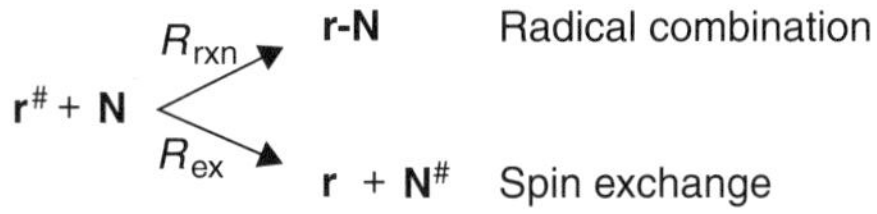

$$r^{\#} + N \quad \begin{array}{c} \xrightarrow{R_{rxn}} \quad \textbf{r-N} \quad \text{Radical combination} \\ \\ \xrightarrow{R_{ex}} \quad \textbf{r} + \textbf{N}^{\#} \quad \text{Spin exchange} \end{array}$$

SCHEME 12.13 Competition between chemical reaction (cross-section R_{rxn}) and spin exchange (cross-section R_{ex}) of a reactive polarized radical $\mathbf{r}^{\#}$ and a nitroxyl radical $\mathbf{N}$.[47]

The diamagnetic molecular combination product **r-N** is invisible by ESR, but we *can state that the radical* $\mathbf{r}^{\#}$*was in proximity of the stable nitroxyl* $\mathbf{N}$, *because the TR ESR spectrum of* $\mathbf{N}^{\#}$*was observed.* Polarization transfer from one radical to another occurs by spin exchange.[48] It is believed that spin exchange does not require physical contact of the reagents, and it can occur at the distance of one or several molecular diameters between radicals.

In the case of a reaction of $\mathbf{r}^{\#}$ with (**N-O-N**) (Scheme 12.12), two polarization transfers occur: by spin exchange with an appearance of (**N-O-N**)$^{\#}$ and by addition of $\mathbf{r}^{\#}$ to (**N-O-N**) with the formation of polarized monoradical see Scheme 12.14.

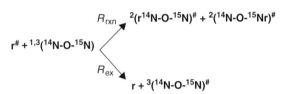

SCHEME 12.14 Competition between chemical reaction (cross-section R_{rxn}) and spin exchange (cross-section R_{ex}) of a reactive polarized radical $\mathbf{r}^{\#}$ and an asymmetric nitroxyl biradical ($^{14}\mathbf{N}$-**O**-$^{15}\mathbf{N}$).[47]

In the case of symmetric ($^{14}\mathbf{N}$-**O**-$^{14}\mathbf{N}$), one observes only one polarized mono-radical (substituted **TEMPO**) and ($^{14}\mathbf{N}$-**O**-$^{14}\mathbf{N}$)$^{\#}$. A relatively simple TR ESR 2D spectrum that was obtained during photolysis of **IRG651** in the presence of ($^{14}\mathbf{N}$-**O**-$^{14}\mathbf{N}$) is demonstrated below.

The key feature of the isotopically asymmetric system is that an interaction of a polarized reactive free radical $\mathbf{r}^{\#}$ with ($^{14}\mathbf{N}$-**O**-$^{15}\mathbf{N}$) results in *three distinct polarized species with different sets of* a_N *for each paramagnetic species and not overlapping spectra* (Scheme 12.14): ($\mathbf{r}^{14}\mathbf{N}$-**O**-$^{15}\mathbf{N}$)$^{\#}$, ($^{14}\mathbf{N}$-**O**-$^{15}\mathbf{N}\mathbf{r}$)$^{\#}$, and ($^{14}\mathbf{N}$-**O**-$^{15}\mathbf{N}$)$^{\#}$. This situation is essentially different from the previously studied case of ($^{14}\mathbf{N}$-**O**-$^{14}\mathbf{N}$), where the three basic components of the ESR spectrum of ($^{14}\mathbf{N}$-**O**-$^{14}\mathbf{N}$)$^{\#}$ coincide with components of the adduct ($^{14}\mathbf{N}$-**O**-$^{14}\mathbf{N}\mathbf{r}$)$^{\#}$ (see Fig. 12.9). The TR ESR spectra obtained with isotopically asymmetric ($^{14}\mathbf{N}$-**O**-$^{15}\mathbf{N}$) are presented in Fig. 12.10.

The relative values of R_{ex} versus R_{rxn} were estimated by employing TR ESR and computer simulation to estimate the concentrations of products of spin exchange and chemical reaction between the same reagents.[25b,47] It was assumed that the exchange is "strong" and chemical reactions are diffusion controlled. The unique feature (Scheme 12.14) of the isotopic asymmetric biradical ($^{14}\mathbf{N}$-**O**-$^{15}\mathbf{N}$) allows the

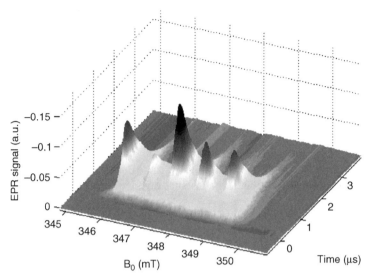

FIGURE 12.9 2D-TR-EPR spectrum produced by photolysis of **IRG651** in the presence of (^{14}N-O-^{14}N) in acetonitrile.[47]

extraction of values of the ratio R_{ex}/R_{rxn} to be ~ 4 from the TR ESR results. Thus, we conclude that spin exchange between a biradical (**N-O-N**) and a polarized organic radical $\mathbf{r}^{\#}$ occurs at a relatively large distance. If k_{rxn} is in fact lower than k_{diff} (rate constant of diffusion-controlled relation), then an estimation of 4–6 is found as an upper limit of R_{ex}/R_{rxn}.[47]

12.4 OPTICAL DETECTION OF TRANSIENTS

12.4.1 UV–vis Spectra of Representative Radicals

Detection of transients by their UV and visible spectra is usually performed by the laser flash photolysis. Such experiments have been performed for more than 35 years, and much of the data on spectra of transients, especially radicals, and on their decay kinetics has been accumulated and has been summarized in reference books.[49,50]

Figures 12.11 and 12.12 demonstrate optical spectra of P-centered radicals. The spectra of benzoyl and 1,1-dimethoxybenzyl radicals have low extinction coefficients. They are presented in Figures 12.13 and 12.14.

12.4.2 Representative Kinetic Data on Reactions of Photoinitiator Free Radicals

Table 12.2 presents rate constants k_{add} for the addition of $\mathbf{r}$ of PIs to acrylates.

In a series of reactions of phosphinoyl free radicals with different HFC a_P a rate constant of addition k_{add} to **NBA** decreases with a decrease in a_P.[59a] Such a correlation

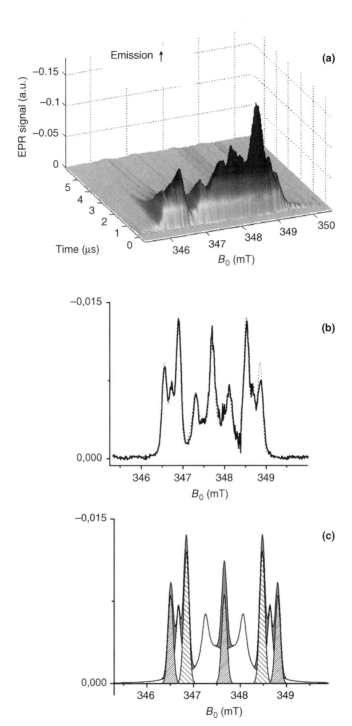

FIGURE 12.10 TR-EPR spectra produced by photolysis of **IRG651** in the presence of (^{14}N-O-^{15}N) in acetonitrile. (a) 2D spectrum, (b) 1D spectrum at 1 μs after the laser flash (solid line) *after* subtraction of the spectrum of $\mathbf{r}^{\#}$ of **IRG651** contribution and its simulation (dotted line), and (c) relative contributions of each of three polarized species into the simulation spectrum, see the text for discussion.[47]

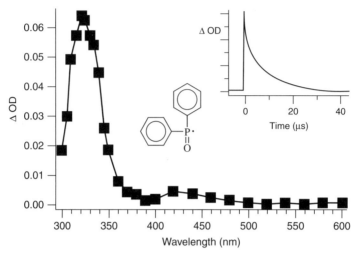

FIGURE 12.11   Transient absorption spectrum obtained under laser flash photolysis of **TPO** in *n*-hexane. The spectrum is ascribed to phospinoyl radical. The inset demonstrates a transient decay measured at $\lambda = 330$ nm.[51]

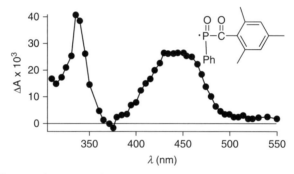

FIGURE 12.12   Transient absorption spectrum obtained under laser flash photolysis of **BAPO** in acetonitrile. The spectrum is ascribed to a phosphinoyl radical.[59a]

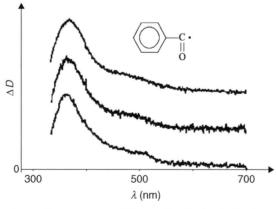

FIGURE 12.13   Absorption spectrum of benzoyl radicals obtained under photolysis of different precursors.[34a]

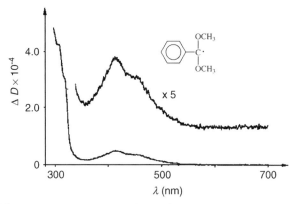

FIGURE 12.14 Absorption spectrum of 1,1-dimethoxybenzyl radicals obtained under photolysis of *tert*-butyl peroxide in the presence of 1,1-dimethoxytoluene in acetonitrile.[34a]

was attributed to a decrease in the *s*-character of the P-atom, indicative of P-centered radicals more planar structure and to a decrease in reactivity.[59a] At the same time, the σ-radical of benzoyl with its high electron density located on the carbonyl C atom[59b] has a much lower k_{add} than the 1,1-dimethoxybenzyl π-radical with its delocalized unpaired electron (see Section 12.3.2). Strictly Generally speaking, there is no correlation between the reactivity of a free radical and electron or spin density on the reactive atom of a radical.

Steric and polar effects on the values of k_{add} have been analyzed in great detail basically exhausting the problem at the present status of chemical kinetics.[61]

A suggestion was made that k_{add} of the 2-hydroxy-2-propyl radicals of **DAR** to **MA** (Schemes 12.1 and 12.3) decreases with the viscosity increase,[3a] which would be a sign of a diffusion-controlled or a diffusion-enhanced reaction.[53] In fact, an increase in viscosity 1200 times leads to a decrease in k_{add} only 5 times and not "by orders of magnitude,"[3a] which does not allow the classification of the addition as almost a diffusion-controlled (diffusion-enhanced) reaction.

Evidently, **r** of PIs can participate in multiple reactions, and the rate constants of many of these reactions have been measured.[41,51,52,58c,60] 2-Hydroxy-2-propyl radical reacts with **TEMPO** with a rate constant of $\sim 10^9\, M^{-1}\, s^{-1}$, a value that is somewhat lower than the rate constant k_{diff} expected for diffusion-controlled reactions.[54] In our opinion, it is quite possible that this reaction is limited by diffusion, but an absolute value is less than k_{diff} due to steric limitations—screening off a reactive oxygen atom by methyl groups of **TEMPO** (pseudodiffusion reactions.[53])

It was demonstrated that the P-centered radical of **BAPO** and its derivatives is oxidized by cationic photoinitiators (onium salts in their ground state).[62a] Phosphonium ion, a **BAPO** fragment, can efficiently initiate cationic polymerization. Thus, under certain conditions, PI used in *free radical polymerization* can initiate a *cationic* polymerization as well.[62a]

Type I PIs that have two aromatic chromophores in their structure and, are dubbed *difunctional* PIs,[62b] demonstrate higher rates of photopolymerization compared to common PIs (excluding phosphine oxides) with benzoyl fragment, in particular to

DAR (Scheme 12.1.) Difunctional PIs usually have higher extinction coefficients than monofunctional PIs.[62b,c]

12.5   IR DETECTION OF FREE RADICALS AND MONITORING THEIR REACTIONS

There are many cases when radical absorption in the UV region overlaps with a tail of absorption of a photoinitiator or other species produced by photolysis. Optical detection may become less sensitive or impossible. Also, a transient radical may have low absorption (extinction coefficient) in the UV–vis region. For example, a benzoyl radical exhibits a weak transition at $\lambda = 368$ nm ($\varepsilon \approx 150\,M^{-1}\,cm^{-1}$ [34a]) (Fig. 12.13). In this case, the IR detection of transients can be a useful alternative. Substituted benzoyl, substituted benzyl and phosphinoyl radicals, often discussed in this chapter, have weak absorption in the UV region. On the contrary, benzoyl radicals have strong absorption in the mid-IR region in the 1780–1880 cm^{-1} due to $>C=O$ vibrations. IR detection of transients is observed by using a diode laser fitted with monochromators and a sample chamber with a fast IR detector. TR IR spectra of several **r** are presented in Fig. 12.15.

Kinetics of reactions of substituted benzoyl radicals of **IRG2959** (Scheme 12.1) is conveniently monitored at their maximum in an IR spectrum at 1805 cm^{-1} (see Fig. 12.1)[20] TR IR measurements allowed determination of the rate constants of elementary reactions of free radicals of PIs with dioxygen, thiophenol, and bromotri-chloromethane.[51,59a] The radicals react with dioxygen in nonviscous solutions with rate constants of $\sim 10^9\,M^{-1}\,s^{-1}$ meaning that reactions are completely or partially controlled by diffusion.[53]

Radicals of **TPO** (Fig. 12.16) participate in self- and cross-termination and second-order kinetics fit is probably a random case.

12.6   CONCLUDING REMARKS

TR ESR allows the identification of the structure and even configuration of free radicals, and helps in the understanding of their reactivity. Polarized and nonpolarized radicals have the same reactivity because their magnetic energy is negligible compared to thermal energy ($k_B T$). Studying reactions in the magnetic field of an X band ESR spectrometer (~ 0.3 T) does not affect the reactivity of radicals toward multiple bonds of molecules. An external moderate magnetic field can slightly decrease the rate constant of bimolecular self- or cross-termination of free radicals.[19f,23] Polarization of radicals is considered a noninvasive labeling of radicals, which allows observation of their fast reactions.

In some exceptional cases, one can detect not only a secondary polarized radical **r-M**$^{\bullet \#}$ but also a tertiary radical **r-M-M**$^{\bullet \#}$.[58b]

Unfortunately, it is difficult to use TR ESR in determining the rates of disappearance of radicals due to a complex interplay of chemical reaction and paramagnetic

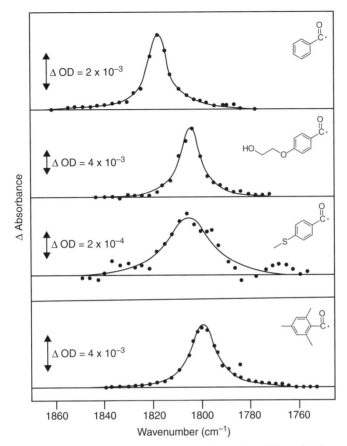

FIGURE 12.15 Diode laser-based TR IR spectra detected at 0.5 μs after laser excitation (λ = 355 nm) of acetonitrile solutions of **DAR**, **IRG2959**, **IRG907**, and **TPO** (Scheme 12.1) as shown from the top to the bottom.[58c]

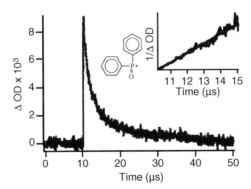

FIGURE 12.16 Decay of a phosphinoyl radical produced under laser flash photolysis of **TPO** (Scheme 12.1) in *n*-heptane. Decay of the radical was monitored at 1805 cm^{-1}. The insert demonstrates the second-order kinetics fit.[51]

relaxation.[41] Radical kinetics can be studied by laser flash photolysis with detection by absorption spectra. Light of the same pulsed laser can be directed into the ESR cavity with a flow flat cell or into an optical cell of a time-resolved spectrophotometer. Both cells may contain the same solution.

During a microsecond time frame, which is the typical paramagnetic relaxation time of free radicals, polarized free radicals can participate in addition to multiple bonds, to dioxygen, in hydrogen (electron) transfer, in addition to polyradicals, etc. Products of these reactions are polarized in most cases, and they demonstrate TR ESR signals.

The same conclusions on the structure and configuration of primary and secondary radicals can be achieved by SS ESR. TR ESR allows the relatively simple observation of highly reactive transient free radicals. The method is especially convenient in the case of transient radicals with a few magnetic nuclei, that is, with a simple ESR spectrum.

Detection of triplet states and photogenerated radicals and in the nanosecond timescale by their optical spectra is probably the most common method used for several decades. However, detection of radicals by their IR spectra becomes valuable in cases of low optical extinction coefficients of transients and overlay of spectra of radicals and parent compounds. In conclusion, laser flash photolysis with optical, IR, and ESR detection provides a powerful arsenal to investigate radical species produced in the photoinitiation of radical polymerization.

ACKNOWLEDGMENTS

The authors are grateful to all their colleagues for contributing to this report and for useful discussions. Their names are presented in the references to joint publications listed below. The authors also thank the National Science Foundation for generous support of this research through grant CHE 07-17518.

REFERENCES

1. (a) Fouassier, J.-P. *Photoinitiation, Photopolymerization, and Photocuring: Fundamentals and Applications*; Hanser: Munich, 1995. (b) Type II photoinitiators participate in bimolecular hydrogen atom/electron abstraction in the excited state with a formation of two radicals. Unfortunately, Type II PIs are often confused with the Norrish process II.

2. Jockusch, S.; Landis, M. S.; Freiermuth, B.; Turro, N. J. *Macromolecules* **2001**, *34*, 1619.

3. (a) Lalevée, J.; Allonas, X.; Jradi, S.; Fouassier, J. P. *Macromol.* **2006**, *39*, 1872; (b) Allonas, X.; Moriet-Savary, F.; Lavalée, J.; Fouassier, J. P. *Photochem. Photobiol.* **2006**, *82*, 88.

4. Turro, N. J. *Modern Molecular Photochemistry*; University Science Books: Mill Valley, CA, 1991.

5. Neckers, D. C.; Jager, W. *Photoinitiation for Photopolymerization: UV and EB at the Millennium*; SITA: London, 1998.

6. Jockusch, S.; Koptyug, I. V.; McGarry, P. F.; Sluggett, G. W.; Turro, N. J.; Watkins, D. M. *J. Am. Chem. Soc.* **1997**, *119*, 11495.

7. Originally cage *effect* meant an *effect of the liquid (condensed) phase* on a reaction studied in the gas phase. Strictly speaking, there is no sense to refer to any *effect* when one deals with reactions in the liquids.

8. The cage effect is observed in the gas phase under high pressures.

9. A question arises, how highly reactive radicals can escape singlet G pairs? Such pairs are formed under thermolysis of initiators, which occurs in the ground singlet state S_0, or under photolysis of PIs occurring is the excited singlet state S_1. The known photoinitiators **HABI** dimers[10] dissociate in a singlet state, but radicals are poorly reactive toward each other. That provides high *f*. Benzoyl peroxide, 2,2'-azobis(2-methylpropionitrile) (**AIBN**) undergo thermolysis or photolysis with a formation of singlet RP, but the reactive radicals are separated by inert molecules of CO_2 or N_2. Such separation most probably facilitates a dissociation of RP. Thermolysis occurs at elevated temperatures (low solvent viscosities) which eases separation. Still, values of *f* are not high.

10. Dietliker, K. *A Compilation of Photoinitiators Commercially Available for UV Today*; SITA: Edinburgh, 2002.

11. Khudyakov, I. V.; Yakobson, B. I. *Russ. J. General Chem.* **1984**, *54*, 3.

12. Khudyakov, I. V.; Kiryukhin, Yu. I.; Yasmenko, A. I. *Chem. Phys Lett.* **1980**, *74*, 462.

13. Levin, P. P.; Khudyakov, I. V.; Kuzmin, V. A. *J. Phys. Chem.* **1989**, *93*, 208.

14. Khudyakov, I. V.; Arsu, N.; Jockusch, S.; Turro, N. J. *Des. Monom. Polym.* **2003**, *6*, 91.

15. Tominaga, K.; Yamauchi, S.; Hirota, N. *Chem. Phys. Lett.* **1991**, *179*, 35.

16. Maliakal, A.; Weber, M.; Turro, N. J.; Green, M. M.; Yang, S. Y.; Pearsall, S.; Lee, M.-J. *Macromolecules* **2002**, *35*, 9151.

17. Savitsky, A. N.; Paul, H.; Shushin, A. I. *J. Phys. Chem. A* **2000**, *104*, 9091.

18. Grossman, A.Yu.; Khokhlov, A. R. *Statistical Physics of Macromolecules*; American Institute of Physics: 2002.

19. For review articles, see (a) Nagakura, S.; Hayashi, H.; Azumi, T. *Dynamic Spin Chemistry*; Kodansha-Wiley: Tokyo, 1998; (b) Woodward, J. R. *Prog. React. Kinet.* **2002**, *27*, 165; (c) Hayashi, H.; Sakaguchi, Y.; Wakasa, M. *Bull. Chem. Soc. Jpn.* **2001**, *74*, 773; (d) Buchachenko, A. L. *Pure Appl. Chem.* **2000**, *72*, 2243; (e) Brocklehurst, B. *Chem. Soc. Rev.* **2002**, *31*, 301; (f) Khudyakov, I. V.; Serebrennikov, Yu. A.; Turro, N. J. *Chem. Rev.* **1993**, *93*, 537; (g) Murai, H. *J. Photochem. Photobiol. C* **2003**, *3*, 183.

20. Vink, C. B.; Woodward, J. R. *J. Am. Chem. Soc.* **2004**, *126*, 16730.

21. Timmel, C. R.; Cintolesi, F.; Brocklehurst, B.; Hore, P. J. *Chem. Phys. Lett.* **2001**, *334*, 387.

22. Suzuki, T.; Maeda, K.; Tatsuo, A. *Nippon Kagak. Koen Yok.* **2003**, *83*, 424.

23. Margulis, L. A.; Khudyakov, I. V.; Kuzmin, V. A. *Chem. Phys. Lett.* **1986**, *124*, 483.

24. Cozens, F. L.; Scaiano, J. C. *J. Am. Chem. Soc.* **1993**, *115*, 5204.

25. For review articles, see (a) Forbes, M. D. E. *Photochem. Photobiol.* **1997**, *65*, 73; (b) Turro, N. J.; Khudyakov, I. V. *Res. Chem. Intermed.* **1999**, *6*, 505; (c) Turro, N. J.; Kleinman, M. H.; Karatekin, E. *Angew. Chem., Int. Ed.* **2000**, *39*, 4436.

26. Weber, M.; Turro, N. J.; Beckert, D. *Phys. Chem. Chem. Phys.* **2002**, *4*, 168.

27. Turro, N. J.; Koptyug, I. V.; van Willigen, H.; McLauchlan, K. A. *J. Magnet. Res. A* **1994**, *109*, 121.

28. Williams, R. M.; Khudyakov, I. V.; Purvis, M. B.; Overton, B. J.; Turro, N. J. *J. Phys. Chem. B* **2000**, *104*, 10437.

29. Weber, M.; Khudyakov, I. V.; Turro, N. J. *J. Phys. Chem. A* **2002**, *106*, 1938.

30. Savitsky, A. N.; Galander, M.; Möbius, K. *Chem. Phys. Lett.* **2001**, *340*, 458.

31. Makarov, T. N.; Savitsky, A. N.; Möbius, K.; Beckert, D.; Paul, H. *J. Phys. Chem. A* **2005**, *109*, 2254.

32. Turro, N. J.; Khudyakov, I. V. *Chem. Phys. Lett.* **1992**, *193*, 546.

33. Wu, C.; Jenks, W. S.; Koptyug, I. V.; Ghatlia, N. D.; Lipson, M.; Tarasov, V. F.; Turro, N. J. *J. Am. Chem. Soc.* **1993**, *115*, 9583.

34. (a) Fischer, H.; Baer, R.; Hany, R.; Verhoolen, I.; Walbiner, M. *J. Chem. Soc. Perkin Trans. 2*, **1990**, 787; (b) Galagan, Y.; Su, W. -F. *J. Photochem. Photobiol.* **2008**, *A195*, 387.

35. Batchelor, S. N.; Shushin, A. I. *J. Phys. Chem. B* **2001**, *105*, 3405.

36. Konkin, A. L.; Roth, H.-K.; Schroedner, M.; Nazmutdinova, G. A.; Aganov, A. V.; Ida, T.; Garipov, R. R. *Chem. Phys.* **2003**, *287*, 377.

37. Caspar, J. V.; Khudyakov, I. V.; Turro, N. J.; Weed, G. C. *Macromolecules* **1995**, *28*, 636.

38. Moad, G.; Solomon, D. H. *The Chemistry of Free Radical Polymerization*; Pergamon: Oxford, 1995.

39. Karatekin, E.; O'Shaugnessy, B.; Turro, N. J. *Macromolecules* **1998**, *31*, 7992.

40. Vacek, K.; Geimer, J.; Beckert, D.; Mehnert, R. *J. Chem. Soc., Perkin Trans. 2*, **1999**, 2469.

41. Forbes, M. D. E.; Yashiro, H. *Macromolecules* **2007**, *40*, 1460.

42. Gatlik, I.; Rzadek, P.; Gerscheidt, G.; Rist, G.; Hellrung, B.; Wirz, J.; Dietliker, K.; Hug, G.; Kunz, M.; Wolf, J.-P. *J. Amer. Chem. Soc.* **1999**, *121*, 8332.

43. Khudyakov, I. V.; Hoyle, C. E. US Patent 6,900,252.

44. Jönsson, E. S.; Lee, T. Y.; Viswanathan, K.; Hoyle, C. E.; Roper, T. M.; Guymon, C. A.; Nason, C.; Khudyakov, I. V. *Prog. Organic Coatings* **2005**, *52*, 63.

45. Fukuda, W.; Nakao, M.; Okumura, K.; Kakiuchi, H. *J. Polym. Sci. A* **1972**, *10*, 237.

46. Beckert, A.; Naumov, S.; Mehnert, R.; Beckert, D. *J. Chem. Soc., Perkin Trans. 2*, **1999**, 1075.

47. Sartori, E.; Khudyakov, I. V.; Lei, X.; Turro, N. J. *J. Amer. Chem. Soc.* **2007**, *129*, 7785.

48. Jenks, W. S.; Turro, N. J. *J. Am. Chem. Soc.* **1990**, *112*, 9009.

49. Fischer, H., Ed.; *Rate and Equilibrium Constants for Reactions of Polyatomic Free Radicals, Biradicals, and Radical Ions in Liquids*; New Series, Springer: Berlin, 1984; Landolt-Börnstein Series, Vol. 13.

50. Scaiano, J. C., Ed.; *CRC Handbook of Organic Photochemistry*; CRC Publishers: 1989.

51. Sluggett, G. W.; Turro, C.; George, M. W.; Koptyug, I. V.; Turro, N. J. *J. Am. Chem. Soc.* **1995**, *117*, 5148.

52. Jockusch, S.; Turro, N. J. *J. Am. Chem. Soc.* **1999**, *121*, 3921.

53. Burshtein, A. I.; Khudyakov, I. V.; Yakobson, B. I. *Prog. React. Kinet.* **1984**, *13*, 221.

54. Sobek, J.; Martschke, R.; Fischer, H. *J. Am. Chem. Soc.* **2001**, *123*, 2849.

55. Koptyug, I. V.; Ghatlia, N. D.; Sluggett, G. W.; Turro, N. J.; Ganapathy, S.; Bentrude, W. G. *J. Am. Chem. Soc.* **1995**, *117*, 9486.

56. Martschke, R.; Farley, R.; Fischer, H. *Helv. Chim. Acta* **1997**, *80*, 1363.

57. Héberger, K.; Fischer, H. *Int. J. Chem. Kinet.* **1993**, *25*, 249.

58. (a) Weber, M. Ph.D. Thesis, University of Zurich, 2000; (b) Weber, M. Unpublished results, 2001; (c) Colley, C. S.; Grills, D. C.; Besley, N. A.; Jockusch, S.; Matousek, P.;

Parker, A. W.; Towrie, M.; Turro, N. J.; Gill, P. M. W.; George, M. W. *J. Am. Chem. Soc.* **2002**, *124*, 14952.

59. (a) Jockusch, S.; Turro, N. J. *J. Am. Chem. Soc.* **1998**, *120*, 11773; (b) Leffler, J. E. *An Introduction to Free Radicals*; Wiley: New York, 1993.

60. Sluggett, G. W.; McGarry, P. F.; Koptyug, I. V.; Turro, N. J. *J. Am. Chem. Soc.* **1996**, *118*, 7367.

61. Fischer, H.; Radom, L. *Angew. Chem., Int. Ed.* **2001**, *40*, 1340.

62. (a) Dursun, C.; Degirmenci, M.; Yagci, Y.; Jockusch, S.; Turro, N. J. *Polym.* **2003**, *44*, 7389; (b) Dietlin, C.; Lavalee, J.; Allonas, X.; Fouassier, J. P.; Visconti, M.; Li Bassi, G.; Norcini, G. *J. Appl. Polym. Science* **2008**, *107*, 246; (c) Allonas, X.; Grotzinger, C.; Lavalee, J.; Fouassier, J. P.; Visconti, M. *Europ. Polym. J.* **2001**, *37*, 897.

APPENDIX: GLOSSARY

PI – photoinitiator

$\mathbf{r}$, $\mathbf{r}^{\#}$ – reactive free radical formed under photolysis of PI

HFC – hyperfine coupling

RP – radical pair

SCRP – spin-correlated radical pair

TM – triplet mechanism

RPM – radical pair mechanism

TR ESR, (CW) TR ESR – time resolved continuous wave ESR

FT ESR – time resolved Fourier transform ESR

SS – steady state

Φ_{diss} – quantum yield of photodissociation inside a solvent cage

Φ – cage effect value

f – cage escape value

k_{diss}, s^{-1} – rate constant of dissociation of molecule of radical into fragments

k_{add}, $M^{-1} \cdot s^{-1}$ – rate constant of addition of $\mathbf{r}$ to a substrate

k_{diff}, $M^{-1} \cdot s^{-1}$ – rate constant of diffusion-controlled relation

G-pair – geminate RP

F-pair – pair of radicals met in the solvent bulk as a result of random wandering (an encounter)

LPF – laser flash photolysis

τ_T – life-time of a triplet state

T_M – phase memory

A, E, A/E, E/A – absorptive, emissive, absorptive/emissive, emissive/absorptive pattern of CIDEP spectra, respectively

MFE – magnetic filed effect (general)

LFE – low (magnetic) field effect

13

DYNAMICS OF RADICAL PAIR PROCESSES IN BULK POLYMERS

CARLOS A. CHESTA[1] AND RICHARD G. WEISS[2]

[1]*Departamento de Química, Universidad Nacional de Río Cuarto, Río Cuarto, Argentina*
[2]*Department of Chemistry, Georgetown University, Washington, DC, USA*

13.1 INTRODUCTION

13.1.1 General Considerations

This chapter presents a critical review and a state-of-the-art assessment of the dynamics of the motions and reactions of geminal radical pairs that are generated by photolyses of guest molecules in glassy or rubbery polymer matrices. By virtue of their mode of creation, these radical pairs must reside initially (that is, at their moment of birth) in a common "cage"[1] within their polymer host. Specifically excluded from this treatment will be processes of biradicals (i.e., single molecular species containing two radical centers separated by a saturated group),[2] radicals that are not spin $^1/_2$, and radical pairs generated from the polymer itself. The dynamics of charge recombination[3] and electron or hole migration[4] processes in bulk polymers are not treated here as well.

A detailed description of bulk polymers as hosts for geminal radical pairs and their precursors is also beyond the scope of this chapter. For general sources of information about photochemical and photophysical processes in bulk polymers, we recommend the classic book by Guillet[5] as well as the book edited by Winnik,[6] the journal *Polymer Degradation and Stability* (incorporating the defunct journal,

Carbon-Centered Free Radical and Radical Cations. Edited by Malcolm D. E. Forbes
Copyright © 2010 John Wiley & Sons, Inc.

Polymer Photochemistry), and a number of other physical chemistry journals. The *Polymer Handbook*[7] is an excellent source of information about the macroproperties of bulk polymers. Because the design of the experiments described in this chapter requires that the incident radiation be absorbed by guest molecules directly, host polymers that are largely transparent in the near UV and visible regions, such as polyethylenes and polyacrylates, will be featured. Two aspects that are important in assessing the kinetics of radical motions in bulk polymers, diffusion[8] and relaxation properties,[9] of bulk polymers have been treated in detail for other purposes. It should be noted that history and hysteresis can complicate analyses of dynamic data in the glassy state.[10–12] In this regard, although probe molecules remain in the same cages for long periods in glassy polymers (such as in poly(alkyl methacrylate)s below their glass transition temperatures, Tg[13]),[14] they migrate much more rapidly in the rubbery states, but still several orders of magnitude more slowly than in low viscosity liquids. More than 50 years ago, Norman and Porter noted the differences in radical pair dynamics in "soft" and "hard" glasses.[15] That work presaged an understanding of the "hole" free volumes in the glassy and rubbery states of polymers. The motional freedom experienced by a radical pair will also depend acutely upon their shapes and volumes, and the "hole" free volumes of the host polymers in their glassy and rubbery states.[10d,16]

The reactions analyzed here (and used to generate the initial radical pairs in the polymer media) can be separated into two distinct categories: those that involve lysis of one molecule into two radicals (such as the Norrish Type I, photo-Fries, and photo-Claisen reactions) and those that require bimolecular processes in which a part of one molecule is abstracted by another (e.g., H-atom abstractions from a phenol or an amine by the lowest energy triplet state of benzophenone). Each reaction produces either singlet or triplet radical pairs and, thus, allows the influence of spin multiplicity on radical pair reaction rates to be separated somewhat from other influences, such as the natures of the polymer matrices and the radical structures. Different methods for extracting rates of processes for the radicals from both static and dynamic data will be discussed.

In principle, geminal radical pairs in polymeric matrices, $\overline{\text{A}\dot{...}\dot{\text{B}}}$, can select among a myriad of processes. The empirically observed major modes of reaction and motion, summarized in Scheme 13.1, include: (1) recombination to reform their precursor; (2) recombination to form a rearranged species, **C**, that is an isomer of the precursor; (3) disproportionation into two non-radical species, **D** and **E**; (4) escape from the cage and react with another entity (**P**) within the polymer matrix; (5) escape from the cage and then recombine discriminately or indiscriminately upon reencounters; (6) fragmentation within the cage to produce one or two different radicals that still exist as a geminal pair and then undergo one or more of the processes in (2), (3), (4), and (5).

The absolute and relative rates at which these processes occur depend on the molecular structures of the radicals in a pair, their initial spin multiplicity and rates of intersystem crossing, their rates of both rotational and translational diffusion[17] within and outside their initial cages, and the nature of the radical centers. In addition, they can be affected enormously by "external factors," such as temperature, the nature of

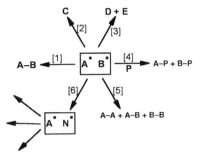

SCHEME 13.1 Important potential processes available to a geminal radical pair in a polymer cage. A unimolecular precursor, **A–B**, is assumed here for purposes of illustrating step [1].

the polymer matrix,[1,7] and applied magnetic fields.[18] Although almost all of these considerations apply to the dynamics of radical pairs in liquid isotropic media of low viscosity, as well, their influence may be very different in bulk polymers, and local anisotropy of the polymer microenvironments can lead to the suppression of processes that dominate in liquid media or to new processes that are not observed there.

13.1.2 Escape Probability of an Isolated, Intimate Radical Pair in Liquids and Bulk Polymers

Estimating the probability that an isolated, intimate radical pair, $\overline{\text{A} \ldots \text{B}}$, will either combine or separate prior to combining has been the subject of a great deal of experimental and theoretical effort.[19]

Diffusion models of geminate pair combination connect the time-dependent pair survival probability, $P(t)$, with the macroscopic properties of the host solvent.[20] Radicals are treated as spherical particles immersed in a uniformly viscous medium. The pair is assumed to undergo random Brownian movements that ultimately lead to either recombination or escape. The expression of $P(t)$ depends on the degree of sophistication of the theory chosen for analyzing the process. In the simplest theory, known as the exponential model, it is assumed that (a) the radical pair $\overline{\text{A} \ldots \text{B}}$ is born and reacts in a sphere of radius σ with a uniform (unimolecular) rate constant k_r and (b) radicals escape the sphere and become free with a time-independent (unimolecular) rate constant k_{dAB}. Under these simplifications, the survival probability is given by Equation 13.1.[19b]

$$P(t) = (1 - F_{cAB}) + F_{cAB} e^{-t/\tau} \qquad (13.1)$$

Here, $\tau = 1/(k_{dAB} + k_r)$ and $F_{cAB} = k_r/(k_{dAB} + k_r)$ represent the fraction of radical pairs that combine in the geminate cage (i.e., the cage factor). According to Equation 13.1, $P(t)$ decreases monoexponentially with time, approaching $(1 - F_{cAB})$. Approximation (a) of this model is acceptable, although most experimental evidence is that radicals are created at an (average) separation distance (r_0) that generally exceeds the contact distance σ. Approximation (b) has received much more

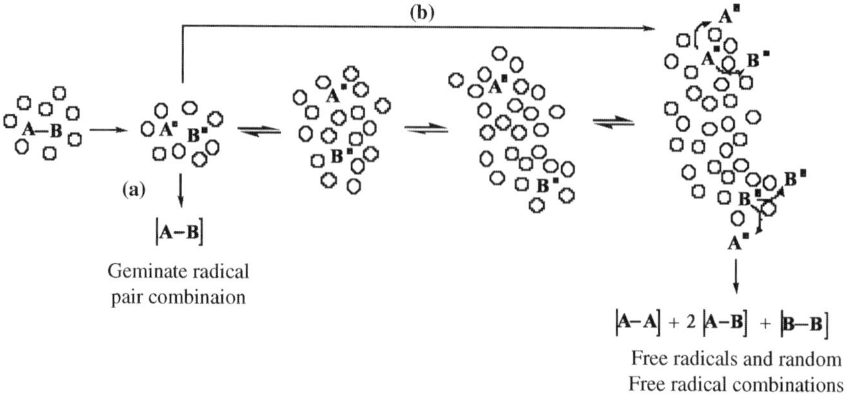

(b)

(a)

$|A–B|$

Geminate radical
pair combinaion

$|A–A| + 2\,|A–B| + |B–B|$

Free radicals and random
Free radical combinations

SCHEME 13.2 Simplified representation of the formation of a geminate radical pair, "in-cage" combination of the radical pair (a), progressive separation (and reencounters) of the radicals, and bimolecular random free radical combination reactions taking place at a longer time scale when the rates of diffusion and reaction of A˙ and B˙ are approximately equal. An important aspect of this scheme is that radicals A˙ and B˙ can reencounter each other and form **A–B** in "out-of-cage" reactions.

criticism because it implies that the separation of a radical pair occurs in a single step, from intimate contact to (operationally) infinite distance.

More rigorous treatments of the geminate combination also take into consideration the probability that the radicals of a pair escape from each other, reencounter in a later event, and finally recombine (Scheme 13.2). This model leads to time-dependent radical pair combination rates and, accordingly, they predict that $P(t)$ does not follow a simple exponential decay. For instance, even for the simple case of a contact-start recombination process ($r_0 = \sigma$), the survival probability is a complex function as shown in Equation 13.2[21]:

$$P(t) = (1 - F_{cAB}) + F_{cAB}\,e^{t/t_0}\,\mathrm{erfc}\,\sqrt{t/t_0} \qquad (13.2)$$

where "*erfc*" is the complementary error function and the relaxation time of the process, t_0 (Eq. 13.3), is expressed in terms of D (the mutual diffusion coefficient of the radicals), σ and $F_{cAB} = k_r/(k_{dAB} + k_r)$. In this model, k_r represents an intrinsic (bimolecular) rate constant as defined in the partially reflecting boundary condition[22] and k_{dAB} is approximated by the Smoluchowski equation[23]: $k_{dAB} = 4\pi\sigma D$. According to Equation 13.2, $P(t)$ undergoes a sharp initial decrease followed by a slow decay that approaches asymptotically $(1 - F_{cAB})$. This behavior clearly differs from that predicted by the exponential model (Eq. 13.1).

$$t_0 = \frac{\sigma^2}{D}(1 - F_{cAB})^2 \qquad (13.3)$$

In principle, the cage factor can be estimated from stationary (F_c) or time-resolved (F_{cAB}) experiments. As will be shown later, only F_{cAB} represents the actual fraction of cage reaction in all cases. Radical pairs generated via thermal or photochemical decomposition of diazenes,[24–26] ketones,[27–29] disulfides,[30] and other substrates[31] have been used to study cage effects in liquids or constrained media. Asymmetrically substituted compounds allow estimation of the fraction of radicals reacting in the geminate cage by analyzing the product distribution according to Equation 13.4. This equation relates product yields to (what are thought to be) the fraction of "in-cage" and "out-of-cage" combinations by the $\dot{A}...\dot{B}$ radical pair.

$$F_c = \frac{[AB] - [AA] - [BB]}{[AB] + [AA] + [BB]} \qquad (13.4)$$

Implicit in the form of Equation 13.4 are the assumptions that **AA**, **AB** and **BB** are the only products, that they are always obtained 1:2:1 ratios of from out-of-cage combinations, and that exclusive formation of **AB** signifies complete in-cage reaction of the radical pairs. The limitations of Equation 13.4, particularly when applied to the evaluation of the cage factor in constrained or microphase-compartmentalized media, have been discussed recently.[29] In symmetrically substituted compounds where the geminate radicals produced upon lysis have the same structures, the addition of radical scavengers[32] or time-resolved detection of the radicals is needed for the estimation of F_{cAB}.

Several studies directed to the estimation of the relevant parameters that control the kinetics of radical pair combination processes using time-resolved techniques have been reported. Results have been interpreted generally using expressions such as Equations 13.1 or Equation 13.2, as well as more sophisticated theoretical treatments. However, obtaining dependable absolute (or even relative) values of k_r and k_{dAB} has been a formidable task.

As pointed out above, the possibility of observing the actual competition between combination and escape processes depends on numerous factors, of which the most important are: the initial multiplicity and separation of the radical pair, the mass and size of the radicals, the viscosity of the medium, and, in some cases, specific interactions between the radicals and their medium. In low viscosity media, only combination and escape processes from singlet radical pairs, occurring on the pico-/femto-second time scale, can be observed.[30,31] As noted, $F_{cAB} = 1/(1 + k_{dAB}/k_r)$ can be obtained from the extrapolated value of $P(t \rightarrow \infty)$. However, this apparently simple procedure has several drawbacks. For example, on the pico-/femto-second time scales, deconvoluting $P(t)$ from the initiating (pump) and interrogating (probe) light pulses becomes difficult, making the estimation of the actual value of the initial concentration of radical pairs inaccurate and, consequently, the verification of the exact time dependence of $P(t)$ and the estimation of the F_{cAB} values derived from such experiments suspect as well. In addition, in order to fit the experimental data to Equation 13.1 or Equation 13.2, the decay curves have to be observed over all time (i.e., until the "plateau-like" region of $P(t)$ is reached). However, this is not a straightforward task because the plateau is reached asymptotically, and what may be considered to be the plateau value of $P(t)$ in a "short" time window is usually not. The scenario can be complicated further by the interference of (bimolecular) free radical combination reactions that become increasingly important in the long time scale. Nevertheless, reasonably good correlations between the values of $(1 - F_{cAB})$ and the inverse of the solvent viscosity $(1/\eta)$ have been found.[33,34]

Few time-resolved studies involving triplet radical pairs have been reported.[29a,35–38] Because the direct interaction of a triplet radical pair to give singlet-state products is spin-forbidden, radicals in low viscosity media can diffuse massively into the bulk, making $F_{cAB} = 0$. Thus, competition between combination and escape processes in geminate triplet radical pairs can be observed only in very viscous media, such as the environments provided by polymers in their in glassy or rubbery states. Although many of the phenomena experienced by radical pairs in liquids of low viscosity and in polymers are (in principle) similar, the anisotropy the polymer microenvironments and the large restraints to the rotational reorientation of the radicals within the "reaction cages" in the polymeric matrices introduce new and intriguing (theoretical and experimental) challenges. For example, when the radii of the diffusing radicals are *not* much larger than that of the solvent, as is the case of a polymer in which motions of chain segments larger than the solutes constitute the dominate relaxation mode, classical diffusion theory (the basis for Equations 13.1 and 13.2) is not entirely suitable and alternative approaches must be considered.[*]

Hence, although following the history of combination/escape processes of radical pairs based either on conventional laser flash photolysis or time-resolved ESR measurements[41,42] is quite tempting, obtaining reliable results from analyses of the experimental kinetic decay curves using diffusion theory (Equations 13.1 and 13.2) is not simple. This statement is particularly true when the methodology is applied to the study of reactions in complex matrices such as a polymer media. Frequently, more reliable estimations of absolute values of k_r, k_{dAB}, and F_{cAB} can be achieved using a different approach that is more directly related to the evolution and loss of singular radical species in mixtures of them. Examples of both types of analyses are shown in this chapter.

13.2 SINGLET-STATE RADICAL PAIRS FROM IRRADIATION OF ARYL ESTERS AND ALKYL ARYL ETHERS

13.2.1 General Mechanistic Considerations From Solution and Gas-Phase Studies

The important mechanistic steps leading to and proceeding from singlet-state radical pairs are shown in Scheme 13.2.[43]

---

[*] The diffusion coefficients associated with translational motions when the radii of the diffusing radicals are *not* much larger than that of the solvent are expressed more accurately by $D = kT/6\pi r \eta_m$ (where r is the radius of the diffusing radical assuming a spherical shape and η_m ($= f x \eta$) is the microviscosity. The value of f, the microfriction factor, can be calculated[39] or taken equal to D_{SE}/D_{exptl}, the ratio between the Stokes–Einstein diffusion coefficient (that considers van der Waals volumes, but not interstitial volumes) and the experimentally measured diffusion coefficient, D_{exptl}. As will be discussed later, these relationships appear to hold even in some polymer matrices.

13.2.1.1 *Photo-Fries Reactions of Aryl Esters*   Meyer and Hammond were the first to show that the first reactive step in the photo-Fries reactions of structurally simple aromatic esters is homolysis, leading to an aryloxy and a bent acyl radical in which the odd electron is somewhat localized in an sp^2-like orbital.[44] Their experiments and subsequent ones by Miranda and coworkers[45] demonstrate unequivocally that the mechanisms leading to the rearranged products, acylphenols, are not concerted. Both experimental data (unsuccessful attempts to sensitize the rearrangements with triplet sensitizers or to quench it with triplet quenchers)[46,47] and calculations[48] eliminate the lowest energy triplet and strongly suggest the first excited singlet as the immediate precursor of the reactive species. The observation that the rate of acetyl/1-naphthoxy radical pair recombination in acetonitrile at room temperature is $>10^{10}\,s^{-1}$[49] is further evidence for the singlet nature of the precursor in this case, 1-naphtyl acetate (**1-NA**).

At the moment of their "birth" (i.e., when the excited singlet states of the aryl esters undergo lysis), the geminal radical pairs must be in positions that make their shape similar to that of their precursor ester: the radical center of the acyl part is very near the oxygen atom of the aryloxy part, and all subsequent diffusion of the two species starts from this orientation. As a result, addition of an acyl radical to its aryloxy partner is favored *spatially* at the nearer adjacent (*ortho*) position(s) than at the more distant *meta* and *para* positions. However, the ability of the acyl radical to add to each of the positions of an aryloxy radical is expected to depend on the energies of the adducts[48,50] (which are keto intermediates that enolize thermally with time to the eventual products[51]; Equation 13.5 shows an example of the keto intermediate for acyl addition to the *ortho* position of phenoxy). The spin densities at the 2-, 4-, and oxy-positions of 1-naphthoxy (and of phenoxy) from ESR measurements[52] and HF/6-31G* level calculations[53] (in parentheses) are shown in Fig. 13.1. Those for the 2-naphthoxy radical by ESR measurements[54] and MNDO-UHF calculations[55] are collected in Table 13.1. These considerations explain why the yields of products from addition at the "*meta*" positions are very low.

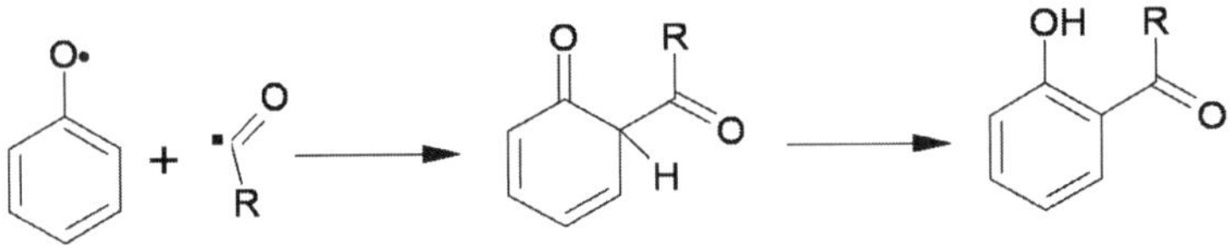

(13.5)

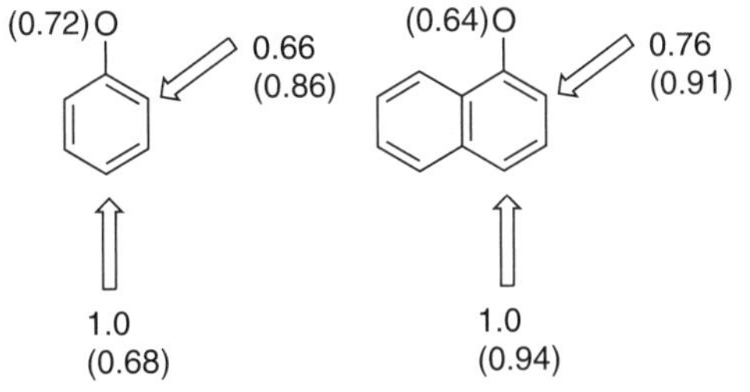

FIGURE 13.1   Spin densities at different positions of phenoxy and 1-naphthoxy radicals. Values at the *meta* and *meta*-like positions are not shown because they are very low. Reproduced with permission from Ref 43. Copyright (2001) Japanese Photochemistry Association.

TABLE 13.1 Spin Densities (Atomic Units) of Carbon Atoms and Oxygen of the 2-naphthoxy Radical from MNDO-UHF Calculations and ESR Data

position	MNDO-UHF	ESR
1	0.399	0.454
3	0.055	0.061
4	0.005	0 (assumed)
5	0.004	0.062
6	0.134	0.228
7	0.000	0.051
8	0.100	0.181
oxygen	0.110	

Decarboxylation of excited singlets of aromatic esters (Eq. 13.6) is a concerted process,[56] and it can account for a significant fraction of the reaction by appropriately substituted aryl esters, especially in bulk polymers and other media that are capable of imposing conformational constraints on guest molecules.[57] Because photoinduced decarboxylation occurs before lysis of the aryl esters, it does not influence the rates at which the $\mathbf{A}^{\bullet}\,\mathbf{B}^{\bullet}$ singlet pair react. For that reason, the relative yields of decarboxylation products need not be considered in analyses of the radical pairs unless their formation precludes sufficient radical pair production for their easy direct or indirect detection.

$$(13.6)$$

As indicated in Scheme 13.1, radicals that escape from their cage of origin can undergo a myriad of reactions, but the dominant one for the aryloxy radicals in media with potential H-atom donors is production of arylols (step 4). Thus, in the absence of radical pair disproportionation (step 3) and reencounters of aryloxy and acyl as in step 5 (or their decarbonylated counterparts, alkyl radicals; see below), the yield of arylol-type product can be related directly to the amount of cage escape suffered by the geminal radical pair. Alternatively, the absence of **A–A** or **B–B** type products in step 5 (or combination of radicals derived from $\mathbf{A}^{\bullet}$ and $\mathbf{B}^{\bullet}$) constitutes evidence against the escape of geminal radical pairs from their initial cages.

Reaction of all of the radical pairs within their initial cages does not preclude one or both of the radicals from undergoing a structural change (N. B., step [6] in Scheme 13.1). In the case of photo-Fries reactions, the most commonly encountered structural change is loss of carbon monoxide (CO) from the acyl radical, leading to formation of an alkyl radical. The rates of decarbonylation of acyl radicals have been measured for a wide variety of acyl structures as a function of the medium viscosity and polarity and at different temperatures. The results indicate that the rate of loss of CO is not affected appreciably by solvent viscosity,[29a,58] but it is profoundly by the acyl structure[59] and somewhat by solvent polarity.[58a,59b,60] Based upon data for decarbonylation of arylacetyl radicals,[59b] the rates for loss of CO from alkanoyl radicals are expected to be a factor of about 3 slower in a very polar solvent, acetonitrile ($\varepsilon \sim 36$) than in the low polarity solvent, n-hexane ($\varepsilon \sim 1.9$). Because the bulk polarities of the polymers discussed in this chapter, polyethylene ($\varepsilon \sim 2.3$), poly(vinyl acetate) ($\varepsilon \sim 3.5$), and poly(methyl methacrylate) ($\varepsilon \sim 2.8$),[7] are much closer to that of hexane or the gas phase ($\varepsilon = 1.0$) than that of acetonitrile, it is reasonable to use the decarbonylation rates in the gas phase or in alkane solutions throughout.

13.2.1.2 *Photo-Claisen Reactions of Alkyl Aryl Ethers*   Photo-Claisen reactions of alkyl aryl ethers[61] proceed via a homolytic cleavage of the aryloxy–alkyl bond, yielding an aryloxy/alkyl geminal radical pair initially[62,63] that can undergo all of the processes shown in Scheme 13.1. Interestingly, photolysis of an aryl ester and subsequent decarbonylation can produce a radical pair that is structurally the same as the pair made directly upon photolysis of the analogous alkyl aryl ether. As will be discussed in some detail, the mechanistic consequences of the immediate (from an alkyl aryl ether) and delayed (from an aryl ester) formation of an aryloxy/alkyl radical pair can yield interesting dynamic information about microdiffusional processes.

The multiplicity of the excited state of the alkyl aryl ether responsible for the direct formation of aryloxy/alkyl geminal radical pairs is not as clearly defined as in the case of the aryl esters. For example, although allyl phenyl ether can react from its excited singlet or excited triplet states, the photoproduct distributions from each are very different,[64] and direct irradiation initiates reaction from only the excited singlet state.[62,64] Although reaction occurs from both the excited singlet and triplet states of dimethylallyl 1-naphthyl ether when it is irradiated directly,[63] very little (if any) lysis from the triplet state is found when benzyl 1-naphthyl ether is irradiated directly.[65] Thus, reaction by the triplet state of benzyl 1-naphthyl ether can be sensitized, but the rate of intersystem crossing from the excited singlet state is much slower than the sum of the rates for its deactivation, including that for lysis.

13.3 PHOTO-REACTIONS OF ARYL ESTERS IN POLYMER MATRICES. KINETIC INFORMATION FROM CONSTANT INTENSITY IRRADIATIONS

Control over the course of the rearrangements of aryl esters can be achieved when the host matrix can interact with the ester and its intermediates through electrostatic or H-bonding interactions. Zeolites[66] and Nafion membranes[67] are two polymeric matrices in which such studies have been conducted. As mentioned above, the positions of the radical centers of the acyl radicals are known to be very near the oxygen atoms of their aryloxy partners at the moment of "birth" of the radical pairs from the excited singlet state of an aryl ester in all media; all subsequent diffusion commences from this arrangement. The amount of acyl migration to positions adjacent to the aryloxy oxygen atom is enhanced and escape of the radicals from their initial cage is greatly suppressed in Zeolites, Nafion membranes, and bulk polymers in their rubbery states. Escape is also suppressed in the glassy state of poly (methyl methacrylate) (**PMMA**), where diffusion of small guest molecules is slowed much more than in the melted (rubbery) state, and irradiation results in the creation of a nonrandom distribution of "quenchers" of additional reaction[13a] that we suggest may be the keto intermediates that form upon recombination of the radical pairs and eventually enolize to yield the isolated acylarylol rearrangement products. Regardless, the inability of the radicals to escape from their initial cages and the inability of the "quenchers" to diffuse over long time periods are manifestations of the difference between bulk polymeric and normal liquid solvent hosts. To understand these differences quantitatively and to exploit them, it is necessary to analyze the kinetics of the processes responsible for the reactions.

13.3.1 Relative Rate Information from Irradiation of Aryl Esters in Which Acyl Radicals Do Not Decarbonylate Rapidly

As mentioned above, the general preference for the *ortho*- and *para*-like rearrangement products can be understood on the bases of the heat of formation and spin density arguments. It is not necessary for the guest aryl ester to be imbedded in a glassy matrix or to be exposed to electrostatic and H-bonding interactions for a polymer matrix to exert control over the motions of an aryloxy/acyl radical pair within a polymeric cage that affect profoundly the preference for *ortho* or *para* combinations.

For example, irradiation of either 2-naphthyl acetate (**2-NA**) or 2-naphthyl myristate (**2-NM**) in the rubbery state of a low density polyethylene (**LDPE**) or a high density polyethylene (**HDPE**) yields a variety of in-cage rearrangement products and 2-naphthol (**2-NOL**), a cage-escape product (Table 13.2).[68,69] Both **LDPE** and **HDPE** consist of amorphous and crystalline regions; **LDPE** has a larger volume

TABLE 13.2   Photoproduct Distributions from Irradiations of 2-NA and 2-NM in Hexane and in PE Films at Room Temperature

substrate	medium	1-AN	3-AN	6-AN	3-AN/6-AN	2-NOL[a]
2-NA	hexane	57 ± 2	16 ± 1	6 ± 3	9.5	21 ± 2
	LDPE(u)	38 ± 2	17 ± 1	22 ± 2	0.77	23 ± 3
	LDPE(s)	31 ± 2	21 ± 1	17 ± 2	1.2	31 ± 5
	HDPE(u)	21 ± 4	29 ± 5	21 ± 3	1.4	29 ± 5
	HDPE(s)	20 ± 4	66 ± 4	2	>33	12
2-NM	hexane	43 ± 2	32 ± 2	8 ± 1	4.0	15 ± 6
	LDPE(u)	0	85 ± 2	10 ± 2	8.9	5
	LDPE(s)	0	>90	1	>90	10
	HDPE(u)	0	83 ± 2	12 ± 2	6.9	5
	HDPE(s)	0	94 ± 3	1	>94	5

[a]Limit of detection was 5%.

fraction of amorphous parts than **HDPE**. In general, guest molecules reside only in the amorphous parts and in "interfacial regions" that are at the borders between the crystallites and amorphous chains. Equations 13.7 and 13.8 summarize the relevant reactions and products.

$$(13.7)$$

$$(13.8)$$

Stretching the **LDPE** film, a *macroscopic* perturbation on the matrix, resulted in an interesting *microscopic* change in the guest cages that causes the some suppression of the formation of the most globularly-shaped product, **1-AN**. Stretching the **HDPE** film also suppresses strongly the formation of **6-AN**, a photoproduct whose formation requires much more translational acetyl radical diffusion within the cage than does formation of **3-AN**, a similarly shaped product. The effects of increasing the size of the acyl radical (and, thereby, decreasing it rate of diffusion), as well as film stretching, are demonstrated by results from irradiation of 2-naphthyl myristate (**2-NM**). Even in unstretched **LDPE** and **HDPE**, neither the cage-escape product **2-NOL** nor the bulky rearrangement product **1-AN** is detected. The absence of **2-NOL** from irradiation of **2-NM** marks the acetyl radical rather than the 2-naphthoxy radical as the species that escapes from the cages when **2-NA** is the guest molecule. Stretching the films increases the relative yields of the more linear product from **2-NM**.

The influence of irradiating 1-naphthyl myristate (**1-NM**) in **LDPE** or **HDPE** is even more dramatic (Table 13.3). In hexane solutions, two photoproducts, 2-myristoyl-1-naphthol and 4-myristoyl-1-naphthol are obtained in a 1.3/1 ratio[68b]; only one photoproduct, 2-myristoyl-1-naphthol (**2-AN**), is obtained in unstretched **LDPE** and **HDPE**.[68b,70]

The irradiations of **1-NA** and **1-NM** have also been investigated above (at 323K) and below (273K) the T_g of two poly(vinyl acetate) films (**PVAC-83** and **PVAC-140**; average molecular weight 83K and 140K, respectively).[71] The results in Table 13.4 are

TABLE 13.3   Photoproduct Distributions from Irradiations of 1-NA and 1-NM in Hexane and Unstretched (u) and Stretched (s) PE Unstretched (u) and Stretched (s) Films at Room Temperature

substrate	medium	2-AN	4-AN	2-AN/4-AN	1-NOL[a]
1-NA	hexane	28.0 ± 0.4	31.2 ± 0.4	0.9	40.8 ± 0.2
	LDPE(u)	77.8 ± 0.5	15.0 ± 0.3	4.2	8.2 ± 0.4
	LDPE(s)	77.8 ± 0.4	15.9 ± 0.5	4.9	6.3 ± 0.4
	HDPE(u)	78.1 ± 0.2	16.1 ± 0.4	4.9	5.8 ± 0.3
	HDPE(s)	76.8 ± 0.2	13.1 ± 0.4	5.9	10.1 ± 0.4
1-NM	hexane	4.0 ± 0.1	3.0 ± 0.1	1.3	<2
	LDPE(u)	100	<1	>100	<2
	LDPE(s)	100	<1	>100	<2
	HDPE(u)	100	<1	>100	<2

[a] Detection limit was 2%.

representative. They indicate that the fate of the radical pairs depends on the state of the **PVAc** as well as the specific aryl ester being irradiated. Thus, the **2-AN/4-AN** ratios for acetyl/1-naphthoxy radical pairs from **1-NA** are very high in the glassy states and decrease significantly in the rubbery states (where polymer chain mobility is increased and acetyl radical motions to the more distal 4-position of the 1-naphthoxy radical are facilitated). The relative yields of **NOL** are surprisingly high from irradiations in both states of the polymer and their formation may be associated with the presence of abstractable H-atoms along the polymer chains. Regardless, the **2-AN/4-AN** ratios for myristoyl/1-naphthoxy radical pairs from **1-NM** exhibit a completely different dependence on the state of the polymer. Their ratios *are higher* in the glassy state than in the rubbery one! A possible explanation for this result is that the glassy polymer matrix controls the ground state conformations of **1-NM** and subsequent motions of myristoyl radicals with respect to their 1-naphthoxy partners. The qualitative observation that the efficiency of the photoreactions of **1-NM** in the glassy states is very low may be important here as well—reaction may be occurring in cages that force **1-NM** molecules into conformations that are not representative of the vast majority of molecules.[13a,29b]

TABLE 13.4   Photoproduct Distributions from Irradiations of 1-NA and 1-NM in PVAc Films

Substrate	PVAc type	T (K)	2-AN	4-AN	NOL	[2-AN]/[4-AN]
1-NA	**PVAc-83**	273	65.5 ± 3.7	2.3 ± 0.5	32.2 ± 3.8	28
		323	80.2 ± 3.2	4.0 ± 0.4	15.8 ± 3.3	20
	PVAc-140	273	90.9 ± 0.1	1.6 ± 0.1	7.5 ± 0.2	57
		323	61.3 ± 5.7	2.5 ± 0.6	36.1 ± 5.1	24
1-NM	**PVAc-83**	273	66.3 ± 2.7	32.0 ± 2.5	1.7 ± 0.2	2.1
		323	83.3 ± 2.4	13.2 ± 2.8	3.5 ± 0.4	6.3
	PVAc-140	273	81.2 ± 1.2	16.9 ± 1.7	1.9 ± 0.6	4.8
		323	87.9 ± 0.3	8.3 ± 0.7	3.8 ± 0.3	12

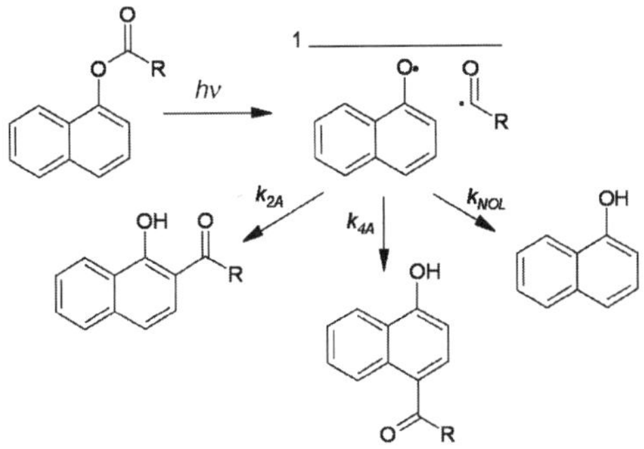

SCHEME 13.3 A simplified mechanism for photo-Fries reactions of 1-naphthyl esters, showing the central role of the initially formed singlet radical pairs.

In each of the aryl esters discussed above, the acyl radical formed upon lysis of the excited singlet state of the ester loses CO very slowly at the temperatures of the irradiations. At 296K, the rates of loss of CO by acetyl and propanoyl radicals in the gas phase are 4.0 and $2.1 \times 10^2 \, s^{-1}$, respectively.[44] As a result, no products from decarbonylation and rearrangement are expected[49] (or have been found) when either of the **NA** or **NM** isomers is irradiated in liquid solvents or bulk polymers, and kinetic information from photoproducts alone is limited to relative rates of radical pair processes (Scheme 13.3). For example, if no Fries products from **1-NA** or **1-NM** emanate from reencounter of radicals that have escaped from their initial cages, $[2\text{-}AN]/[4\text{-}AN]/[1\text{-}NOL] = k_{2A}/k_{4A}/k_{NOL}$.

13.3.2 Absolute and Relative Rate Information from Constant Intensity Irradiation of Aryl Esters in Which Acyl Radicals Do Decarbonylate Rapidly

When the rate of decarbonylation by an acyl radical is comparable to the rates of in-cage combination by the acyl/aryloxy radical pair and no Fries- or Claisen-like (from in-cage combination of alkyl/aryloxy radical pairs after decarbonylation) rearrangement products are derived from radicals that reencounter after escaping from their initial cages, a great deal of additional kinetic information can be obtained. Because no acyl radicals of which we are aware lose CO at rates that are comparable to the rates at which geminal singlet acyl/aryloxy radicals combine in low-viscosity liquid solutions[44,49] and the rates of decarbonylation cannot be enhanced significantly by solvent effects, the lifetime of the radical pairs in their cages must be prolonged in order to allow decarbonylation to precede radical pair combination and formation of the needed Claisen-like products (Eq. 13.9). Irradiating aryl esters such as **1a** and **1b** in bulk polymers is a convenient method to accomplish this goal; placing the esters in a polymer matrix does not affect the "clock," the rate of loss of CO, but placing the radical pairs in "viscous space" does slow their rates of in-cage combination by a factor of 10^2–10^3. Using this approach and aryl esters whose lysis yields substituted arylacetyl radicals with rates of decarbonylation from $\sim 10^6$ to $>10^8 \, s^{-1}$ at 296K in low-viscosity alkane solutions,[44]

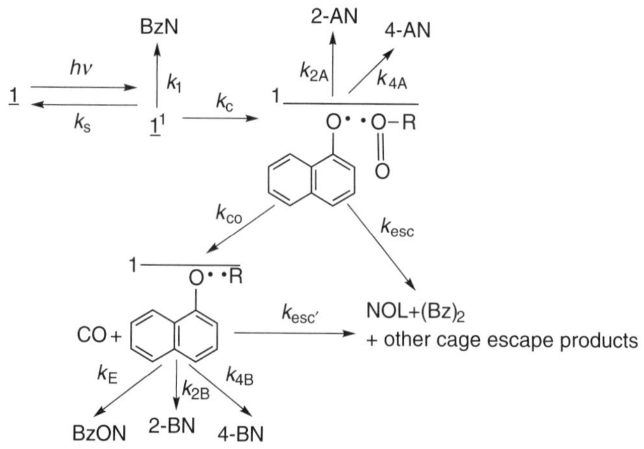

SCHEME 13.4 Reprinted with permission from Ref 47. Copyright (2000) Elsevier Science Ltd.

it is possible to form Claisen-like products can be formed.[47]

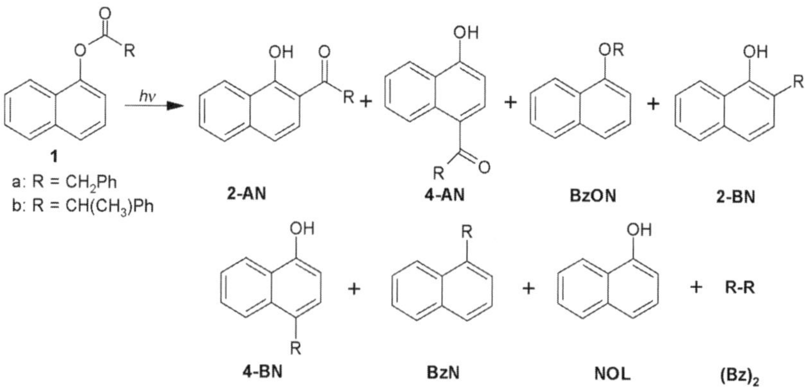

$$(13.9)$$

Starting from the mechanistic Scheme 13.4, it is possible to apply Equations 13.10 and 13.11 (below), provided some verifiable assumptions hold:

1. The fraction of the acyl/naphthoxy radical pairs that escape from their initial cages (k_{esc}) should be known. That fraction can be approximated from the yields of $(Bz)_2$, the dimerization product of the decarbonylated radical. In practice, the fractions of **NOL** and $(Bz)_2$ arising from the k_{esc} and $k_{esc'}$ pathways cannot be determined easily. As long as the relative yield of $(Bz)_2$ is very small, the values for k_{2A} and k_{4A} calculated from Equations 13.10 and 13.11 are overestimated by amounts that are usually less than the experimental error in determining the relative yields of the **2-AN**, **4-AN**, **2-BN**, **4-BN**, **BzON**, and

NOL. The **NOL**$'$ in Equations 13.10 and 13.11 is the naphthol formed from the $k_{\text{esc}'}$ step.

2. There should be no return of CO to add to an alkyl radical (i.e., the reverse of $k_{-\text{CO}}$). Ryu and coworkers[72] have demonstrated that the activation barrier for such a process is prohibitively high—6.0 kcal/mol in benzene solution for a primary radical, and it must be higher for a benzylic or aryloxy radical—to allow it to compete with the other in-cage processes in Scheme 13.4, even when the medium is a polymer matrix. In addition, it is possible to estimate the translational distance (d) traveled by a molecule of expelled CO during the periods required for a singlet radical pair to combine. Using the simple expression, $d = (2Dt)^{1/2}$,[73] and the diffusion coefficient D for CO in polyethylene,[74] the distances are <5 Å in the first ns and 100 Å during 100 ns. If the radius of a cage is approximately the same as the van der Waals radius of a simple aryl ester (such as those discussed here), <10 Å, a large fraction of the CO should be able to escape before a benzylic radical combines with an aryloxy partner.[75]

3. The absolute rate constant for loss of CO from the acyl radical in the cage occupied by the other radical, $k_{-\text{CO}}$, must be known. Laser flash photolyses carried out in both liquid solutions[58–60] and bulk polymers[29a] on dibenzyl ketone and derivatives of it, yielding triplet benzylic/arylacetyl radical pairs, indicate that the presence of a benzylic radical within a cage has no discernible effect on the rate of loss of CO from the arylacetyl partner. Thus, where values of $k_{-\text{CO}}$ are not known in a polymer, rate constants measured in liquid media of similar polarity may be used.

Rate constants for radical combinations for several arylacetyl/aryloxy radical pairs in unstretched (u) and stretched (s) polyethylene films of different crystallinities[43] and in **PVAc-83** above Tg[76] have been calculated in this way over a range of temperatures.

In polyethylene, the values for k_{2A} are in the range of 10^8 s^{-1} and those for k_{4A} are about one order of magnitude lower, ca. 10^7 s^{-1}; rate constants at the lowest and highest temperatures examined and activation energies (E_{2A} and E_{4A}) for the radical pair combinations from irradiations of **1b** in an **LDPE** and an **HDPE** film are collected in Table 13.5. The preexponential factors associated with the activation energies were not reported because the rate constants were not as precise as is necessary for the relatively small temperature ranges over which the data were collected.

$$k_{2A} \approx k_{-\text{CO}}[\text{2-AN}]/\{[\text{2-BN}] + [\text{4-BN}] + [\text{BzON}] + [\text{NOL}']\} \qquad (13.10)$$

$$k_{4A} \approx k_{-\text{CO}}[\text{4-AN}]/\{[\text{2-BN}] + [\text{4-BN}] + [\text{BzON}] + [\text{NOL}']\} \qquad (13.11)$$

At room temperature, film stretching reduces both k_{2A} and k_{4A} of the radical pair from **1b** by a factor of 2–3. The consequences of film stretching on the microscopic properties of polymer films containing both amorphous and crystalline parts, such as **LDPE** and **HDPE**, explain why the rate constants decrease. Although the relative

TABLE 13.5   Rate Constants and Activation Energies for Combination of 2-phe-nylpropanoyl/1-naphthoxy Radical Pairs and Relative Rate Constants (at 295 K) for Combination of 1-phenylethyl/1-naphthoxy Radical Pairs from Irradiation of 1b in Unstretched (u) and Stretched (s) Polyethylene Films

polyethylene	T (K)	$10^8 k_{2A}$ (s^{-1})	$10^7 k_{4A}$ (s^{-1})	k_{2A}/k_{4A}	E_{2A} (kcal mol^{-1})	E_{4A} (kcal mol^{-1})	$k_{2B}/k_{4B}/k_E$
LDPE (u)	278	0.51	0.13	39.2	5.1 ± 0.2	8.3 ± 0.7	1.0/1.9/1.4
	333	14.3	24.8	5.8			
LDPE (s)	278	0.23	<0.02	>115	10.2 ± 1.4	12.8 ± 0.7	1.0/3.0/1.1
	333	10.4	16.9	6.2			
HDPE(u)	265	0.22	0.24	9.2	10.7 ± 1.5	11.5 ± 0.9	1.0/4.5/2.6
	333	15.5	25.2	6.2			
HDPE(s)	265	0.11	0.15	7.3	13.1 ± 1.7	11.6 ± 2.5	1.0/2.5/1.1
	333	1.3	1.1	11.8			

yields of **NOL** decrease somewhat as a function of increasing polyethylene crystal-linity or upon stretching, no distinct trend has emerged—the relative yields are probably controlled more by conformational constraints placed on the radical pair within their cage than by the ability of a radical to escape from it.[77]

Film stretching causes a net translocation of guest molecules from less constraining cages in amorphous regions to cages in more constraining interfacial ones (at the borders between crystallite lateral surfaces and amorphous chains)[78] in a cage and a decrease in the average "hole" free volume.[68b] Both factors reduce the ability of guest molecules (or radicals derived from them) to diffuse and are instrumental in explaining the reductions in both the rate constants for in-cage radical pair combinations and the relative yields of **NOL** that attend stretching the polymer host. Other changes that are germane to reactions of radical pairs—breaking the spherulitic crystal assemblies into rod-like pieces that become oriented along the direction of the film extension and increasing in the degree of crystallinity, especially for the **LDPE**[78]—also occur when a polymer is stretched. As a result, stretching **LDPE** has a much greater influence on the activation energy for combination of the 2-phenylpropanoyl/1-naphthoxy radical pair than does stretching **HDPE**. Additionally, the activation energies in unstretched **LDPE** are significantly lower than in unstretched **HDPE** because a larger fraction of **1** molecules reside in the more constraining interfacial sites in the more crystalline **HDPE**,

Contrary to the much higher rate constants for the formation of **2-AN** than **4-AN**, the *relative* rate constants for formation of **2-BN**, k_{2B}, are *lower* than those for formation of **4-BN**, k_{4B}! In addition, a third combination process, leading to **BzON**, must be considered. It is analogous to recombination by the phenylacyl/1-naphthoxy radical pairs leading to reformation of **1**; the rate of that process has not been determined. Regardless, the relative yields (and, therefore, the relative rate constants in Table 13.5) for formation of **2-BN**, **4-BN**, and **BzON** mirror the relative spin densities of 1-naphthoxy (Figure 13.1). This is the result expected if the 1-phenylethyl/1-naphthoxy radical pairs are able to explore completely the space within their cages

TABLE 13.6 Rate Constant Ratios (k_{2B}/k_{4B}) from Irradiations of 1a, 3a, 2a and 3c and k_{2A}/k_{4A} from 2a (in parentheses) in Unstretched (u) and Stretched (s) PE Films Under a Nitrogen Atmosphere at 295 K

polyethylene	from **1a**	from **3a**	from **2a**	from **3c**
LDPE (u)	0.75	2.1	1.8 (1.1)	0.9
LDPE (s)	0.79	2.0	1.3 (1.3)	1.1
HDPE (u)	0.70	2.2	0.7 (0.9)	0.6
HDPE (s)	0.90	2.1	0.5 (0.9)	0.8

before combining. The k_{2B}/k_{4B} rate constant ratios from irradiations of **1a** in unstretched and stretched **LDPE** and **HDPE** are also less than one (Table 13.6).[65] However, the relative yields of **BzON** are <0.02%. They indicate that the greater reactivity of the benzyl radical (from **1a**) than the 1-phenylethyl radical (from **1b**) is less selective in its reactions with 1-naphthoxy and that the longer period for decarbonylation of a phenylacetyl radical (than a 2-phenylpropanoyl radical) places the benzyl radical farther than the 1-phenylethyl radical from the oxygen atom of their 1-naphthoxy partners at the moment of their birth.

The k_{2A} and k_{4A} from **1b** in **PVAc-83** at 323 K (i.e., well above Tg) as calculated from Equations 13.10 and 13.11 are $(0.8–1.3) \times 10^9$ and $(2.0–3.3) \times 10^7\,s^{-1}$, respectively;[76] the range limits are determined by assuming that either all or none of the **NOL** emanates from cage escape at the 1-phenylethyl/1-naphthoxy stage.[47] Although it is possible to make similar calculations from data in the glass phase at 278K, the validity of the numbers is questionable because the efficiency of the irradiations is very low, indicating that the photochemistry may be occurring in sites that do not represent the bulk.[13a,29b] The k_{2A}/k_{4A} ratio, ~40, is much larger than the 5.8 ratio for unstretched **LDPE** and the 6.2 ratio for unstretched **HDPE** at 333K.[47] Also, the $k_{2B}/k_{4B}/k_E$ ratios in **PVAc-83** at 323K are 1/0.35/2.4.[76] The most dramatic difference between these ratios and those reported in Table 13.5 for irradiations in the polyethylene films is the much lower relative rate of formation of **4-BN**. The most economical explanation for these observations is that translational motions of the 2-phenylpropanoyl and 1-phenylethyl radicals (as well as of their 1-naphthoxy partner) are impeded by attractive interactions of the radical centers with ester groups of the polymer.

13.4 RATE INFORMATION FROM CONSTANT INTENSITY IRRADIATION OF ALKYL ARYL ETHERS

At room temperature, the rate constant for loss of CO from a 2-phenylpropanoyl or a phenyacetyl radical in a low polarity solvent, isooctane, is 4.2×10^{7} [59a] or $8.1 \times 10^6\,s^{-1}$,[60d] corresponding to a lifetime of 24 or 120 ns, respectively. Those periods are sufficient in a polyethylene host for the acyl/1-naphthoxy radicals from both **1a** and **1b** that eventually lose a molecule of CO (i.e., that neither regenerate the initial ester nor combine to form one of the rearrangement products, **2-AN** or **4-AN**) to diffuse within their cages over distances that encompass the cavity dimensions (see above). As a result, the positions of the geminal benzylic and 1-naphthoxy radicals with respect to each other are not known at the moment of their "birth" from their acyl/1-naphthoxy radical parents. The general uncertainty associated with the initial relative positions of an alkyl/aryloxy geminal radical pair generated upon irradiation of an aryl ester can be largely eliminated by using alkyl aryl ethers as the precursors. Lysis of the alkyl-aryloxy bond results in a radical pair that is chemically equivalent to the one obtained after decarbonylation of an acyl radical in an analogous acyl/aryloxy pair. In addition, as in the case of the acyl/aryloxy pair, the *initial* position of the alkyl radical center with respect to its aryloxy radical partner is known by virtue of the nature of the chemical process leading to the alkyl fragment; its radical center must be very near the oxygen atom of aryloxy. Thus, a comparison of the relative rate constants, k_{2B} and k_{4B}, for radical pairs generated *indirectly* from aryl esters and *directly* from analogous alkyl aryl ethers can provide insights into the motions of the radicals within their cages over very short periods of time.

      2a   **3a**   **3b**   **3c**

The k_{2B}/k_{4B} ratios from **1a** and benzyl 1-naphthyl ether (**3a**) are compared in Table 13.6.[43] Although the ratios from **3a** may be affected slightly by the very small amounts of out-of-cage reactions indicated by the presence of very low relative yields of $(\mathbf{Bz})_2$ from irradiations in **LDPE**, the pattern is clear. Addition of benzyl radicals to the closer 2-position of 1-naphthyl is favored when the benzyl/1-naphthoxy radical pair is produced directly and, therefore, the benzyl radical center is closer to the 2-position than to the 4-position of 1-naphthoxy.

Another interesting comparison is between the k_{2B}/k_{4B} ratios from the ester, phenyl phenylacetate (**2a**), and benzyl phenyl ether (**3c**), in which the aryl part is phenyl instead of naphthyl.[79] Because the phenyl ring has two *ortho* and one *para* sites at which a radical can add, the rate constant ratios for **2a** and **3c** in Table 13.6 have been statistically adjusted. Although all of the ratios from the esters **1a** and **2a** are close to unity, the differences between an unstretched and a stretched film or between a **LDPE** and a **HDPE** (un)stretched film do not follow a consistent order. Similarly, the ratios from **2a** and **3c** do not exhibit an obvious trend.

Volume differences between the guest molecules containing naphthoxy and phenoxy and the hole free volumes of the polyethylenes do not appear to be the reason for these results.[68b] An indication of what may be responsible is found in the k_{2A}/k_{4A} ratios from **2a**; regardless of the polyethylene type or its unstretched/stretched state, the ratios remain near unity. If the rotational motion of the coin-shaped phenoxy moiety of a radical pair is faster than (or comparable to) to the rate combination of phenylacetyl to phenoxy, the overall effect on the relative rates would be the same as if there were translational motions between the two radicals.[80]

13.4.1 Rate Information from an Optically Active Ether

More detailed kinetic information about the motions and reactions of the alkyl/aryloxy radical pairs can be obtained if the alkyl carbon atom making an ether bond is chiral and the alkyl aryl ether is optically active.[81] Because the radical centers of the alkyl fragments derived from irradiation of such ethers are either planar or invert very rapidly, they exist initially with their aryloxy partners as prochiral pairs. Their ability to retain the memory of the chirality of their parent ethers and impart it to the rearrangement products depends on the ratio between the in-cage rates of radical tumbling and radical pair combination.

The corresponding esters are much less informative because the centers of chirality in their acyl radicals are structurally protected from racemization like that experienced by translational or rotational motions of prochiral alkyl radicals. In addition, the decarbonylated radicals derived from them are formed long after their acyl precursors have moved to orientations with respect to their aryloxy partners that result in a loss of the memory of their host stereochemistry within a cage; see above. Thus, of the Claisen-like photoproducts from irradiation of (R)-**1b**, only the **BzON** (i.e., **3b**) retains a measurable amount of optical activity even in the solid phases of long n-alkane.[82] However, in polyethylene films, *all* of the Claisen products from irradiation of (R)-**1b**—**2-BN**, **4-BN**, and **3b**—exhibit significant ee values.[83] In the same media, the photo-Fries products from **1b** retain virtually all of the enantiomeric purity of the starting ester.[84]

A more convenient and sensitive probe of the tumbling (rotational) and translational motions within a cage is the 1-phenylethyl/1-naphthoxy radical pair *formed directly* upon photolysis of optically active (R)-**3b**. Even in hexane solutions, the **2-BN** and **4-BN** photoproducts retain a large enantiomeric excess (ee) of the parent (R) configuration, and the ee values increase when the irradiations are conducted in solid n-alkane phases.[82] To be directly comparable, ee values of photoproducts must be extraplolated to 0% conversions of (R)-**3b** because return of the 1-phenylethyl/1-naphthoxy radical pair to (S)-**3b** (k_3 in Scheme 13.5) is a viable alternative to in-cage formation of **2-BN** and **4-BN** or out-of-cage formation of **1-NOL**. As the degree of conversion (and, therefore, formation of (S)-**3b**) increases, the *ee*s of the rearrangement products will decrease, regardless of the mechanism of their formation. For that reason, the ee values discussed here are extrapolated to 0% conversion of (R)-**3b**.

13.4.1.1 Results from Irradiation in n-Alkane Solutions
To obtain useful kinetic information based upon the detailed mechanism outlined in Scheme 13.5 is a daunting task. By simplifying it, useful kinetic data can be extracted from irradiations of (R)-**3b** in polymer films. Before doing so, information obtained in isotropic media, n-alkanes of differing viscosities, will be analyzed.[85]

As shown in Figure 13.2, the selectivities of the radical pair combinations, as measured by the [**2-BN**]/[**4-BN**] ratios, increase with n-alkane viscosity and reach a plateau value at $\eta \geq 2.5$ cP.[85] At the same time, the ee%$_{total}$ values (i.e., ee% values for products formed from both in-cage and out-of-cage combinations and extrapolated to 0% conversion of **3b**) for **2-BN** and **4-BN** follow a very similar dependence on

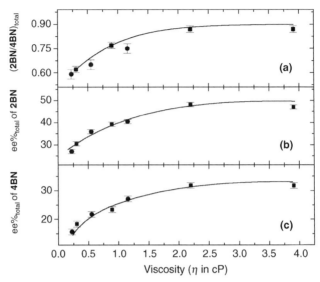

FIGURE 13.2   Plots of (a) **2BN/4BN**, (b) ee%$_{total}$ of **2BN**, and (c) ee%$_{total}$ of **4BN** from irradiations of (R)-**3b** at 296 (■) or 333K (●) versus bulk viscosity (η) of various n-alkanes. The curves are visual aids only; they are not based on a physical model. Reprinted with permission from Ref [85]. Copyright (2005) American Chemical Society.

viscosity. The yield of out-of-cage diagnostic product in this case, 2,3-diphenylbutane, is very small in the most viscous solvent, n-heptadecane ($\eta = 3.91$ cP at 296K), examined; almost all of the radical pairs in this solvent combine in cage. Yet the ee%$_{total}$ values in n-heptadecane are only 47 and 32% for **2-BN** and **4-BN**, respectively; the rates of tumbling and translational diffusion within a cage are comparable (i.e., k_t and the sum, $k_{2B} + k_{4B} + k_3$, must be the same order of magnitude). In the much lower viscosity solvent, n-hexane ($\eta = 0.304$ at 296K), the ee%$_{total}$ values are lower, 31% and 18%, for the two photoproducts. The fact that they are not close to 0% is a clear indication that most of the radical pairs combine in cage even in this solvent.

As the viscosity of the n-alkanes increases, radical movements are attenuated and so are the rates of the three possible in-cage recombination processes, k_{2B}, k_{4B}, and k_3. However, there is no a priori reason to believe that the three rate constants are equally dependent on viscosity. As mentioned above, if translational diffusion of a radical from its initial solvent cage is the most sensitive motion, the reciprocal of the fraction of in cage combination of a radical pair can be expressed as a linear function of $(1/\eta)^n$.[33] Although this expression does not always hold in solvents of low viscosity, like hexane,[86] it was observed that $(2 \times (S)$-**3b** + **2-BN** + **4-BN**)$_{in\text{-}cage}$, the relative yield of in-cage 1-phenylethyl/1-naphthoxy combination products from irradiations of (R)-**3b**, does increase qualitatively with increasing viscosity.[83]

In hexane, the sum of the diffusion coefficients for a 1-phenylethyl and a 1-naphthoxy radical is calculated from Equation 13.2 to be 8.7×10^{-5} cm^2 s^{-1} at 296K and 1.3×10^{-4} cm^2 s^{-1} at 333K. Using them and Equation 13.2, $k_{diff} = 3.6 \times 10^{10}$ M^{-1} s^{-1} at 296K and 5.4×10^{10} M^{-1} s^{-1} at 333K. Based on

Scheme 13.5, the ratio between the percent of relative product yields from *out-of-cage* (100-[2 × (S)-**3b** + **2-BN** + **4-BN**]$_{\text{in-cage}}$) and *in-cage* ([2 × (S)-**3b** + **2-BN** + **4-BN**]$_{\text{in-cage}}$) is given by Equation 13.12 in terms of four rate constants. Assuming that ca. 10 ps is needed at 296K in hexane for the radicals in the pair from (R)-**3b** to move from their initial positions and combine at the 4-position of 1-naphthoxy[49] and an initial separation of 5.1 Å[87] between the two immediately after lysis of the C–O bond,[88] an *in-cage* diffusion coefficient, $D_{\text{in-cage}} \approx 1.3 \times 10^{-4} \, \text{cm}^2 \, \text{s}^{-1}$, that is very near the value calculated from Equation 13.2, is calculated from Equation 13.13.[34,89]

$$k_{\text{esc}}/(2k_3 + k_{2\text{B}} + k_{4\text{B}}) = 100/(2 \times (S)\text{-}\mathbf{3b} + \mathbf{2\text{-}BN} + \mathbf{4\text{-}BN})_{\text{in-cage}} - 1 \qquad (13.12)$$

$$d \approx (2D_{\text{in-cage}}t)^{1/2} \qquad (13.13)$$

Because k_{esc} is directly proportional to k_{diff}, the *in-cage* combination product yield can be related directly to the solvent microviscosity, η_{m}. Thus, plots of ln [100/(2 × (S)-**3b** + **2-BN** + **4-BN**)$_{\text{in-cage}}$-1] versus ln (1/η_{m}) using data from irradiations of (R)-**3b** in the *n*-alkanes are linear, with positive slopes, 0.47 ± 0.10 at 296K and 0.6 ± 0.1 at 333K. Unfortunately, the number of data points is too small to determine whether the two slopes are truly different. Regardless, the positive slopes indicate that the sum of the *in-cage* combination rate constants, $k_3 + k_{2\text{B}} + k_{4\text{B}}$, is less sensitive to microviscosity than is k_{esc} (and k_{diff}).

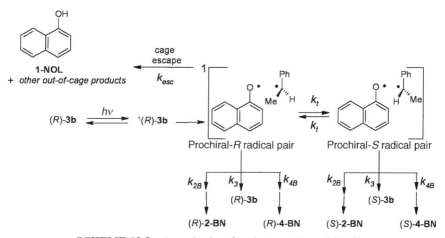

SCHEME 13.5   A mechanism for photoreactions of (R)-**3b**.

    The efficiency of photoinduced racemization (S) of (R)-**3b**, due to recombination of 1-phenylethyl/1-naphthyl radical pairs is given by Equation 13.14.[90] In this equation, the conversion fraction is the part of **3b** that has become photoproducts. The values of **S**, as calculated from the slopes of the plots in Fig. 13.3, are nearly constant, 0.13–0.16,

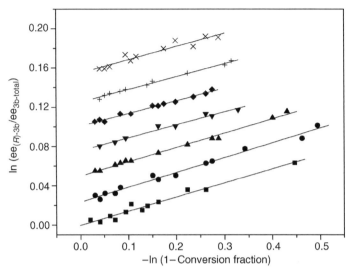

FIGURE 13.3 Plots of ln ($ee_{(R)-3b}/ee_{3b\text{-total}}$) versus $-\ln(1 - \text{conversion fraction})$ for irradiations of (R)-**3b** in n-hexane at 296K (■) and 333K (●), in n-decane at 296 (▲) and 333K (▼), in n-tetradecane at 296 (◆) and 333K (+), and in n-heptadecane at 333K (×). Points for each data set are vertically offset by 0.025 increments for clarity. Reprinted with permission from Ref. 85. Copyright (2005) American Chemical Society.

for irradiations in a series of n-alkanes of different viscosities and at 296 and 333K.

$$\mathbf{S} = \frac{\ln\left(\frac{ee_{3b\text{-total}}}{ee_{(R)-3b}}\right)}{\ln(1 - \text{conversion fraction})} \tag{13.14}$$

Because the energies of the enantiomeric (prochiral) 1-phenylethyl/1-naphthoxy radical pairs from (R)-**3b** and their rate constants leading to (R)-**3b** and (S)-**3b** are the same, $\mathbf{S} = 2P_{inv}/(1 - P_{inv} - P_{ret})$, where P_{inv} and P_{ret} are the probabilities that a radical pair will form the (S)- and (R)-enantiomers of **3b**, respectively.[90] The expressions for P_{inv}, P_{ret}, and $\mathbf{S}$ based on Scheme 13.5 are very complex,[85] and even it does not describe all of the processes involved in the tumbling of the 1-phenylethyl radicals because k_t really should not be described by one rate constant. To do so requires the introduction of k_{t-3}, k_{t-2B}, and k_{t-4B}, defined as the specific tumbling rate constants inside a cage for 1-phenylethyl radicals that combine with 1-naphthoxy to form, respectively, **3b** and the keto tautomers of **2-BN** and **4-BN**; taking their sum as k_t is convenient but not accurate. The necessity to invoke separate tumbling rate constants for formation of **3b**, **2-BN**, and **4-BN** is indicated by the observation that the ee values for **2-BN** and **4-BN** extrapolated to 0% conversion of (R)-**3b** are different. Were one rate constant for tumbling sufficient, the limiting ee values should be the same. Mechanistically, $\mathbf{S}$ is a function of radical motions from tumbling (k_{t-3}, k_{t-2B}, and k_{t-4B}) and translation (k_3, k_{2B}, k_{4B}, and k_{esc}) as shown in Equation 13.15. More information can be extracted from schemes and resulting kinetic expressions that concentrate on specific aspects of the processes experienced by 1-phenylethyl/1-naphthoxy radical pairs.

$$S = \frac{2k_3 k_{t\text{-}3}}{[(k_{2B} + k_{4B} + k_t + k_{esc})^2 + k_3(k_{2B} + k_{4B} + k_{t\text{-}2B} + k_{t\text{-}4B} + k_{esc})]} \tag{13.15}$$

In simplified Scheme 13.6, both prochiral radical pairs are combined implicitly into one, so that $P_{inv} = k_i/(k_i + k_r + k_o)$, $P_{ret} = k_r/(k_i + k_r + k_o)$. Thus, $k_i \approx 0.07 k_o$ regardless of the n-alkane or temperature among those investigated. Relating the two schemes, k_o is approximately $k_{esc} + k_{2B} + k_{4B}$, and k_i equals k_3 multiplied by the fraction of prochiral-(R) radical pairs that become prochiral-(S) radical pairs. Since **S** is essentially constant as well, viscosity- or temperature-induced changes of k_i and the sum or the dominant term of $(k_{esc} + k_{2B} + k_{4B})$ must be proportional.

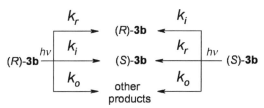

SCHEME 13.6 A simplified mechanism for photoracemization of (R)-**3b**.

Whereas Scheme 13.6 focuses on the kinetics of *stereochemical* changes to (R)-**3b** induced by its photolyses, Scheme 13.7 focuses on *regiochemical* considerations of product formation. Scheme 13.7 links in-cage formation of **2-BN** and **4-BN** to specific, regio-isomeric radical pairs, [radical pair]$_{2B}$ and [radical pair]$_{4B}$, and Equation 13.16 can be derived from it.

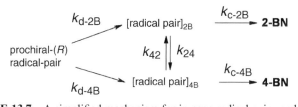

SCHEME 13.7 A simplified mechanism for in-cage radical pair combinations.

$$(\textbf{2-BN}/\textbf{4-BN})_{in\text{-}cage} =$$
$$[k_{c\text{-}2B}/k_{c\text{-}4B}] \times [(k_{42} + k_{c\text{-}4B})k_{d\text{-}2B} + k_{42}k_{d\text{-}4B}]/[(k_{24} + k_{c\text{-}2B})k_{d\text{-}4B} + k_{24}k_{d\text{-}2B}] \tag{13.16}$$

Because the radical pairs in this case are singlets, no spin change is needed for their combination and their rate constants for combination after the 1-phenylethyl radical center reaches the proximity of C(2) or C(4) of 1-naphthoxy, $k_{c\text{-}2B}$ and $k_{c\text{-}4B}$, should be $>10^{11}$ s^{-1},[49] regardless of the microviscosity. Thus, they should be much faster than the viscosity-dependent rates, $k_{d\text{-}2B}$ and $k_{d\text{-}4B}$, at which the radicals of the pair move from their initial positions to the orientations appropriate for bond formation, [radical pair]$_{2B}$ and [radical pair]$_{4B}$. Also, although $k_{24} \neq k_{42}$, both are expected to be similar in magnitude to the k_{d-2B} and k_{d-4B} rate constants. Thus, the rate-limiting steps of the combination process are expected to involve formation and interconversion of [radical pair]$_{2B}$ and [radical pair]$_{4B}$, but not the actual bond-formation steps in Scheme 13.17. Because the slope of a plot of ln (**2-BN/4-BN**)$_{in\text{-}cage}$ versus ln $(1/\eta_m)$ is slightly negative, microviscosity influences the in-cage combination process involving the longer pathway to **4-BN** slightly more than the shorter pathway to **2-BN**.

The extent of tumbling by 1-phenylethyl radicals along the pathways to the 2- and 4-positions of their 1-naphthoxy partners are related to the ee values of **2-BN** and **4-BN**. For the purpose of simplifying the kinetic treatment, we assume that no tumbling occurs in those cases where the (R)-enantiomer of a product forms and that tumbling is the only route to an (S)-enantiomer. The activation energy of another possible route, internal rotation about the phenyl-benzylic C–C bond of 1-phenylethyl is ca. 54 kJ/mol[91] and, therefore, too high to compete with combinations within a cage; typically, the activation energy for a diffusion-controlled radical–radical combination leading to a C–O bond is ca. 12 kJ/mol.[81] Thus, the slopes of plots of ln (S)-**2-BN**/(R)-**2-BN**)$_{in\text{-}cage}$ and ln (S)-**4-BN**/(R)-**4-BN**)$_{in\text{-}cage}$ versus ln $(1/\eta_m)$ at 296K are both 0.2 ± 0.1.[85] These values indicate that microviscosity influences in a similar fashion the k_t (Scheme 13.5) associated with the pathways to both **BN** products. However and unfortunately,[33b–d,40a–d,86] no simple function appears to describe adequately the relationship between the fates of *in-cage* radical pairs and the properties of their host solvents, even in "simple" media such as *n*-alkanes!

13.4.1.2 Results from irradiations in polyethylene films.

Given these analyses, the data from irradiations of (R)-**3b** in polyethylene films will now be discussed (Table 13.7).[77] As mentioned above, the ability of the radical pairs from (R)-**3b** to *diffuse translationally* within a cage is related *qualitatively* to the **2-BN/4-BN** ratios whereas the ability of the 1-phenylethyl radical to *tumble* along the translational course that brings it to combine at either the 2- or 4-position of its 1-naphthoxy partner is related to %ee$_{2B}$ or %ee$_{4B}$. Although a 1-phenylethyl radical center is attracted intrinsically more to the 4-position than the 2-position of 1-naphthoxy (Fig. 13.1), it is

TABLE 13.7 Distribution Ratios and Extrapolated Enantiomeric Excesses (*ees*) of 2-BN and 4-BN Photoproducts (*ee*$_{2B}$ and *ee*$_{4B}$, Respectively) from Irradiations of 99.5% Enantiomerically Pure (R)-3b in *n*-hexane, *n*-heptadecane, and Unstretched (u) and Stretched (s) Completely Amorphous (PE0) and LDPE Films at 296 K

medium	[2-BN]/[4-BN]	%ee$_{2B}$	%ee$_{4B}$
Hexane	0.62	31	18
Heptadecane	0.87	47	32
PE0 (u)	1.2	56	42
PE0 (s)	1.41	57	43
PE46 (u)	1.5	50	33
PE46 (s)	1.45	42	26

reasonable to assume a proportionality between the times and distances required for the radical center of 1-phenylethyl to traverse the vectoral distances to the closer 2-position and farther 4-position.[73] Viewing the data in Table 13.7 in this way, both the much higher **2-BN/4-BN** ratios in polyethylene than in liquid *n*-alkane solutions and the >1 values for the ee_{2B}/ee_{4B} ratios in the polyethylene matrices are expected; the general conclusion that can be reached from inspection of the data in Table 13.7 is that the **PE** reaction cages exert much more control over the regio- and stereochemistry of 1-phenylethyl/1-naphthoxy radical pairs than the cages of *n*-alkanes. It can be attributed to a combination of "templating" and microviscosity effects that are related to the much longer relaxation times of the chains constituting the walls of the **PE** cages. In partially crystalline polyethylenes, such as **LDPE**, analyses of the effects is complicated further by the aforementioned presence of two distinct cage types— those within amorphous regions and those within interfacial regions. In the completely amorphous polyethylene, **PE0**, only one general site type exists and an unknown, the fraction of (*R*)-**3b** molecules residing and reacting within each site type, is eliminated.[92]

Because amorphous and interfacial cages influence translational and tumbling motions differently, some being attenuated more than others, radical pair combinations are predicted and found to be less regio- and stereoselective in interfacial sites than in the amorphous ones:[83] the higher ee values for **2-BN** and **4-BN** are found upon irradiation of (*R*)-**3b** in the completely amorphous **PE0** (i.e., a polymer without crystalline parts) than in a mixture of amorphous and *more highly ordered cages* in the partially crystalline **LDPE**. In addition, from the slopes of the lines drawn in Fig. 13.4, the values of **S** (Eq. 13.15)[90] are 0.23 and 0.24 for irradiations of (*R*)-**3b** in unstretched and stretched **PE0**, respectively, and 0.27 and 0.44 in **LDPE** before and after stretching.[77] All of these are much higher than the **S** found in liquid a *n*-alkanes, 0.13–0.16,[85] and only in the partially crystalline polyethylene, **LDPE**, does stretching have a discernible effect. The net result is that (*R*)-**3b** is more easily photoracemized in partially crystalline **LDPE** than in completely amorphous **PE0** and least easily in the much less viscous environments offered by liquid *n*-alkanes! These results also suggest that most of the (*R*)-**3b** reside and react in amorphous cages of unstretched **LDPE**; stretching increases the fraction of reaction occurring in interfacial cages, where the probability that a 1-phenylethyl radical will tumble before combining is lower than in an amorphous cage. This conclusion is supported by (1) the lack of change in the $\%ee_{2B}$ and $\%ee_{2B}$ induced by stretching **PE0**, (2) the decreases in both $\%ee_{2B}$ and $\%ee_{2B}$ caused by stretching **LDPE**, and (3) the lower values of $\%ee_{2B}$ and $\%ee_{2B}$ obtained in unstretched **LDPE** than in unstretched **PE0**.

Comparisons between the results from irradiations of (*R*)-**3b** in liquid *n*-alkanes and in polyethylene films lead to several other interesting conclusions. Foremost among them is that *lower* tumbling probabilities are found for radical pair combinations leading to reformation of an enantiomer of **3b** in liquid *n*-alkanes than in polethylene films (N. B., **S** values), but the tumbling probabilities are *higher* for combinations leading to **2-BN** and **4-BN** in the liquid *n*-alkanes (N. B., %ee values)! An economical explanation for these and a myriad of other results from both thermal and photochemical reactions in polymer matrices is that the cages act as templates as well as diffusion-

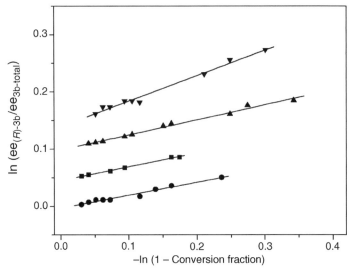

FIGURE 13.4 Plots of $\ln(ee_{(R)\text{-}3b}/ee_{3b\text{-total}})$ versus $-\ln(1 - \text{conversion fraction})$ for irradiations of (R)-**3b** in unstretched ($\bullet$) and stretched ($\blacksquare$) **PE0** films, and in unstretched ($\blacktriangle$) and stretched ($\blacktriangledown$) **LDPE** films at 296K. Points at each plot from bottom to top are vertically offset by 0.05 increments. Reproduced from Ref. 77 by permission of the Royal Society of Chemistry on behalf of the European Society for Photobiology and the European Photochemistry Association.

modulating environments for the durations of the lifetimes of the reactive species[1] (which are radical pairs in the examples of this chapter). Especially when reaction occurs in an interfacial cage of **PE46**, the proximity of the radical center of 1-phenylethyl to the oxygen atom of the 1-naphthoxy radical immediately after lysis of (R)-**3b** makes $k_{t\text{-}3}$ more impervious to medium anisotropy (and microviscosity) than are k_{esc}, k_3, k_{2B}, and k_{4B}. Clearly, the "templating" effects of interfacial cavities impose additional anisotropic constraints to some motions and may be accelerating others. Those radical pairs that follow the more difficult *and longer* routes in **PE** films (Fig. 13.5) experience greater resistance to tumbling than to translational motions (i.e., lower $k_{t\text{-}2B}$ and $k_{t\text{-}4B}$ with respect to k_{2B} and k_{4B} in the same medium), allowing higher ee values of **2-BN** and **4-BN** in **PE** films. A corollary of this qualitative analysis is that the fraction of the radical pairs that returns to **3b** rather than escaping from their cages or forming one of the **BN** rearrangement products is larger in **PE** films than in liquid *n*-alkanes.

13.5   COMPARISON OF CALCULATED RATES TO OTHER METHODS FOR POLYETHYLENE FILMS

The diffusion coefficients of *N,N*-dialkylanilines (where alkyl was varied from methyl to dodecyl) in unstretched or stretched **PE** films have been measured using methods

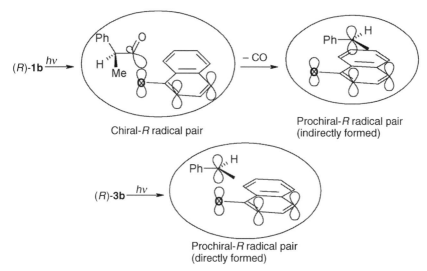

FIGURE 13.5 Radical pairs from irradiation of (R)-**1b** and (R)-**3b** showing their relative positions at their moments of formation. Adapted from Ref. 77 by permission of the Royal Society of Chemistry on behalf of the European Society for Photobiology and the European Photochemistry Association.

based on static and dynamic fluorescence.[78,93] In general, they indicate that molecules move slower in **LDPE** than in the completely amorphous **PE0**. Two diffusion coefficients, attributed to the movement of χ_a and χ_i mole fractions of dialkylanilines in amorphous and interfacial regions, respectively, were found for diffusion within **LDPE**; in **PE0**, $\chi_a = 1$ and $\chi_i = 0$ in terms of diffusing species, occupancy, and even for photoreactions of (R)-**3b**. As in the case of diffusion, the photochemical data in **LDPE** are based on weighted averages of the fractions of reaction occurring in amorphous and interfacial reaction cages. Because the dialkylanilines are similar in size to the radicals in the pairs derived from **3b**, the diffusion coefficients obtained from the fluorescence measurements can provide interesting comparisons with the rate data obtained from the irradiations. These comparisons are especially useful for **PE0** because there is no ambiguity about the value of the mole fractions for occupancy or reaction in the cages—all are amorphous. Then, taking the values obtained from this film and applying them to the **LDPE** may shed additional light on the differences between the dynamics of the reactions occurring in the amorphous and interfacial cages.

Using the D values of the dialkylanilines at 298K and Equation 13.13, the time required for 1-phenylethyl and 1-naphthoxy radicals to move to the locations amenable to combine at the 2- and 4-positions of 1-naphthoxy, ca. 3.1 and 5.0 Å, respectively,[87] correspond to ca. 2.4–24 and 6.3–63 ns in unstretched **PE0** and ca. 24–240 and 63–630 ns in unstretched **PE46**, depending on the specific diffusion coefficient employed.[93] The times required for formation of the keto intermediates of **2-AN** and **4-AN** from **1b** by the "radical clock" method described above[43,47] are ca. 3 and 14 ns, respectively, in unstretched *or* stretched **PE0** and ca. 2–3 and 26–33 ns, respectively, in unstretched **LDPE**. In stretched **LDPE**, where translocation increases

χ_i, the times are longer, ca. 5–8 and 59–91 ns, respectively. Whereas the "radical clock" technique considers the effects of templating by a cage that may affect translational motions (and, therefore, rates of combinations) by a radical pair, Equation 13.13 does not. The intracage radical pair combinations are influenced less by surrounding polymeric chains than are bulk diffusional processes (that are germane to Eq. 13.13) and, as a consequence, are faster than those calculated from bulk diffusion constants.[93] Good agreement between the two methods in **PE0** may be ascribed to its lack of interfacial reaction cavities.

It is important to note that a 2-phenylpropanoyl radical (from which the radical clock based rate constants were derived) should be more reactive than a 1-phenylethyl radical toward an electron-rich 1-naphthoxy radical since the odd electron density of the former radical is more localized.[85,94] Therefore, absolute comparisons between k_3, k_{2B}, and k_{4B} rate constants from irradiations of (R)-**3b** and k_{2A} and k_{4A} from irradiations of **1b**, even in the same medium, are not very useful.

13.6 TRIPLET-STATE RADICAL PAIRS

13.6.1 Triplet-State Radical Pairs from the Photoreduction of Benzophenone by Hydrogen Donors

The photoreduction of benzophenone (**B**) in solution is one of the most extensively investigated photochemical reactions, having been studied using a wide range of reductants, solvents, reactant concentrations, and irradiation conditions.[95] In the presence of H-donating substrates, the triplet excited state of **B** is able to abstract an H-atom directly (Equation 13.17) or *via* a sequential electron/H$^+$ transfer mechanism (Equation 13.18) to produce triplet radical pairs, $^3\overline{\text{BH}^\bullet\text{R}^\bullet}$.

$$^3\text{B}^* + \text{HR} \rightarrow {}^3\overline{\text{BH}^\bullet\text{R}^\bullet} \tag{13.17}$$

$$^3\text{B}^* + \text{HR} \rightarrow {}^3\overline{\text{B}^{-\bullet}\text{HR}^{+\bullet}} \rightarrow {}^3\overline{\text{BH}^\bullet\text{R}^\bullet} \tag{13.18}$$

Because direct interaction of the triplet radical pair to give products in the ground singlet state is spin-forbidden, the radicals massively diffuse into the bulk and ultimately react to give a variety of products in fluid solvents. Hence, the rates of cage-escape and in-cage combination of the $^3\overline{\text{BH}^\bullet\text{R}^\bullet}$ become competitive only in high-viscosity media.

The kinetics of combination of radical pairs formed by quenching of the triplet excited state of **B** by 2,4,6-trimethylphenol (**HR**) in films of rigid and plasticized polyvinylchloride (**PVC**) were studied by the method of laser flash photolysis by Levin et al.[35] At high concentrations of the phenol quencher ($\sim$15 wt%), most of the triplet excited state of the ketone is quenched and the formation and disappearance of the ketyl and phenoxy radicals can be followed from changes in their characteristic transient absorptions at 540 and 390 nm, respectively.

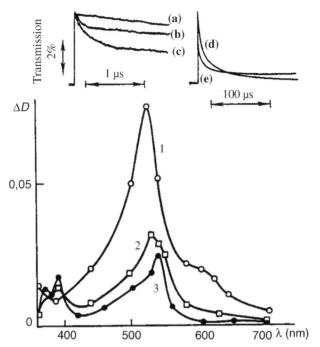

FIGURE 13.6 Transient absorption spectra obtained from laser flash photolysis experiments in **PVC** films containing 6% **B** (1), 6% **B** and 15% **HR** (2), 6% **B**, 15% **HR**, and 59% plasticizer (3). Kinetic curves are for the disappearance of the 2,4,6-trimethylphenoxyl radicals (390 nm) in **PVC** films containing 6% **B** and 15% **HR**, and 49% [(a) and (d)], 29%, (b) and 0% [(c) and (e)] of plasticizer. Taken from Ref. 35.

In a liquid plasticizer solution comprised of a polychlorinated decane or in films containing the plasticizer with less than 30 wt% of **PVC**, the combination of the radicals follows a second-order rate law (Fig. 13.6). The observed reaction rate constant (k_{r2}) in neat plasticizer is $\sim 3 \times 10^8\,M^{-1}\,s^{-1}$ and decreases slightly with increasing concentration of **PVC** in the films. This observation suggests that the radical combination reactions occur in segments of the films with higher molecular mobility, where molecules of the plasticizer are aggregated.

With 30–70 wt% **PVC**, the kinetic curves show an initial sharp decrease followed by a slow component that satisfactorily obeys a second-order rate law (Fig. 13.6). Levin et al. proposed that the initial change in the absorption corresponds to the "in-cage" combination of the radicals pairs, while the slow component represents the combination of the radicals that have diffused into the (polymer) bulk. Interestingly, at >70 wt% **PVC**, only the fast "in-cage" combination of the geminate radical pairs can be observed. From the relative amplitudes of the fast and slow components of the kinetic curves, approximated values for the cage factor F_{cAB} were calculated. F_{cAB} increase from zero in films with <30 wt% **PVC** to near unity (>0.95) in films with >70 wt % **PVC**.

The value of k_{r2} calculated from the slow component of the kinetic curves decreases by more than one order of magnitude on going from the liquid plasticizer to films with 70 wt% of **PVC**. In this range of **PVC** concentrations in the films, the activation energy associated with k_{r2}, ~30–40 kJ/mol, is consistent with the activation barriers expected for the forward diffusion of the radicals in the viscous "liquid" phases of the polymers.

The first-order combination rate constants from in-cage processes, $k_{r1} \sim 3 \times 10^6 \, s^{-1}$, were determined from the initial slopes of the fast components of the kinetic curves. They remain nearly constant in films with 30–100 wt% of **PCV**. The activation energy of k_{r1} is very small, ~8 kJ/mol, suggesting that geminal recombination in the **PVC** films is limited only by the rotational mobility of the radicals in their cages. Furthermore, k_{r1} was insensitive to the application of a strong external magnetic field (0.3 T), as well as to the influence of an internal heavy atom.[37] These facts strongly support a model in which geminal combinations are controlled by molecular (rather than spin) dynamics.

Kinetic curves of films containing >70 wt% **PVC** and in glassy **PMMA** films were clearly nonexponential. Figure 13.7 shows the results obtained in **PMMA** at room temperature. Because the characteristic half-lives are independent of the initial concentration of the radical pairs (as changed by varying the laser pulse energy), the radical pairs appear to combine according to a unimolecular mechanism despite the fact that the decays cannot be fitted to a first-order rate law. Above 333 K, the glass transition temperatures of the **PVC** film containing 15 wt% 2,4,6-trimethylphenol and 6 wt% **B**, results in a qualitative change in the kinetics in the 100% **PVC** film. The decay curve becomes similar to that observed in films with 40–60% wt **PVC** at room temperature; that is, in addition to the fast geminate combination, a slow component due to in-bulk combination of the free radicals also appears.

The observed nonexponential character of the decays observed in films containing >70 wt% **PVC** (or in the **PMMA** films) was ascribed to the existence of a distribution of unimolecular combination rates, apparently caused by the dependence of the rate constants on the properties of different local environments in the polymer films. This effect was quantified according to a kinetic dispersive model,[36,38,96] which predicts

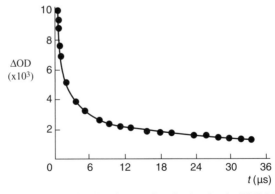

FIGURE 13.7 Kinetic curve for the decay of radical pairs in **PMMA** at 20°C. Points represent the experimental data and the line is the fitting to Eq. 13.9. Taken from Ref. 36.

that the survival probability $P(t)$ of the radical pairs follows Equation 13.19:

$$P(t) = [1 + nk_{av}t]^{-1/n} \tag{13.19}$$

where k_{av} is the initial rate constant of the process and n characterizes the distribution width of unimolecular rate constants. Kutsenova et al.[36] showed that n is independent of temperature and that it approaches 2 in neat **PMMA** and **PVC** films. As noted by the same authors, this result is somewhat surprising because it suggests that the distribution of unimolecular rate constants is independent of the nature of the polymer matrix and temperature. The values of k_{av} measured at $20\,°C$ in the **PMMA** and **PVC** films are $0.84 \times 10^6\,s^{-1}$ and $2.4 \times 10^6\,s^{-1}$, respectively, and they increase with increasing temperature. The activation energies associated with k_{av} in both type of films are small, $\sim 13\,kJ/mol$, giving further support to the contention that the combination processes are controlled primarily by molecular dynamics.

The data in Fig. 13.7 is amenable to other interpretations. For instance, it can be fitted well to Equation 13.20 which is Equation 13.2 solved in the limit, $F_{cAB} \to 1$.[19b]

$$P(t) = \exp\left(-\sqrt{\frac{4t}{\pi t_0}}\right) \tag{13.20}$$

In this case, the relaxation time of the process is given approximately by $t_0 \cong \left(\frac{\sigma^2}{D}\right)\left(\frac{k_{dAB}}{k_r}\right)^2$. The t_0 calculated in this way from the data in Fig. 13.7 is $\sim 2 \times 10^{-6}\,s$. Thus, taking reasonable values for the contact distance σ ($\sim 5–7\,Å$) and assuming F_{cAB} to be in the range, $\sim 0.95–0.99$, an approximate value of $D \sim 10^{-11}\,cm^2/s$ can be estimated. In turn, D can be used to recalculate k_{dAB} ($= 4\pi\sigma D$) and k_r, which are $\sim 10^4$ and $\sim 10^6\,M^{-1}\,s^{-1}$, respectively. Not surprisingly, k_r is of the same order of magnitude as k_{av}. However, according to this approach (Eq. 13.20), the nonexponential behavior of the decay curve in Fig. 13.7 is caused by transient effects on the combination rate. As shown in Scheme 13.2, the radicals can escape intimate contact and explore (the geminate and) the nearest reaction cages before combining, even in the case of $F_{cAB} \sim 1$. In principle, both models (Eqs 13.19 and 13.20) explain reasonably well the pseudo-first-order kinetic curve observed in Fig. 13.7. However, while the kinetic dispersive model attributes the phenomenon to the heterogeneity of the polymeric media, the diffusion-based theory attributes the nonexponential decay to the probability that, after several escape/reencounters of the radicals, an effective collision leads to product formation.

13.6.2 Triplet-State Radical Pairs from Norrish Type I Processes

Asymmetrically substituted dibenzyl ketones (**ACOB**) have been employed frequently as precursors of geminal triplet radical pairs in studies of cage effects in constrained or microphase-compartmentalized media.[27,28]

Irradiation of **ACOB** results in a classic Norrish Type I process. The first homolytic step from the triplet state leads to an arylacetyl/benzylic radical pair,[97,98] as shown in Scheme 13.8. In high viscosity media, loss of carbon monoxide from the arylacetyl part is sufficiently rapid to compete with cage-escape by one of the radicals. Thus, a large fraction of the resulting pairs of benzylic radicals $\dot{\text{A}} \ldots \dot{\text{B}}$ can be formed in the reaction cavity of their origin while retaining the triplet spin multiplicity of the precursor ketone excited state. The triplet $\dot{\text{A}} \ldots \dot{\text{B}}$ pairs can either combine within their cage of origin, yielding **AB** or escape and react after reencounters (if no trapping species are present in the medium), yielding principally **AA**, **AB**, and **BB**.

ACOB$_0$ **ACOB$_1$**   **ACOB$_{16}$**

Photochemical reactions of the dibenzyl ketones in shown above have been studied by Bhattacharjee et al. in fluid solutions and in **PE** films of different degrees of crystallinity: **PE0**, **LDPE**, and **HDPE**.[29] Irradiation of **ACOB$_1$** and **ACOB$_{16}$** at >300 nm in fluid media (such as n-hexane, pentane, etc.) yields **AA**, **AB**, and **BB** as the only detected products in a $1:2:1$ distribution (i.e., implying a zero cage effect: all geminal radical pairs escape from their cages into the solvent bulk before combining). The F_c values (Eq. 13.4) calculated from steady-state irradiations of the **ACOB$_n$** in the fluid media are collected in Table 13.8. In contrast, photolyses of **ACOB$_1$** and **ACOB$_{16}$** in the **PE** films produce large excesses of **AB**, suggesting (*but not requiring*[29a]) that a significant component of reaction by the geminal radical pairs involves in-cage combinations in the polymeric matrices. In the particular case of **ACOB$_{16}$**, the values of F_c are ~ 1 in all the **PE** films employed. This result was unexpected because even assuming that the mobility of the long-tailed benzyl radical (**B$_{16}$**$\dot{}$) can be severely attenuated by the polymer matrices, the diffusion of the lighter benzyl radical (**A**$\dot{}$) should be relatively rapid. Hence, the observed $F_c \sim 1$ cannot be entirely ascribed to a decrease in k_{dAB}; instead, it is attributed to persistent radical effects (see below).

Laser flash photolyses of **ACOB$_0$**, **ACOB$_1$**, and **ACOB$_{16}$** in the **PE** films produced transient absorptions with maxima around 320–324 nm that were assigned to the benzyl and p-alkylbenzyl radicals. Typical kinetic curves are shown in Fig. 13.8. The profiles show three different time domains: an initial "instantaneous" rise (1), a time-resolved rise (2) and (3) a time-resolved decay. The initial rise was attributed to the fast homolysis of dibenzyl ketone triplets yielding a benzylic and arylacetyl radical pair. The time-resolved rise is from formation of additional benzylic radicals as decarbonylation of the arylacetyl radical occurs. In agreement with the mechanism in Scheme 13.8, the ratio of the change of optical density (ΔOD) for the initial "instantaneous" rise and the protracted rise is ~ 1. The decay of the ΔOD signal in the <20 μs time window was ascribed to in-cage combination of the triplet radical pairs (see below).

TABLE 13.8 Rate Constants and F_{cAB} Values at 290 K Calculated from Laser Flash (LF) Photolysis Experiments According to the Model and F_c Values from Steady-state (SS) Photolyses Conducted at 295 K.[a] *Error Limits are the t-test Confidence Intervals Calculated at the 95% Confidence Level*

polyethylene		ACOB$_0$	ACOB$_1$	ACOB$_{16}$
PE0[b]	$k_{-CO}/10^6\,\mathrm{s}^{-1}$	4.6 ± 0.5	6.0 ± 0.3	6.2 ± 0.3
		4.7 ± 0.2		
		4.7 ± 0.3^c		
	$k_{dAB}/10^4\,\mathrm{s}^{-1}$	7.4 ± 0.9	7.5 ± 0.9	11.0 ± 0.5
	$k_r/10^4\,\mathrm{s}^{-1}$	4.7 ± 0.2	5.0 ± 0.2	5.4 ± 0.2
	F_{cAB}	0.38 ± 0.04	0.40 ± 0.04	0.33 ± 0.01
LDPE	$k_{-CO}/10^6\,\mathrm{s}^{-1}$	4.7 ± 0.7	5.5 ± 0.2	$>5^d$
	$k_{dAB}/10^4\,\mathrm{s}^{-1}$	8.6 ± 0.9	9.4 ± 0.3	16.0 ± 0.2
	$k_r/10^4\,\mathrm{s}^{-1}$	11.0 ± 0.2	11.0 ± 0.1	14.0 ± 0.1
	F_{cAB}	0.55 ± 0.04	0.55 ± 0.05	0.46 ± 0.02
	F_c	–	0.51 ± 0.03	1.00 ± 0.04
HDPE	$k_{-CO}/10^6\,\mathrm{s}^{-1}$	4.8 ± 0.6	5.6 ± 0.5	$>5^e$
	$k_{dAB}/10^4\,\mathrm{s}^{-1}$	7.8 ± 0.8	7.4 ± 0.4	13.0 ± 0.4
	$k_r/10^4\,\mathrm{s}^{-1}$	9 ± 1	12 ± 1	11.0 ± 0.1
	F_{cAB}	0.53 ± 0.01	0.61 ± 0.07	0.44 ± 0.06
	F_c	–	0.67 ± 0.04	1.00 ± 0.04
n-hexane	$k_{-CO}/10^6\,\mathrm{s}^{-1}$	6.4		
	F_c		0.01 ± 0.03	-0.01 ± 0.04

[a] Except as indicated, all laser flash data are averages from 10 laser shots on one spot of an aerated film. In the fitting routines using the model, k_{-CO} was fixed to the values derived from the rise portions of transient absorption traces.
[b] For **ACOB$_0$** from one shot-averaged decay trace in each of two different films.
[c] Stored for 1 month under an argon atmosphere prior to irradiation.
[d] For **ACOB$_1$** average from four separate shot-average decay traces from four different spots of the film surface.
[e] No accurate measurement possible due to strong scattering and fast decays.

Experiments performed on a 4 ms time scale in **PE** films that were either equilibrated with air or (partially) deoxygenated demonstrated that molecular oxygen can efficiently trap the benzyl radicals during the longer time periods. A manifestation of the incomplete (and systematic) elimination of oxygen (and other possible radical scavengers intrinsic to the polymer films) is the poor reproducibility of the intensities and characteristic decay times of the transient absorption signals at the longer times. However, the decay kinetics during the first 20–30 μs were reproducible; molecular oxygen or other quenchers within the **PE** matrices do not interfere with the processes shown in the faster events box in Scheme 13.8 (Fig. 13.8). The decay characteristics were also independent of the laser-pulse energies (in the 40–120 mJ range). These facts are consistent with a pseudo-unimolecular in-cage combination cage-escape of the triplet radical pairs and in the 20–30 μs timescale. Simple analyses of the decay curves using a first-order model always produced consistently better fits than when second-order fits (i.e., indicative of bimolecular out-of-cage combination processes) were used.[99]

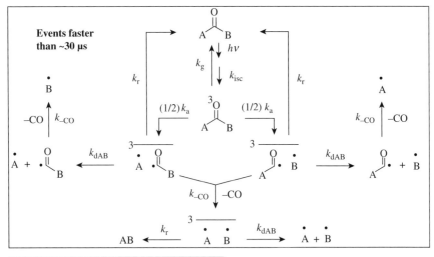

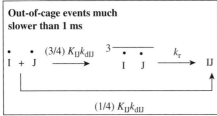

SCHEME 13.8 Basic processes in a classic Norrish Type I reaction of **ACOB** in a constraining medium. Events faster that 30 μs and much slower (bimolecular) free radical combination processes are separated for simplicity. $I^{\bullet}$ and $J^{\bullet}$ represent any of the radicals shown in the top panel. The scheme involves a number of important assumptions. For details, see Ref. 29a. Reprinted with permission from Ref. 29a. Copyright (2007) American Chemical Society.

The values of k_{-CO} reported in Table 13.8 were obtained from analyses of the ΔOD traces recorded in a much shorter time period, 2–8 μs, and they were calculated from the rise portions of the curves. These k_{-CO} actually represent weighted averages for the decarbonylation of the phenylacetyl and p-alkylphenylacetyl radicals that are known to be very similar.[58,60,100] The k_{-CO} measured in the **PE** films are very similar to those found in alkane solvents, supporting earlier suggestions that the decarbonylation process is independent of the viscosity of the medium.[100]

Due to the complex nature of the systems inspected, modeling the decay portions of the kinetic curves according to Equation 13.2 (or similar expressions) becomes very difficult. Fitting the experimental data to Equation 13.2 requires treatment of the ΔOD over all time and, as was commented above, the kinetic curves obtained in the millisecond time window are affected by dissolved oxygen and other possible scavengers. Also, the ΔOD signals in the early time domains (0–2 μs) are convoluted due to the simultaneous decarbonylation of the phenylacetyl radicals and in-cage combination processes; rigorous determination of the initial concentration of the triplet radical pairs

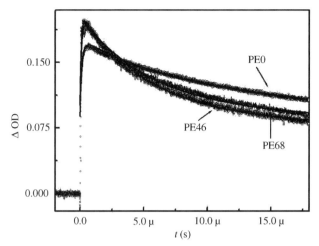

FIGURE 13.8 Transient absorption profiles (black) monitored at 320 nm for laser-pulsed photolyses of **ACOB$_0$** in **PE** films at 293K and best fits according to the fitting model. Numbers after **PE** refer to % crystallinity. Reprinted with permission from Ref. 29a. Copyright (2007) American Chemical Society.

(and, therefore, the exact form of $P(t)$) is not possible. Hence, experimental kinetic curves such as those in Fig. 13.8 were analyzed using a relatively simple approach based on the exponential model in Equation 13.1. The proposed fitting model solves numerically eight simultaneous differential equations and allows the calculation of approximated values of k_r, k_{dAB}, and F_{cAB}. The best fits to the experimental data achieved using the proposed model are shown in Fig. 13.8. The values of the rate constants and the corresponding cage factors are collected in Table 13.8. Note that the model ignores the possible consequences of secondary cage effects and it is applicable only to systems in which the diffusion is Fickian.[101,102]

The cage factors calculated from either the steady-state (F_c) or time-resolved (F_{cAB}) photolysis experiments for **ACOB$_1$** in the different **PE** films are, within experimental uncertainties, the same. However, in the case of **ACOB$_{16}$**, the F_c are always larger than the F_{cAB}. This apparent disagreement can be attributed to the fact that the diffusion coefficients for the benzyl and p-methylbenzyl radicals are nearly the same, but that for the p-hexadecylbenzyl radical is much smaller, and is rationalized in terms of the persistent radical effect.[58a,60] The cage factor F_c, as calculated from photoproduct distributions and Equation 13.4, can be related to the rates constants shown in Scheme 13.8 by Equation 13.21. In this equation, F_{cAB} is given by Equation 13.22 and R_- and R_+ are complex functions that depends on the relative diffusion coefficients of radicals A$^{\bullet}$ and B$^{\bullet}$. F_{cAB} represents the "true" cage effect of the Norrish Type I reactions.

$$F_c = \frac{F_{cAB} + R_-(1 - F_{cAB})}{F_{cAB} + R_+(1 - F_{cAB})} \tag{13.21}$$

$$F_{cAB} = \left(\frac{k_r}{k_r + k_{dAB}} \right) \left(\frac{k_{-CO}}{k_{-CO} + k_{dAB}} \right) \qquad (13.22)$$

F_{cAB} depends not only on k_r and k_{dAB} but also on k_{-CO}. Central to the results from the steady-state and laser-pulsed determinations of the cage effect, Equation 13.21 predicts that the "true" cage effect (F_{cAB}) and that obtained from photoproduct distributions (F_c) will be the same only when special conditions are met. In principle, there are two practical limiting conditions for F_c:

(1) If radical A$^\bullet$ is the same as radical B$^\bullet$ (as in the case of irradiations of **ACOB$_0$**), $k_{dAA} = k_{dBB}$ (the same diffusion rate), $R_- = 0$, $R_+ = 1$, and $F_c = F_{cAB}$. It is not possible to measure F_c experimentally from the product yields of **ACOB$_0$** because **AA**, **AB**, and **BB** are the same molecules. However, due to the similar structures and diffusivities of A$^\bullet$ and B$^\bullet$ from **ACOB$_1$**, the approximation that $k_{dAA} \cong k_{dBB}$ should be valid and F_{cAB} values from laser-flash experiments and F_c from steady-state irradiations should be (and are confirmed experimentally to be) almost the same in this case.

(2) However, when A$^\bullet$ and B$^\bullet$ have very different rates of diffusion, the value of F_c can be evaluated in the limit, $\frac{k_{dBB}}{k_{dAA}} \to 0$ (taking arbitrarily $k_{dAA} \gg k_{dBB}$, in keeping with the nature of B$^\bullet$ from **ACOB$_n$** when n is a large number). In such a case, $\mathrm{Lim}_{\frac{k_{dBB}}{k_{dAA}} \to 0}(F_c) = 1$.

Case (2) indicates that when $k_{dAA} \gg k_{dBB}$, F_c approaches unity independent of the absolute values of k_{dAA} and F_{cAB}. Hence, the observed value $F_c \sim 1$ measured for **ACOB$_{16}$** in the **PE** films can be ascribed to a large difference in the mobilities of the two benzylic radicals and the real cage effect must be obtained via analyses of transient absorption profiles such as those in Fig. 13.8. The observed excess of **AB** in the polymer films in the case of **ACOB$_{16}$** is *not* (or not exclusively) a result of the cage effect, but rather to the enhancement of "out-of-cage" formation of **AB**. Conceptually, the low diffusivity of **B$^\bullet$** makes the probability of it finding another radical **B$^\bullet$** very small. As a result, the concentration of **B$^\bullet$** in the photostationary state will exceed that of radical **A$^\bullet$**. Also, the diffusion rate constant associated with the formation of **AB** (k_{dAB}) is not dramatically affected when $k_{dAA} \gg k_{dBB}$: although a radical B$_n$$^\bullet$ can be considered to be "static" during its lifetime in a very high-viscosity medium, the high mobility of radical A$^\bullet$ allows it to diffuse to sites occupied by radicals B$^\bullet$, producing **AB** preferentially because $k_{dAB} = k_{dAA}/2$. Hence, although the large accumulation of B$^\bullet$ in the photostationary state of a constant intensity irradiation experiment favors the formation of **BB**, it is offset by the relatively small value of k_{dBB}; although k_{dAA} is large, the formation of **AA** is offset by the small $[A^\bullet]$. As a result, the efficiency of out-of-cage formation of **AB** is greatly enhanced, independent of the value of the true cage effect factor, F_{cAB}. This situation is tantamount to "the persistent radical effect".[103]

The data in Table 13.8 are reasonable and self-consistent. Note that k_{CO} is not influenced by the degree of crystallinity of the **PE** matrices and matches the values found in fluid solutions of similar polarity.[100] Interestingly, in all cases, $k_{-CO} \gg k_{dAB}$, confirming that in high-viscosity media, loss of carbon monoxide from the arylacetyl part is fast enough to compete with cage-escape by one of the radicals.

The k_{dAB} in Table 13.8 are approximately five orders of magnitude lower than in fluid alkane solutions, but are comparable to that measured for the combination of triplets radical pairs formed upon quenching the triplet excited state of **B** by 2,4,6-trimethylphenol in **PMMA** or **PVC** (Fig. 13.7). The larger value of k_{dAB} for **ACOB$_{16}$** than for **ACOB$_0$** or **ACOB$_1$** in the three **PE** films may be indicative of more than one type of cage even within the amorphous regions. It has been conjectured that the **ACOB$_{16}$**, because of strong London dispersive interactions with long, unbranched segments of the polymer chains, are able to reside preferentially in cages which differ somewhat from those preferred by **ACOB$_0$** and **ACOB$_1$**.[29] If that is correct, what appears to be faster random diffusional motion over an average distance by a *p*-hexadecylbenzyl radical may, in fact, be slower directed motion over shorter distances.

The k_r from one of the **ACOB$_n$** did increase between the completely amorphous **PE0** and partially crystalline (**LDPE** and **HDPE**) films, but the differences among the k_r values in one **PE** film for the three **ACOB$_n$** are nearly the same. At first glance, a tentative conclusion from these data is that the in-cage radical combination rates may increase slightly in more confining environments. However, other interpretations are also possible (see below). Crystallinity is only one of several factors that contribute to the nature of the sites of **PE** in which radical pairs combine. The k_r in Table 13.8 are three to four orders of magnitude lower than the analogous values obtained for combination of singlet radical pairs from photo-Fries lyses of aromatic esters in the same media.[77,79] Because the singlet radical pairs are similar in size to the A•/B• from the **ACOB$_n$**, and the steric requirements for combination of the singlet pairs are more severe than those of the A•/B• triplets, the rate-limiting process for in-cage combination of the triplets seems to be intersystem crossing. In principle, k_r is the rate for hyperfine-coupling-induced intersystem crossing followed by fast, spin-allowed bond formation. Because intersystem crossing usually requires some spatial separation of a radical pair (so that exchange interactions are minimized and hyperfine coupling interactions become comparable to electron exchange interactions), k_r should represent a convolution of spin evolution and radical diffusion processes.[29b,104,105] The experimental data in Table 13.8 show that an increase of k_r is always accompanied by an increase of k_{dAB}. However, given the large number of approximations involved in the model used for the interpretation of the decay curves (Fig. 13.8), this conclusion will require additional testing before it can be accepted fully.

Finally, the values of k_r (or k_{av}) estimated for the combination of benzophenone/2,4,6-trimethylphenol (Fig. 13.7) in the **PMMA** and **PVS** polymer matrices are approximately two orders of magnitude larger than the k_r calculated for the ^{3} Ȧ Ḃ from the **ACOB$_n$** in the **PE** films. While the combination of the ^{3} BḢ Ṙ is apparently controlled by molecular dynamics, spin dynamics seems to be mostly responsible in the case of triplet benzyl radical pairs. The hetero atoms in the ^{3} BḢ Ṙ pairs are probably responsible for the fast intersystem crossing process by enhancing spin-orbital coupling (for example, by allowing electrons in singly occupied orbitals to move from p_x to p_y as the electron spin changes). Considering that the estimated values of k_{dAB} are similar in both cases, the substantially different k_r have practical consequences. Although most of ^{3} B$\dot{\text{H}}$ $\dot{\text{R}}$ pairs combine "in-cage" in the first 5 µs, an important fraction of the benzyl radicals from the **ACOB**$_n$ in the **PE** films are able to escape from their cages of origin and, therefore, the laser-flash decay traces observed at $t > 30$ µs are extremely sensitive to the presence of oxygen and other possible scavenges in the polymeric matrices.

13.7   CONCLUDING REMARKS

Although aspects of the processes suffered by radical pairs in polymeric media have been investigated for many years, and information about those processes is of fundamental importance to industry and new technologies, the field is still in its infancy. Few studies have delved into the detailed mechanisms by which radical pairs move and interact in the confining environments of bulk polymers.

Clearly, there are many benefits to investigating radical processes in bulk polymers because the media constitute a "viscous space" in which the processes suffered by radicals (and, especially, pairs of radicals) are slowed significantly, allowing them to be observed more easily. We have provided an example in which decarbonylation can be used as a "clock" over time domains that are appropriate in polymeric media, but would be too slow in many fluid solvents. A great deal of basic scientific information about the dynamics of polymeric radicals can be derived as well.

The information presented in this chapter is intended to provide a platform for future studies. It has attempted to point readers toward challenges and possible approaches for those studies. The challenges and opportunities are manifold.

ACKNOWLEDGMENTS

The authors thank the U.S. National Science Foundation and the Petroleum Research Fund of the American Chemical Society for their support of the research from Georgetown University that is described here. The authors are also grateful to their colleagues, Weiqiang Gu, Jinqi Xu, Changxing Cui, Xiaoqiang Wang, Tadashi Mori, Jyotirmayee Mohanty, Werner M. Nau, Urbashi Bhattacharjee, Philippe Passin, Pavela Meakin, Anita J. Hill, and Steven Pas, for their contributions to the research presented.

REFERENCES

1. (a) Weiss, R. G.; Ramamurthy, V.; Hammond, G. S. *Acc. Chem. Res.* **1993**, *26*, 530. (b) Ramamurthy, V.; Weiss, R. G.; Hammond, G. S. *Adv. Photochem.* **1993**, *18*, 67.
2. A classic example is the first intermediates along the reaction pathway of Norrish-Yang (Type II) reactions. See for example: (a) Wagner, P. J.; Wagner, P. J.; Park, B. -S. *Org.*

Photochem. **1991**, *11*, 227. (b) *Acc. Chem. Res.* **1989**, *22*, 83. (c) Wagner, P. J. *Top. Curr. Chem.* **1976**, *6*, 1. (d) Turro, N. J.; Dalton, J. C.; Dawes, K.; Farrington, G.; Hautala, D.; Morton, D.; Niemczyk, M.; Schore, N. *Acc. Chem. Res.* **1972**, *5*, 92. (e) Yang, C.; Elliott, S. P. *J. Am. Chem. Soc.* **1969**, *91*, 7550. (f) Stephenson, L. M.; Cavigli, P. R.; Parlett, J. L. *J. Am. Chem. Soc.* **1971**, *93*, 1984.

3. See for example: Zhang, G. -H.; Thomas, J. K. *J. Phys. Chem. A* **1998**, *102*, 5465.

4. See for example: Miyasaka, H.; Khan, S. R.; Itaya, A. *J. Photochem. Photobiol. C Photochem. Rev.* **2003**, *4*, 195.

5. Guillet, J. *Polymer Photophysics and Photochemistry: An Introduction to the Study of Photoprocesses in Macromolecules*; Cambridge University Press: New York, 1985.

6. Winnik, M. A. Ed.; *Photophysical and Photochemical Tools in Polymer Science: Conformation, Dynamics, Morphology*; Reidel Pub. Co.: Hingham, MA, 1986.

7. Brandrup, J.; Immergut, E. H.; Grulke, E. A., Eds.; Polymer Handbook; Wiley: New York, 1999.

8. (a) Vieth, W. R. *Diffusion in and Through Polymers*; Hanser: New York, 1991. (b) Comyn, J., Ed.; *Polymer Permeability*; Elsevier: New York, 1985. (c) Neogi, P., Ed.; *Diffusion in Polymers*; Marcel Dekker: New York, 1996.

9. (a) Matsuoka, S., Ed.; *Relaxation Phenomena in Polymers*; Hanser: New York, 1992. (b) Bartenev, G. M.; Zelenev, Yu. V., Eds.; *Relaxation Phenomena in Polymers*; Keterpress: Jerusalem, 1974.

10. (a) Angell, C. A.; Ngai, K. L.; McKenna, G. B.; McMillan, P. F.; Martin, S. W. *J. Appl. Phys.* **2000**, *66*, 3113. (b) Florey, A.; McKenna, G. B. *Macromolecules* **2005**, *38*, 1760. (c) Florey, A. L.; McKenna, G. B. *Polymer* **2005**, *46*, 5211. (d) Alcoutlabi, M.; McKenna, G. B. *J. Phys. Condens. Matter* **2005**, *17*, R461. (e) Alcoutlabi, M.; Banda, L.; McKenna, G. B. *Polymer* **2004**, *45*, 5629.

11. (a) Royal, S. J.; Victor, J. G.; Torkelson, J. M. *Macromolecules* **1992**, *25*, 729. (b) Hall, D. B.; Dhinojwala, A.; Torkelson, J. M. *Phys. Rev. Lett.* **1997**, *79*, 103. (c) Priestley, R. D.; Ellison, C. J.; Broadbelt, L. J.; Torkelson, J. M. *Science* **2005**, *309*, 456.

12. Pekarski, P.; Hampe, J.; Bohm, I.; Brion, H. G.; Kirchheim, R. *Macromolecules* **2000**, *33*, 2192–2199.

13. (a) Wang, Z.; Holden, D. A.; McCourt, F. R. W. *Macromolecules* **1990**, *23*, 3773. (b) Yoshii, K.; Machida, S.; Horie, K. *J. Polym. Sci.: Part B: Polym. Phys.* **2000**, *38*, 3098.

14. Vrentas, J. S.; Vrentas, C. M. *J. Polym. Sci.: Part B: Polym. Phys.* **2003**, *41*, 501.

15. Norman, I.; Porter, G. *Proc. R. Soc. Lond. A Math. Phys. Sci.* **1955**, *230*, 399.

16. Duda, J. L.; Zielinski, J. M.; In *Diffusion in Polymers*, Neogli, P., Ed.; Marcel Dekker: New York, 1996; Chapter 3.

17. Especially in polar media where dipole–dipole and H-bonding interactions with the solvent may be important, radicals are known to diffuse translationally more slowly than their parent molecules (in which an H-atom has been added to the radical center). These effects are usually small, and we are not aware of studies in bulk polymers which they have been quantified. (a) Terazima, M. *Acc. Chem. Res.* **2000**, *33*, 687.

18. Steiner, U. E.; Wolff, H. -J. In *Photochemistry and Photophysics*; Rabek, J. F., Ed.; CRC Press: Boca Raton, 1991; Vol. 4, Chapter 1.

19. See for instance: (a) Rice, S. A.; In *Diffusion-Limited Reactions. Comprehensive Chemical Kinetics*; Bamford, C. H.; Tipper, C. F. H.; Compton, R. G., Eds.; Elsevier: Amsterdam,

1985; Vol. 25, pp 1–400. (b) Burshtein, A. I. in *Unified Theory of Photochemical Charge Separation. Advances in Chemical Physics* Prigogine, I.; Rice, S. A., Eds.; Wiley: New York, 2000; Vol. 114, pp 419–587.

20. The survival probability relates the concentration of the radical pairs, $[RP]$, at a given elapsed time t, $[RP]_{t=t}$, to the initial concentration of geminate pairs, $[RP]_{t=0}$ according to $P(t) = [RP]_{t=1}/[RP]_{t=0}$.

21. Shin, K. J.; Kapral, R. *J. Chem. Phys.* **1978**, *69*, 3685.

22. Collins, F. C.; Kimball, G. E. *J. Colloid Sci.* **1949**, *4*, 425.

23. Smoluchowski M. V. *Z. Phys. Chem.* **1917**, *92*, 129.

24. Kelly, D. P.; Serelis, A. K.; Solomon, D. H.; Thompson, P. E. *Australian J. Chem.* **1987**, *40*, 1631.

25. Kopeckyp, K.; Pope, P. M.; Lopez-Sastre, *J. Can. J. Chem.* **1976**, *54*, 639.

26. Engel, P. S. *Chem. Rev.* **1980**, *80*, 99.

27. Hrovat, D. A.; Liu, J. H.; Turro, N. J.; Weiss, R. G. *J. Am. Chem. Soc.* **1984**, *106*, 5291.

28. (a) Ruzicka, R.; Barakova, L.; Klan, P. *J. Phys. Chem. B* **2005**, *109*, 9346–9353. (b) Veerman, M.; Resendiz, M. J. E.; Garcia-Garibay, M. A. *Org. Lett.* **2006**, *8*, 2615. (c) Petrova, S. S.; Kruppa, A. I.; Leshina, T. V. *Chem. Phys. Lett.* **2004**, *385*, 40. (d) Aspée, A.; Maretti, L.; Scaiano, J. C. *Photochem. Photobiol. Sci.* **2003**, *2*, 1125–1129. (e) Lipson, M.; Noh, T. H.; Doubleday, C. E.; Zaleski, J. M.; Turro, N. J. *J. Phys. Chem.* **1994**, *98*, 8844. (f) Roberts, C. B.; Zhang, J.; Brennecke, J. F.; Chateauneuf, J. E. *J. Phys. Chem.* **1993**, *97*, 5618.

29. (a) Chesta, C. A.; Mohanty, J.; Nau, W. M.; Bhattacharjee, U.; Weiss, R. G. *J. Am. Chem. Soc.* **2007**, *129*, 5012. (b) Bhattacharjee, U.; Chesta, C.; Weiss, R. G. *Photochem. Photobiol Sci.* **2004**, *3*, 287.

30. (a) Scott, T. W.; Liu, S. N. *J. Phys. Chem.* **1989**, *93*, 1393. (b) Hirata, Y.; Niga, Y.; Makita, S.; Okada, T. *J. Phys. Chem.* **1997**, *101*, 561. (c) Autrey, T.; Devdoss, C.; Sauerwein, B.; Franz, J. A.; Shuster, G. B. *J. Phys. Chem.* **1995**, *99*, 869.

31. (a) Elles, C. G.; Cox, M. J.; Barnes, G. L.; Crim, F. F. *J. Phys. Chem. A* **2004**, *108*, 10973. (b) Lipton, M.; Deniz, A. A.; Peters, K. S. *J. Phys. Chem.* **1996**, *100*, 3580. (c) Oelkers, A. B.; Scatena, L. F.; Tyler, D. R. *J. Phys. Chem.* **2007**, *111*, 5353.

32. (a) Hammond, G. S.; Waits, H. P. *J. Am. Chem. Soc.* **1964**, *86*, 1911 and references therein. (b) Kiefer, H.; Traylor, T. *J. Am. Chem. Soc.* **1967**, *89*, 6667. (c) Ghibaudi, E.; Colussi, A. J. *Chem. Phys. Lett.* **1982**, *94*, 121.

33. (a) Glasstone, S.; Laidler, K.; Eyring, H. *Theory of Raze Processes*; McGraw-Hill: New York, **1941**; Chapter 9. (b) Koening, T. *J. Am. Chem. Soc.* **1969**, *91*, 2558. (c) Olea, A. F.; Thomas, J. K. *J. Am. Chem. Soc.* **1988**, *110*, 4494. (d) Niki, E.; Kamiya, Y. *J. Am. Chem. Soc.* **1974**, *96*, 2129.

34. Einstein, A. *Investigations on the Theory of Brownian Motion*; Dover: New York, **1956**.

35. Levin, P. P.; Ivanov, V. B.; Selikhov, V. V.; Kuźmin, V. A. *Ivz. Akad. Nauk. SSSR, Ser. Khim.* **1988**, *8*, 1742.

36. Kutsenova, A. V.; Kutyrkin, V. A.; Levin, P. P.; Ivanov, V. B. *Kinet. Catal.* **1998**, *39*, 335.

37. Kutsenova, A. V.; Levin, P. P.; Ivanov, V. B. *Polymer Sci. Ser. B* **2002**, *44*, 124.

38. Levin, P. P.; Kutyrkin, V. A.; Kutsenova, A. V. *Khim. Fiz.* **1989**, *8*, 1338.

39. Spernol, A.; Wirtz, K. *Z. Naturforsch, A* **1953**, *8*, 522. $f = (0.16 + 0.4r_A/r_B)(0.9 + 0.4T_{rB} - 0.25T_{rA})$. In this equation, r_A and r_B are the radii of solute and solvent molecules,

respectively, and can be estimated from their van der Waals volumes (V; $\mathring{A}^3$) assuming the molecules are spherical ($r = (3V\chi/4\pi)^{1/3}$, $\chi = 0.74$ is the space filling factor for closest packed spheres, and T_{rA} and T_{rB} are reduced temperatures of solute and solvent, respectively, and can be calculated from the equation, $T_r = (T - T_{mp})/(T_{bp} - T_{mp})$, where T_{mp} and T_{bp} are melting and boiling points, respectively, of each molecule.

40. (a) Kramers, H. A. *Physica* **1940**, *7*, 284. (b) Akesson, E.; Hakkarainen, A.; Laitinen, E.; Helenius, V.; Gillbro, T.; Korppi-Tommola, J.; Sundstrom, V. *J. Chem. Phys.* **1991**, *95*, 6508. (c) Ben-Amotz, D.; Drake, J. M. *J. Chem. Phys.* **1988**, *89*, 1019. (d) Dote, J. L.; Kivelson, D.; Schwartz, R. N. *J. Phys. Chem.* **1981**, *85*, 2169. (e) Sun, Y.; Saltiel, J. *J. Phys. Chem.* **1989**, *93*, 8310. (f) Terazima, M.; Okamoto, K.; Hirota, N. *J. Chem. Phys.* **1995**, *102*, 2506. (g) Kim, S. H.; Kim, S. K. *Bull. Korean Chem. Soc.* **1996**, *17*, 365.

41. (a) Bagryanskaya, E.; Fedin, M.; Forbes, M. D. E. *J. Phys. Chem. A* **2005**, *109*, 5064. (b) McCaffrey, V. P.; Harbron, E. J.; Forbes, M. D. E. *J. Phys. Chem. B* **2005**, *109*, 10686.

42. Lei, X. G.; Jockusch, S.; Ottaviani, M. F.; Turro, N. J. *Photochem. Photobiol. Sci.* **2003**, *2*, 1095.

43. Gu, W.; Weiss, R. G. *J. Photochem. Photobiol. C* **2001**, *2*, 117.

44. Chatgilialoglu, C.; Crich, D.; Komatsu, M.; Ryu, I. *Chem. Rev.* **1991**, *99*, 1991 and references cited therein.

45. Jimenez, M. C.; Leal, P.; Miranda, M. A.; Tormos, R. *J. Chem. Soc., Chem. Commun.* **1995**, 2009.

46. (a) Miranda, M. A. In *CRC Handbook of Organic Photochemistry and Photobiology*; Horspool, W. M.; Song, P. -S., Eds.; CRC Press: Boca Raton, 1995; Chapter 47. (b) Cui, C.; Wang, X.; Weiss, R. G. *J. Org. Chem.* **1996**, *61*, 1962–1974. (c) Miranda, M. A.; Galindo, F. In *Photochemistry of Organic Molecules in Isotropic and Anisotropic Media*; Ramamurthy, V.; Schanze, K. S., Eds.; Marcel Dekker: New York, 2003; Chapter 2.

47. Gu, W.; Weiss, R. G. *Tetrahedron* **2000**, *56*, 6913.

48. Cui, C.; Wang, X.; Weiss, R. G. *J. Org. Chem.* **1996**, *61*, 1962.

49. From extrapolation of data by: Nakagaki, R.; Hiramatsu, M.; Watanabe, T.; Tanimoto, Y. *J. Phys. Chem.* **1985**, *89*, 3222.

50. (a) Hammond, G. S. *J. Am. Chem. Soc.* **1955**, *77*, 334. (b) Agmon, N. *J. Chem. Soc., Faraday Trans. 2* **1978**, *74*, 388. (c) Arteca, G. A.; Mezey, P. G. *J. Phys. Chem.* **1989**, *93*, 4746.

51. Mori, T.; Takamoto, M.; Saito, H.; Furo, T.; Wada, T.; Inoue, Y. *Chem. Lett.* **2004**, *33*, 254.

52. (a) Dixon, W. T.; Moghimi, M.; Murphy, D. *J. Chem. Soc., Faraday Trans. II* **1974**, *70*, 1713. (b) Dixon, W. T.; Foster, W. E. J.; Murphy, D. *J. Chem. Soc., Perkin Trans, II* **1973**, *15*, 2124.

53. Gu, W. Ph.D. Thesis, Georgetown University, Washington, DC, 2000.

54. (a) Dixon, W. T.; Foster, W. E. J.; Murphy, D. *J. Chem. Soc., Perkin Trans.* **1972**, *2*, 2124. (b) Stone, T. J.; Waters, W. A. *J. Chem. Soc.* **1962**, 253.

55. (a) Ghibaudi, E.; Colussi, A. *Chem. Phys. Lett.* **1983**, *94*, 121. (b) Plank, E. D. A.; Ph.D. Thesis, Purdue University, 1966, as cited in: Sander, M. R.; Hedaya, E.; Trecker, D. J. *J. Am. Chem. Soc.* **1968**, *90*, 7249.

56. (a) Finnegan, R. A.; Knutson, D. *J. Am Chem. Soc.* **1967**, *89*, 1971. (b) Finnegan, R. A.; Knutson, D. *Chem. Ind.* **1965**, 1837.

57. (a) Gu, W.; Abdallah, D. J.; Weiss, R. G. *J. Photochem. Photobiol. A Chem.* **2001**, *139*, 79. (b) Mori, T.; Inoue, Y.; Weiss, R. G. *Org. Lett.* **2003**, *5*, 4661. (c) Mori, T.; Weiss, R. G.; Inoue, Y. *J. Am. Chem. Soc.* **2004**, *126*, 8961.

58. (a) Tsentalovich, Y. P.; Fischer, H. *J. Chem. Soc., Perkin Trans.* **1994**, *2*, 729. (b) Braun, W.; Rajbenbach, L.; Eirich, F. R. *J. Phys. Chem.* **1962**, *66*, 1591.

59. (a) Turro, N. J.; Gould, I. R.; Baretz, B. H. *J. Phys. Chem.* **1983**, *87*, 531. (b) Zhang, X. Y.; Nau, W. M. *J. Phys. Org. Chem.* **2000**, *13*, 634.

60. (a) Kurnysheva, O. A.; Gritsan, N. P.; Tsentalovich, Y. P. *Phys. Chem. Chem. Phys.* **2001**, *3*, 3677. (b) Lunazzi, L.; Ingold, K. U.; Scaiano, J. C. *J. Phys. Chem.* **1983**, *87*, 529.

61. In this chapter, we define an "alkyl" group as one that is bonded to an ether or ester group via a saturated carbon atom. The total structure may contain unsaturation.

62. (a) Waespe, H. R.; Hansen, H. J.; Paul, H.; Fischer, H. *Helv. Chim. Acta* **1978**, *61*, 401. (b) Adam, W.; Fischer, H.; Hansen, H. J.; Heimgartner, H.; Schmid, H.; Waespe, H. R. *Angew. Chem., Int. Ed. Engl.* **1973**, *12*, 662.

63. (a) Pohlers, G.; Grimme, S.; Dreeskamp, H. *J. Photochem. Photobiol. A Chem.* **1994**, *79*, 153. (b) Grimme, S.; Dreeskamp, H. *J. Photochem. Photobiol. A Chem.* **1992**, *65*, 371.

64. Shimamura, N.; Sugimori, A. *Bull. Chem. Soc. Jpn.* **1971**, *44*, 281.

65. Gu, W.; Warrier, M.; Schoon, B.; Ramamurthy, V.; Weiss, R. G. *Langmuir* **2000**, *16*, 6977.

66. Pichumani, K.; Warrier, M.; Weiss, R. G.; Ramamurthy, V. *Tetrahedron Lett.* **1996**, *37*, 6251.

67. Tung, C. -H.; Xu, X. -H. *Tetrahedron Lett.* **1999**, *40*, 127.

68. (a) Wang, C.; Xu, J.; Weiss, R. G. *J. Phys. Chem. B* **2003**, *107*, 7015. (b) Gu, W.; Hill, A. J.; Wang, X. C.; Cui, C. X.; Weiss, R. G. *Macromolecules* **2000**, *33*, 7801. (c) Luo, C.; Atvars, T. D. Z.; Meakin, P.; Hill, A. J.; Weiss, R. G. *J. Am. Chem. Soc.* **2003**, *125*, 11879.

69. Cui, C.; Weiss, R. G. *J. Am. Chem. Soc.* **1993**, *115*, 9820.

70. Luo, C.; Passin, P.; Weiss, R. G. *Photochem. Photobiol.* **2006**, *82*, 163.

71. Gu, W.; Bi, S.; Weiss, R. G. *Photochem. Photobiol. Sci.* **2002**, *1*, 52.

72. Nagahara, K.; Ryu, I.; Kambe, N.; Komatsu, M.; Sonoda, N. *J. Org. Chem.* **1995**, *60*, 7384.

73. Birks, J. B. In *Organic Molecular Photophysics*; Birks, J. B., Ed.; Wiley: New York, 1973; Vol. 1, p 403.

74. Michaels, A. S.; Bixler, H. J. *J. Polym. Sci.* **1961**, *50*, 413.

75. Given the greater stability of a benzylic radical than an acyl radical, the rates of combination of benzylic/aryloxy pairs should be slower than those of phenylacyl/aryloxy pairs.

76. Xu, J.; George, M.; Weiss, R. G. *Ann. Brazil. Acad. Sci.* **2006**, *78*, 31.

77. Xu, J.; Weiss, R. G. *Photochem. Photobiol. Sci.* **2005**, *4*, 348.

78. (a) Phillips, P. J. *Chem. Rev.* **1990**, *90*, 425. (b) Jang, Y. T.; Phillips, P. J.; Thulstrup, E. W. *Chem. Phys. Lett.* **1982**, *93*, 66. (c) Meirovitch, E. *J. Phys. Chem.* **1984**, *88*, 2629. (d) Naciri, J.; Weiss, R. G. *Macromolecules* **1989**, *22*, 3928. (e) He, Z.; Hammond, G. S.; Weiss, R. G. *Macromolecules* **1992**, *25*, 1568. (f) Jenkins, R. M.; Hammond, G. S.; Weiss, R. G. *J. Phys. Chem.* **1992**, *96*, 496.

79. Gu, W.; Weiss, R. G. *J. Org. Chem.* **2001**, *66*, 1775.

80. Attempts to investigate this hypothesis further were foiled when it was found that irradiation of 5,6,7,8-tetrahydro-1-naphthyl phenylacetate, an ester that yields an aryloxy fragment similar in size and shape to 1-naphthoxy but whose electronic properties are more similar to those of phenoxy, did not result in detectable amounts of photoproducts analogous to **4-AN** or **4-BN**, apparently because of steric factors near the 4-position of the aryloxy group.[75]

81. For reviews on the stereoselectivity of radical pair combinations, see (a) Curran, D. P.; Porter, N. A.; Giese, B. *Stereochemistry of Radical Reactions: Concepts, Guidelines, and Synthetic Applications*; VCH: Weinheim, 1995; p 242, and references cited therein. (b) John, L. E. *An Introduction to Free Radicals*; Wiley: New York, 1993; p 56 and references cited therein.

82. Xu, J.; Weiss, R. G. *Photochem. Photobiol. Sci.* **2005**, *4*, 210.

83. Xu, J.; Weiss, R. G. *Org. Lett.* **2003**, *5*, 3077.

84. In the most constraining cages of solid *n*-nonadecane or stretched **LDPE** and **HDPE**, the ee of **2-AN** is somewhat lower because of a secondary photoprocess, analogous to the first step in the Norrish-Yang reaction,[2] that its keto precursor (Equation 13.5) under goes. (a) Mori, T.; Takamoto, M.; Saito, H.; Furo, T.; Wada, T.; Inoue, Y. *Chem. Lett.* **2004**, *33*, 256. (b) Jiménez, M. C.; Miranda, M. A.; Scaiano, J. C.; Tormos, R. *Chem. Commun.* **1997**, 1487.

85. Xu, J.; Weiss, R. G. *J. Org. Chem.* **2005**, *70*, 1243.

86. Kiefer, H.; Traylor, T. G. *J. Am. Chem. Soc.* **1967**, *89*, 6667.

87. The distance between the radical center in 1-phenylethyl and the oxygen atom of 1-naphthoxy immediately after lysis of (*R*)-**3b** is assumed to be the sum of their van der Waals radii, 3.22 Å. Then, the minimum distance traveled by the reactive radical center of 1-phenylethyl to reach the 4-position of 1-naphthoxy (without consideration of the orientation of approach) is estimated to be 5.1 Å based on a length for the newly formed C–C bond of 1.5 Å. (a) Allen, F. H.; Kennard, O.; Watson, D. G.; Brammer, L.; Orpen, A. G.; Taylor, R. *J. Chem. Soc., Perkin Trans. 2* **1987**, S1.

88. The distance between the radical center of 1-phenylethyl needed to form a bond at the 2-position of 1-naphthoxy immediately after lysis of **3b** is estimated to be $\sim$3 Å.[77]

89. Alwattar, A. H.; Lumb, M. D.; Birks, J. B. In *Organic Molecular Photophysics*; Birks, J. B., Ed.; Wiley: New York, 1973; Vol. 1, pp 403–456.

90. (a) Tarasov, V. F.; Shkrob, I. A.; Step, E. N.; Buchachenko, A. L. *Chem. Phys.* **1989**, *135*, 391. (b) Tarasov, V. F.; Ghatlia, N. D.; Buchachenko, A.; Turro, N. J. *J. Phys. Chem.* **1991**, *95*, 10220.

91. Conradi, M. S.; Zeldes, H.; Livingston, R. *J. Phys. Chem.* **1979**, *83*, 2160.

92. In addition to not addressing the need for individual tumbling rate constants for radical pair combinations leading to **3b**, **2-BN**, and **4-BN**, Scheme 13.5 does not take into account the fact the values of k_t probably differ and the translational rate constants definitely differ[83] within amorphous and interfacial cages of either **LDPE** or **HDPE**.

93. (a) Schurr, O. Ph.D Thesis, Georgetown University, Washington, DC, 2002. (b) Schurr, O.; Weiss, R. G. *Polymer* **2004**, *45*, 5713. (c) Taraszka, J. A.; Weiss, R. G. *Macromolecules* **1997**, *30*, 2467.

94. Few examples of direct comparisons of rates of reaction of different radicals with a common species are in the literature. In one, the (nondelocalized) *tert*-butyl radical was found to react more rapidly than pivaloyl radical with an electron-deficient partner, acrylonitrile, in 2-propanol.[a] This is not a good analogy to the comparison between 1-phenylethyl and 2-phenylpropanoyl being made here because we suspect that 1-naphthoxy is more electron-rich than acrylonitrile, polyethylene is much less polar than 2-propanol, and the odd-electron in a 1-phenylethyl radical is delocalized. (a) Jent, F.; Paul, H.; Roduner, E.; Heming, M.; Fischer, H. *Int. J. Chem. Kinet.* **1986**, *18*, 1113.

95. See for instance: (a) Pitts, J. N.; Lentsinger, R. L.; Taylor, R. P.; Patterson, J. M.; Recktenwald, G.; Martin, R. B. *J. Am. Chem. Soc.* **1959**, *81*, 1068. (b) Weiner, S. A. *J.*

Am. Chem. Soc. **1971**, *93*, 425. (c) Filipescu, N.; Minn, F. L. *J. Am. Chem. Soc.* **1968**, *90*, 1544. (d) Viltres Costa, C.; Grela, M. A.; Churio, M. S. *J. Photochem. Photobiol. A: Chem.* **1996**, *99*, 51 and references therein.

96. Emanuel', N. M.; Buchachenko, A. L. In *Chemical Physics of Molecular Degradation and Stabilization of Polymers*; Nauka: Moscow, 1988.

97. (a) Robbins, W. K.; Eastman, R. H. *J. Am. Chem. Soc* **1970**, *92*, 6076–6077. (b) Robbins, W. K.; Eastman, R. H. *J. Am. Chem. Soc* **1970**, *92*, 6077.

98. Engel, P. S. *J. Am. Chem. Soc.* **1970**, *92*, 6074.

99. The magnitudes of the calculated second-order rate constants (k_{r2}) were unreasonably large ($(1–5) \times 10^9 \, M^{-1}s^{-1}$) given the known rates of self-diffusion of molecules of similar size and shape in polyethylene films.

100. (a) Tsentalovich, Y. P.; Kurnysheva, O. A.; Gritsan, N. P. *Russ. Chem. Bull. Intern. Ed.* **2001**, *50*, 237. (b) Zhang, X.; Nau, W. M. *J. Phys. Org. Chem.* **2000**, *13*, 634. (c) Claridge, R. F. C.; Fischer, H. *J. Phys. Chem.* **1983**, *87*, 1960. (d) Tokumura, K.; Ozaki, T.; Nosaka, H.; Saigusa, Y.; Itoh, M. *J. Am. Chem. Soc.* **1991**, *113*, 4974. (e) Maouf, A.; Lemmetyinen, H.; Koskikallio, *J. Acta Chem. Scand.* **1990**, *44*, 36. (f) Turro, N. J.; Gould, I. R.; Baretz, B. H. *J. Phys. Chem.* **1983**, *87*, 531.

101. Crank, J. *The Mathematics of Diffusion*; 2nd ed., Oxford Press: Oxford, UK, 1975.

102. (a) Neogi, P., Ed.; *Diffusion in Polymers*; Marcel Dekker: New York, 1996. (b) Vieth, W. R. *Diffusion in and Through Polymers*; Hanser: Munich, 1991.

103. (a) Bachmann, W. E.; Wiselogle, F. Y. *J. Org. Chem.* **1936**, *1*, 354. (b) Perkins, M. J. *J. Chem. Soc.* **1964**, 5932. (c) Fischer, H. *Chem. Rev.* **2001**, *101*, 3581–3610. (d) Studer, A. *Chem. Eur. J.* **2001**, *7*, 1159.

104. Bohne, C.; Alnajjar, S.; Griller, D.; Scaiano, J. C. *J. Am. Chem. Soc.* **1991**, *113*, 1444.

105. Tarasov, V. F.; Ghatlia, N. D.; Buchachenko, A. L.; Turro, N. J. *J. Am. Chem. Soc.* **1992**, *114*, 9517.

14

ACRYLIC POLYMER RADICALS: STRUCTURAL CHARACTERIZATION AND DYNAMICS

MALCOLM D. E. FORBES AND NATALIA V. LEBEDEVA

Department of Chemistry, University of North Carolina, Chapel Hill, NC, USA

14.1 INTRODUCTION

Polymer degradation is a subject of intense interest to chemists, physicists, and materials scientists.[1–3] Mechanistic information about degradation processes is highly sought after in fields such as environmental science,[4] aerospace,[5] and medicine.[6] Acrylic polymers are used in many applications where exposure to intense light sources is common, such as bendable light pipes in the automobile industry,[7] architectural coatings,[8] and lithography.[9,10] The degradation mechanisms of acrylates often involve free radical reactions,[11–13] therefore knowledge of the structure, molecular dynamics, and chemical reactivity of these reactive intermediates is highly desirable. Electron paramagnetic resonance (EPR) spectroscopy is a powerful tool for such studies and has contributed much to our understanding of polymer degradation and dynamics over the past 60 years.[14,15] This chapter will focus on the use of time-resolved (CW) EPR spectroscopy (TREPR) experiments to investigate free radicals produced via acrylic polymer photodegradation in liquid solution. These studies have been carried out in the authors' laboratory for the last decade.[16–20] The unique ability of TREPR to record spectra of radicals created from the *primary* photochemical events after an acrylic polymer absorbs UV light has provided much insight into macromolecular free radical structure and dynamics.

Carbon-Centered Free Radicals and Radical Cations, Edited by Malcolm D. E. Forbes
Copyright © 2010 John Wiley & Sons, Inc.

325

14.2   THE PHOTODEGRADATION MECHANISM

The photodegradation of acrylic polymers, illustrated in Scheme 14.1 for a generic structure, has been extensively studied by photochemists and polymer scientists for many years.[14,15] Despite some contrary interpretations,[21] there is a general consensus that Norrish I α-cleavage of the ester side chain is the first bond breaking event after absorption of UV light by the ester chromophore. The result of side-chain cleavage is a main-chain polymeric radical **a** and a smaller oxo-acyl radical **b** as a primary radical pair. Radical **a** has available to it a unimolecular decomposition pathway, namely, β-scission,[22] to give a terminal alkene and the so-called propagating radical **c**.[23,24] This reaction is analogous to the reverse of the free radical polymerization reaction, that is, free radical addition to the alkene monomer.[25]

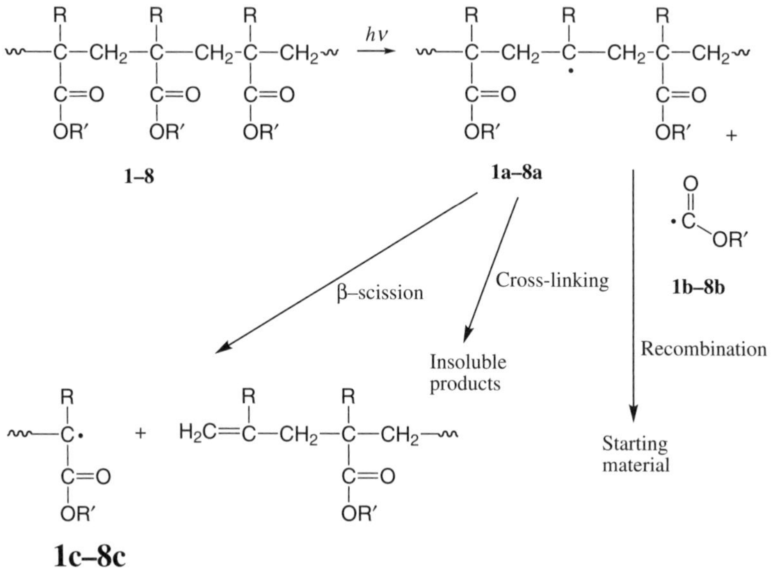

SCHEME 14.1   The general structure of an acrylic polymer and the established photodegradation mechanism via Norrish I α-cleavage of the carbonyl side chain, leading to main-chain polymeric radical **a** and oxo-acyl radical **b**. The secondary β-scission rearrangement reaction leading to the propagating radical **c** is also shown.

Because of the rapid β-scission reaction, steady-state electron paramagnetic resonance (SSEPR) investigations carried out since 1951 on acrylate and methacrylate degradation were unable to confirm the chemistry shown in Scheme 14.1 as the first step. In 1982, Liang et al. suggested the existence of a main-chain acrylate radical in a cold SSEPR experiment,[26] but the spectrum contained many overlapping lines from other radicals, so a definitive conclusion regarding its structure could not be reached. In fact, main-chain polymeric radical such as **a** had not been unambiguously characterized in any work except in that reported here. The TREPR experiments described below will provide unambiguous spectral assignment of both radicals **a** and **b** from many acrylic polymers, confirming the mechanism shown in Scheme 14.1. In addition, our results show novel features beyond the original mechanistic problem, in particular regarding nuclear spin symmetry relationships in these polymeric free radicals, their chain dynamics and spin relaxation processes in solution, and the anomalous intensities associated with chemically induced electron spin polarization (CIDEP) mechanisms[27] that almost always accompany TREPR spectra.

14.3 POLYMER STRUCTURES

The photochemistry and ensuing free radical chemistry of the eight acrylic polymers shown in Chart 14.1 have been studied intensely over the past decade in our laboratory.[16–20] Structural modification of acrylates can be achieved by modifying either the substituent on the polymer backbone (α-substitution), or the ester side chain (β-substitution). Poly(methyl methacrylate) (PMMA), **1**, was the starting point for our investigations as it is the most ubiquitous acrylic polymer. Common commercial names for PMMA include Lucite, Perspex, and Plexiglass; in many applications, PMMA is exposed to UV light where radiation damage can occur. In EPR spectroscopy, it is often desirable to confirm spectral assignments by isotopic labeling, and therefore the deuterated analog of PMMA was also studied. Here, the deuterium substitution is only on the backbone methyl group and the polymer is abbreviated as d_3-PMMA, structure **2** in Chart 14.1.

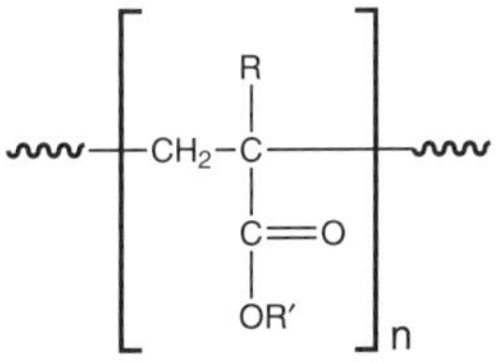

Polymer	Acronym	R	R′
1 Poly(methyl methacrylate)	PMMA	CH_3	CH_3
2 Poly(methyl d_3-methacrylate)	d_3-PMMA	CD_3	CH_3
3 Poly(ethyl methacrylate)	PEMA	CH_3	CH_2CH_3
4 Poly(ethyl cyanoacrylate)	PECA	CN	CH_2CH_3
5 Poly(ethyl acrylate)	PEA	H	CH_2CH_3
6 Poly(fluorooctyl methacrylate)	PFOMA	CH_3	$CH_2(CF_2)_xCF_3$
7 Poly(acrylic acid)	PAA	H	H
8 Poly(methacrylic acid)	PMAA	CH_3	H

CHART 14.1 Structures of all polymers investigated in this work, numbered with their common acronyms from the polymer literature.

The next three polymers in this series are all ethyl acrylates, meaning that while the backbone (α) substituent is different in all three structures, the ester side-chain group (β) is the same for all of them ($-CH_2CH_3$). Polymer **3** is poly(ethyl methacrylate) (PEMA), and **4** is poly(ethyl cyanoacrylate) (PECA), which may be recognizable as a primary component of the so-called "superglues." Polymer **5** is poly(ethyl acrylate) (PEA), with H on the backbone α-position. From structure **3** to **4** to **5**, the α-substituent becomes simpler in structure and this will be reflected in the observed TREPR spectra below in terms of the number of observed transitions, and in some cases the linewidths as well.

Poly(fluorooctyl methacrylate) (PFOMA, structure **6**) has very different physical properties from the other four methacrylates shown in Chart 14.1. Primarily this is due to the bulky (and more rigid) β-substituent, which is actually a mixture of straight chain and branched perfluorooctyl groups. The rigidity of the main chain enforced by these bulky groups has drastic consequences on the appearance of the TREPR spectra of the resulting radicals. PFOMA was the first polymer to be synthesized in liquid CO_2 and was the basis of several new surfactant technologies developed for recyclable solvents.

The last two polymer structures listed in Chart 14.1 are poly(acrylic acid) (PAA, **7**) and poly(methacrylic acid) (PMAA, **8**), which exhibit interesting properties due to their water solubility and ionic strengths. From a structural perspective, PAA and PMAA are the simplest polyelectrolytes, and they are of significant interest as biodegradable materials for wound dressings[28] and other biomedical and bioanalytical[29] applications. The main difference between the polyacrylic acids and their ester counterparts is the presence of charges on the carboxyl side chains in solution. The degree of ionization of these functional groups can influence the morphology of the polymer chains in solution (coiled versus stretched),[30] the redox properties of the carboxylate groups,[31] and the nature of the excited states involved in the photodegradation process.[32] The water solubility of polymers **7** and **8** has some experimental ramifications: EPR in aqueous solution presents some technical challenges due to the high dielectric constant of the sample. In addition, there is the possibility of pH and/or ionic strength dependences of their photochemical reactivity and their TREPR spectral features.

The polymers selected for these studies are structurally similar in that the primary photophysics and chemistry are not expected to change; that is, we do not expect a major deviation from the mechanism shown in Scheme 14.1 in terms of UV photodegradation for these macromolecules. However, the small structural variations from one polymer to another have been chosen to allow structural, dynamic, solvent, and pH effects in the ensuing free radicals to be probed as systematically as possible. It should be noted that all of the polymers under investigation are, to the best of our knowledge, homopolymers of high molecular weight ($M_w > 10,000$), high purity ($>95\%$ by NMR and GPC), and unless otherwise indicated, atactic in terms of macromolecular stereochemistry. Tacticity is an important factor in magnetic resonance of polymers, a fact long recognized in the NMR research community[33] but, as we will show below, it is an issue that was somewhat underappreciated by EPR spectroscopists because of the lack of clean, well-resolved spectra of main-chain macromolecular free radicals.

14.4 THE TIME-RESOLVED EPR EXPERIMENT

In routine steady-state EPR spectroscopy, the transitions are detected by sweeping an external magnetic field B_0 through each resonance at a constant microwave frequency ω_0. The external field is modulated, usually at a frequency of 100 kHz, so that phase-sensitive detection can be used to increase the signal to noise (S/N) ratio and the spectral resolution.[34] The resulting spectra have first-derivative line shapes, and care is usually taken to keep the amplitude of the field modulation smaller than the linewidth to avoid line shape distortions. A consequence of the 100 kHz field modulation is that the time response of the spectrometer becomes limited to, at best, the inverse of the modulation frequency. Practically, for good S/N, three or four cycles of modulation are necessary, which means that species with chemical lifetimes less than about 40 μs become difficult to detect. Since most organic radicals have lifetimes in solution on the order of 10–100 μs, their detection can be problematic in SSEPR. Lower temperatures (below $-50°C$) can help with this problem. Continuous, intense light or heat can be used to generate large steady-state concentrations of radicals.[35]

The CIDEP enhancements of 1–100 above the Boltzmann population differences are common to TREPR spectra and decay with T_1 values on the order of 1 to 10 μs. This makes CIDEP difficult (but not impossible if strong enough) to observe at steady state. Historically, it was quickly recognized that a significant amount of kinetic and magnetic information could be obtained by studying the CIDEP mechanisms, and therefore an EPR experiment with submicrosecond time resolution and response became highly desirable. The earliest attempts to build such an apparatus were coupled to pulse radiolysis instrumentation by Avery and Smaller[36] and by Fessenden,[37] who also made seminal contributions to the analysis of TREPR data.[38] The apparatus and methodology used in the authors' laboratory couples laser flash photolysis to EPR, which was first developed by Trifunac and coworkers[39] and used widely by others such as McLauchlan,[40] Levanon,[41] van Willigen,[42] and Hirota.[43] The experiment found great utility in photosynthesis research, most notably in the research groups of Hoff,[44] Mobius,[45] Lubitz,[46] and Norris.[47]

In our X-band TREPR apparatus, temporal resolution is achieved by pulsing the production of the radicals, and then gating the detection system. Pulsed production of radicals is typically accomplished using an excimer or YAG laser. The EPR signal from the microwave bridge preamplifier is passed directly to a boxcar gated integrator or to a transient digitizer, which allows the signal to be trapped and stored on the submicrosecond timescale. Figure 14.1a shows how the apparatus is connected, and Fig. 14.1b shows the timing sequence. The phase-sensitive detection system of the SSEPR system is bypassed (hence, the experiment is sometimes called "direct detection"). The only additional components to the TREPR spectrometer are the timing electronics (usually a Stanford DG535 delay generator or equivalent), a boxcar signal averager, and the laser (excimer or Nd^{3+}:YAG). A computer is used to store and display the data as an xy array (B_0, TREPR intensity). Quartz flow cells are used to avoid heating and sample depletion, and either holes or slots are cut into the sides of the resonators to allow light access. The microwave excitation is continuous wave throughout the experiment, even during the production of the radicals, as

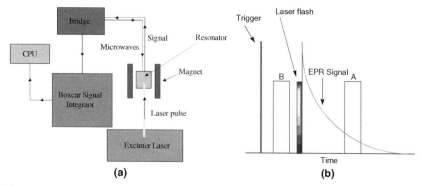

FIGURE 14.1 The Time–resolved EPR experiment. (a) The apparatus. (b) The timing sequence (see text for scale and definitions).

opposed to pulsed microwave methods such as electron spin echo or Fourier transform (FT) EPR. Significant advantages in sensitivity with similar time response are available with FT-EPR,[48] but there are also disadvantages in terms of spectral width of excitation that limit the application of this technique. The TREPR (CW) method is the most facile and cost-effective method for the observation of field-swept EPR spectra of organic radicals on the submicrosecond timescale.

There are two modes in which the experiment can be run. The preferred mode is to fix the delay time between the two boxcar gates. The first gate opens before the laser flash to sample the dark signal and provide a baseline; the second gate opens at a set delay time τ after the flash and samples both light and dark signals. Gate widths are 100 to 300 ns, which defines the time resolution of the experiment. Typical delay times (τ) are 0.1 to 10 μs. A processor in the boxcar unit provides the difference between the gates (the light-induced EPR signal) as an output voltage, which is passed to a computer for storage at each field point. Time constants in the boxcar charging circuit are adjusted to give an exponentially averaged output after 5 to 10 laser flashes at a single magnetic field value. The external magnetic field sweep is usually divided into 1000 data points. Clock pulses are generated during the field sweep to control the rate at which data are downloaded to the computer. In many spectrometers, there exists an option to program the data collection software to provide a DC voltage ramp that can sweep the external magnetic field externally, providing complete computer control of the scan. The laser repetition rate ranges from 10 to 100 Hz, with 60 Hz being nominal. The microwave power in most TREPR experiments is 2–20 mW, but it is essential to vary this parameter during the experiment to investigate whether the line shapes and/or intensities change with it.

The second mode is to run the experiment at a fixed magnetic field and sweep the second boxcar gate over time to collect kinetic information. There are two problems with this approach. First, the experiment must be repeated several times with a slow scan rate in order to get satisfactory S/N. To extract the EPR kinetic curve, the experiment is repeated off resonance and the two curves subtracted. Kinetics are more easily obtained using a high-bandwidth transient digitizer instead of a boxcar, and many researchers perform TREPR in this fashion.[49,50] It is important to note here two disadvantages of the TREPR technique. It is not generally possible to observe a Boltzmann (equilibrium) population of spin states using the boxcar method because of $1/f$ noise. Also, lifetime broadening effects are observed when the second boxcar gate is placed close in time to the laser flash. This is a consequence of having the microwave excitation running continuously. Near the laser flash, and during the photochemical events that produce radicals, the apparatus is attempting to excite spin states that are still in the process of forming. In other words, small interactions such as hyperfine couplings take time to evolve and may not be visible in the TREPR spectrum for several hundreds of nanoseconds after the laser flash.

Since a Boltzmann population is not generally detectable using the boxcar method, it is then logical to ask why the experiment works at all? The answer lies in the fact that in most, if not all, photochemical reactions that produce radicals, radical ions, or biradicals, CIDEP phenomena are observed. It is here that much of the sensitivity is gained back that was lost in bypassing the phase-sensitive detection unit (100 kHz field modulation). A smaller improvement in sensitivity comes from the use of the boxcar to signal average. In all of the TREPR spectra shown in this proposal, transitions below the baseline are in emission (E), while those above the baseline are in enhanced absorption (A). This is different from most conventional EPR spectra that are displayed as first derivative curves representing the change in detected intensity with the external field.

A detailed description of CIDEP mechanisms is outside the scope of this chapter. Several monographs[27,51,52] and reviews[53,54] are available that describe the spin physics and chemistry. Briefly, the radical pair mechanism (RPM) arises from singlet–triplet electron spin wave function evolution during the first few nanoseconds of the diffusive radical pair lifetime. For excited-state triplet precursors, the phase of the resulting TREPR spectrum is low-field E, high-field A. The triplet mechanism (TM) is a net polarization arising from anisotropic intersystem crossing in the molecular excited states. For the polymers under study here, the TM is net E in all cases, which is unusual for aliphatic carbonyls and will be discussed in more detail in a later section. Other CIDEP mechanisms, such as the radical–triplet pair mechanism[55] and spin-correlated radical pair mechanism,[56] are excluded from this discussion, as they do not appear in any of the systems presented here.

The photochemistry taking place in these polymers is destructive. It is therefore essential to flow or recirculate samples during the experiment. Flowing through the microwave resonator also prevents excessive heating of the samples by the laser flashes. To obtain high-temperature TREPR spectra of polymer radicals, a special insulated flow apparatus has been constructed in our laboratory that provides stable laminar flow of liquids through the EPR resonator at temperatures up to 150°C. The choice of solvent is critical for the success of high-temperature experiments. The solvent must (1) dissolve the polymer to concentrations of several grams per 100 mL, (2) have a high enough boiling point that it can withstand our reservoir temperature without evaporating or decomposing, and (3) be optically transparent at 248 nm. To date, we have found only one solvent, propylene carbonate, shown below, which fits all of these criteria. It is an excellent solvent for PMMA at room temperature and above, boils at 240°C, and shows no TREPR signal when run as a blank under 248 nm irradiation. It is used for all acrylates with ester side chains except for PFOMA, where specialized fluorinated solvents must be used. For the polyacids PAA and PMAA, aqueous solutions (neutral or basic) are appropriate, but if temperatures above 100°C are desired, small amounts of ethylene glycol should be added to raise the boiling point.

propylene carbonate

14.5 TACTICITY AND TEMPERATURE DEPENDENCE OF ACRYLATE RADICALS

Acrylic polymers have stereogenic centers on every other carbon atom. As a result, the polymers can be classified as atactic (random stereochemistry), isotactic (the same configuration at each stereogenic center), or syndiotactic (alternating configurations at each stereogenic center).[57] For PMMA, these are denoted as a-PMMA, i-PMMA, and s-PMMA, respectively. The tacticity of the polymers can be controlled during synthesis to some extent. While highly syndiotactic or isotactic acrylic polymers are reasonably straightforward to synthesize, it is rather difficult to generate highly atactic material. In general, samples of acrylic polymers purporting to be completely atactic have large sections of the polymer chain that are syndiotactic. We will see below that polymer tacticity plays a large role in the appearance of the TREPR spectra of acrylic main-chain radicals.

Figure 14.2 shows the temperature dependence of the TREPR spectra obtained upon photodegradation of all three tacticities of PMMA in propylene carbonate. Near room temperature, an alternating linewidth pattern indicative of conformationally induced hyperfine modulation is observed (Fig. 14.2, top). Upon heating, a large transformation takes place. At high temperatures (at or over 100°C), all three samples reveal motionally narrowed spectra, simple in appearance and with sharp linewidths (<1 G). Even more remarkable is that at the highest temperature recorded for each sample, all three spectra are quite similar in appearance. The convergence temperature to the fast motion spectrum is different for radicals from all three tacticities of PMMA, with the i-PMMA radical converging at the lowest temperature. This result is expected as i-PMMA has been found in many studies to be the least rigid of the three polymers.[58–60] Computer simulations of the fully converged, high-temperature limit spectra are shown at the bottom of each data set in Fig. 14.2. The hyperfine coupling constants used for each simulation are nearly identical and will be discussed in the next section.

The fast motion spectrum of the i-PMMA radical consists of 21 lines attributed to three separate isotropic hyperfine coupling constants. There is coupling to the methyl group to form a quartet (22.9 G) that is then split further into a triplet from one set of β-methylene protons (16.4 G) and another triplet from the other set (11.7 G). Theoretically, this should lead to 36 lines (4 × 3 × 3), but a fortuitous degeneracy exists because one of the fast motion β-methylene coupling constants is almost exactly

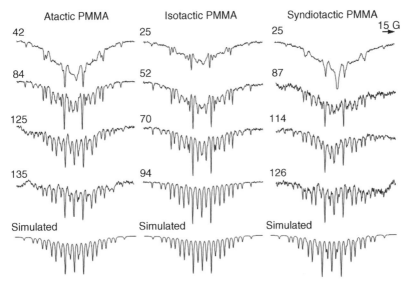

FIGURE 14.2 X-band TREPR spectra of main-chain polymer radical **1a** produced from 248 nm laser flash photolysis of atactic, isotactic, and syndiotactic PMMA in propylene carbonate at 0.8 μs delay time. The temperature for each spectrum is shown in °C, and the magnetic field sweep width is 150 G for all spectra, which exhibit net E CIDEP in all cases. Simulations of each fast motion spectrum (highest temperature) are shown at the bottom of each data set. Hyperfine values for each simulation are 3 $a_H(CH_3) = 22.9$ G, $2a_H(CH_2) = 16.4$ G, $2a_H(CH_2) = 11.7$ G for isotactic PMMA; $3a_H(CH_3) = 22.9$ G, $2a_H(CH_2) = 16.2$ G, $2a_H(CH_2) = 11.7$ G for syndiotactic PMMA; and $3a_H(CH_3) = 23.0$ G, $2a_H(CH_2) = 16.4$ G, $2a_H(CH_2) = 11.3$ G for atactic PMMA.

half the value of the methyl proton coupling constant. The syndiotactic and atactic polymers give rise to radicals with 27 lines, due to a lifting of this degeneracy (or perhaps an incomplete high-temperature averaging process). The slight changes in coupling constants from 16.4 to 16.2 G for s-PMMA, and from 11.7 to 11.3 G for a-PMMA, are minor but clearly observable at 0.8 μs delay time.

A special consequence of the stereoregularity of these polymers is the pseudo-symmetry relationships between the β-methylene protons of the main-chain radicals. The concept is briefly reviewed here by introducing Fig. 14.3, which shows the possible radicals formed by loss of the ester side-chain moiety from PMMA by the Norrish I α-cleavage reaction from the first excited triplet state. Because of the repetition pattern of stereogenic centers in isotactic and syndiotactic materials, these two tacticities are required to lead to the same free radical, which has a mirror plane of symmetry. We call this the "meso" radical. The mirror plane establishes magnetic equivalence in each of the two sets of β-methylene protons and is the reason why these protons show a triplet of triplets in the TREPR spectrum. Note that on each nonstereogenic center, the two protons are chemically and magnetically inequivalent because of the adjacent stereogenic centers.

In the "meso" and "racemic" radical structures, the hyperfine coupling constants to the β-methylene protons are not perturbed greatly by the change in stereochemistry on

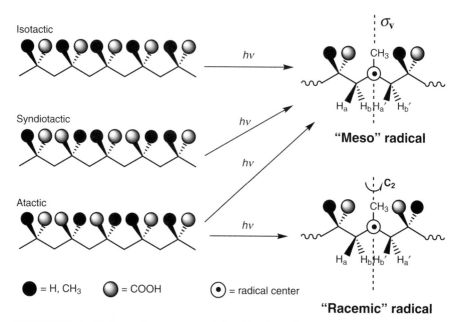

FIGURE 14.3   Nuclear spin symmetry relationships for main-chain radical **1a** from PMMA, as a function of polymer tacticity. The two radicals shown have σ_v mirror plane ("meso") or C_2 axis ("racemic") symmetry elements establishing magnetic equivalence of each set of β-methylene protons.

the next carbon further down the chain. For this reason, the observed spectra for each type of radical are very similar. The difference in coupling constant for a methylene proton pair in either the meso or the racemic radical can be quite large. In PMMA, for instance, the methylene couplings on the same carbon atom differ by about 5 G. In contrast, for an acrylic ester such as PEA, the difference is less than 1 G (see below). The methylene proton inequivalencies are a function of polymer stereochemistry and therefore cannot be removed by fast rotation.

14.6   STRUCTURAL DEPENDENCE

The left side of Fig. 14.4 shows TREPR spectra obtained 1.0 μs after 248 nm laser flash photolysis of eight acrylic polymers, with computer simulations on the right side.

   All these spectra were acquired at elevated temperatures (~100°C), that is, where the observation of fast motion spectra is expected. In Fig. 14.4A, the TREPR spectrum of the main-chain polymer radical from photolysis of *i*-PMMA is repeated from the bottom left side of Fig. 14.2, as it is the starting point for comparisons of spectral features such as linewidths and hyperfine coupling constants. The nomenclature used throughout this section is derived using the notations indicated in Scheme 14.1 and Chart 14.1. For example, a main-chain radical from PMMA will be denoted **1a**, whereas the oxo-acyl radical from PFOMA will be designated as radical **6b**, and so on. For all radicals simulated, the parameters used are listed in Table 14.1.

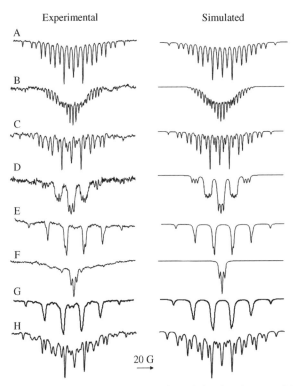

Experimental   Simulated

FIGURE 14.4 Experimental TREPR spectra (left) and simulated spectra (right) for radicals observed at 0.8 μs after 248 nm laser flash photolysis of the following polymers (see Chart 1 for acronym definitions): (A) PMMA, (B), d_3–PMMA, (C) PEMA, (D) PECA, (E) PEA, (F) PFOMA, (G) PAA, (H) PMAA. Except for the spectrum obtained from PFOMA (**6**), the simulation unambiguously assigns the signal carrier to main-chain polymer radical **a**. In the case of PFOMA, oxo-acyl radical **6b** is the dominant signal carrier. For PMMA (spectrum A), the material is isotactic (91% by NMR), but all other polymer samples are atactic material. Simulation parameters are listed along with the radical structures in Table 14.1.

14.6.1 d_3-Poly(methyl methacrylate), d_3-PMMA

Figure 14.4B shows the experimental and simulated TREPR spectra acquired during the photolysis of d_3-PMMA (radical **2a**) in propylene carbonate at 120°C. The spectral width is quite narrow, and there are more transitions observed here than for the protonated analog. There are many overlapping lines and only the outermost lines of the spectrum are clearly resolved. In Table 14.1, it is seen that the value of the hyperfine coupling constant for the α-methyl protons has decreased from 22.9 to 3.5 G, exactly as expected for this isotopic substitution.[61] The values for the β-methylene protons used in the simulation are 16.3 and 10.9 G, which are close to the values used in the simulations of the nondeuterated polymer. The fact that they are slightly different suggests that deuteration of the polymer backbone substitutent has a small but observable effect on the conformational energies of this polymer in solution.

TABLE 14.1 Magnetic Parameters Used in the Simulations in Fig. 14.4

Acronym	Structure	Hyperfine Constant
PMMA radical 1a		$CH_3 = 22.9\,G$ $H_\beta = 16.7\,G$ $H_{\beta'} = 11.2\,G$
d_3-PMMA radical 2a		$CD_3 = 3.5\,G$ $H_\beta = 16.3\,G$ $H_{\beta'} = 10.9\,G$
PEMA radical 3a		$CH_3 = 22.9\,G$ $H_\beta = 15.8\,G$ $H_{\beta'} = 11.2\,G$
PECA radical 4a		$N = 3.3\,G$ $H_\beta = 16.3\,G$ $H_{\beta'} = 14.8\,G$
PEA radical 5a		$H_\alpha = 21.7\,G$ $H_\beta = 23.5\,G$ $H_{\beta'} = 23.8\,G$
PFOMA radical oxo-acyl 6b		$CH_2 = 3.2\,G$ $CF_2 = 0.8\,G$
PAA radical 7a		$H_\alpha = 21.5\,G$ $H_\beta = 23.7\,G$ $H_{\beta'} = 23.8\,G$ $H_\gamma = 0.9\,G$
PMAA radical 8a		$CH_3 = 23.1\,G$ $H_\beta = 27.3\,G$ $H_{\beta'} = 11.0\,G$

14.6.2 Poly(ethyl methacrylate), PEMA

The spectrum of radical **3a** no longer consists of 21 lines (Fig. 14.4C). In a similar fashion to the tacticity dependence described earlier, one or more of the accidental degeneracies observed in *i*-PMMA radical **1a** have been lifted to give a total of 27 lines. Also, the overall spectral width has decreased slightly. The new lines are assumed to arise from the main-chain radical and not a new signal carrier, because the kinetic decay curve for this line is identical to all the other lines and the intensity ratios of all of the transitions in Fig. 14.4C are constant with temperature. These differences in hyperfine coupling constants are somewhat unexpected because the only structural difference between these radicals is the identity of the β-substituent in which the ester R group changes from methyl to ethyl.

From the simulation parameters in Table 14.1, we see that the hyperfine values for the α-methyl protons remain the same for both methacrylate polymers. The methyl group has free rotation about the C−C bond from the backbone to the methyl group, even at temperatures as low as −100°C.[62] The additional splittings in the TREPR spectrum of radical **3a** can be simulated by lowering the hyperfine values for one set of the β-methylene protons (the largest coupling constant) by 1 G compared to that from PMMA. This suggests that making the side-chain ester group bulkier changes the relative energies of the conformations of the polymer near the radical center.

14.6.3 Poly(ethyl cyanoacrylate), PECA

Figure 14.4D shows the experimental high-temperature TREPR spectrum obtained during photolysis of PECA (radical **4a**) in propylene carbonate solution, along with a computer simulation. There are five packets of lines from coupling to the β-methylene protons. Each of these packets of lines is split into three additional lines from the γ-nitrogen ($I = 1$) in the nitrile group. The larger width of the second and the fourth packet of lines can be explained by the diastereotopicity of the β-methylene protons. The hyperfine values used in the simulation are 3.3 G for the γ-nitrogen and 16.3 and 14.8 G for the β-methylene protons (Table 14.1). The difference between the hyperfine coupling constants for each of the β-methylene protons is much smaller for this polymer than for the other acrylic polymers. This may be related to the increased conformational flexibility of this polymer because of the small backbone substituent.

14.6.4 Poly(ethyl acrylate), PEA

The TREPR spectrum obtained after photolysis of PEA (radical **5a**) in solution is shown in Fig. 14.4E. There are six major transitions observed in β-methylene hyperfine couplings of 23.0 and 24.7 G, and one α-hyperfine coupling of 21.5 G. These coupling constants are comparable to values reported in the literature for radicals of similar chemical structure.[63–65] At room temperature, the β-methylene couplings are not resolved due the increased natural linewidth, and at such temperatures the spectrum appears to have five nearly equivalent coupling constants lead to a six-line spectrum, with transitions in a 1:5:10:10:5:1 intensity ratio. It is only at higher temperatures such as in Fig. 14.4E that the difference in the β-methylene coupling and the α-hyperfine coupling are resolved. The β-hyperfine interactions in radical **5a** are close in value at all temperatures, but the fast motion limit appears to be achieved at lower temperatures for acrylates versus methacrylates. This is because there is more conformational flexibility in this polymer.

14.6.5 Poly(fluorooctyl methacrylate), PFOMA

Because of the importance of fluorinated polymers in a wide range of new industrial applications, from dry cleaning to microlithography,[66,67] the photochemistry of PFOMA (**6**) was investigated. Fluorination of the alkyl tail of the ester group increases the rigidity of the polymer chain, decreasing the internal motion of the polymer.[68,69] It should be expected that the TREPR spectra of radical **6a** from photolysis of PFOMA will be much broader, at least at room temperature. To obtain the high-temperature TREPR spectrum of PFOMA, a solvent mixture consisting of different fluorinated hydrocarbons (FC-70, 3M Corporation) was used. This solvent system has a boiling point of 201°C, dissolves PFOMA easily, and is transparent at 248 nm.

The high-temperature TREPR spectrum observed during photolysis of PFOMA is shown in Fig. 14.4F. For all polymeric radicals previously presented and discussed, the polymeric main-chain radical generally exhibits a very intense signal, and the oxo-acyl signal is weak or absent from the TREPR spectrum at high temperatures. The opposite relative intensities are observed here. The well-resolved TREPR spectrum of fluorinated oxo-acyl radical **6b** is superimposed on a wide signal that has some similar features to the signal from the polymeric radical of PMMA (radical **1a**) at lower temperatures (see top left of Fig. 14.2). The broad signal is most likely due to the main-chain polymeric radical from PFOMA (**6a**). Slow dynamics of the PFOMA main-chain radical is reflected in the TREPR spectra by lines that are broadened due to slower motion around the C_α–C_β bond.

We suggest two reasons for this anomaly in the intensities of the two radicals from PFOMA: one is that the relaxation time of radical **6b** is longer than usual for an acyl radical. This is to be expected because the major relaxation mechanism in such radicals is from the spin–rotation interaction.[70] This process will be less effective in structures such as **6b** because the radical is larger in size than typical acyl radicals observed in organic photochemical reactions of carbonyl compounds—it will certainly undergo slower molecular rotations than the oxo-acyl radical **1b**, observed at room temperature from side-chain cleavage of the ester group in PMMA (data not shown). Another reason for the strong intensity is based on the number of total transitions for the two radicals. Since both radicals acquire the same initial spin polarization from the triplet mechanism, more intensity has to be packed into fewer lines for radical **6b** than the polymeric radical. We will return to the TREPR spectroscopy of PFOMA radicals **6a** and **6b** in later section on oxo-acyl radicals.

We have observed similar effects in the TREPR spectra of the polymeric radical of poly(adamantyl methacrylate) (PAMA, spectrum not shown). Large ester side chains such as the adamantyl and fluorinated alkyl groups experience a larger amount of hydrodynamic friction than do smaller side chains such as methyl and ethyl groups.[71]

The polymer then undergoes slower internal rotations, and the TREPR spectrum of the main-chain radical is broadened. The similarity in the spectra from PFOMA and PAMA suggests that the conformational mobility of the polymeric radical in solution plays a major role in the intensity and spectral shape of the TREPR signal from these polymeric radicals and that side chain size and structure can completely prevent access to the fast motion limit, at least at temperatures below 135°C. Higher temperatures are not currently available to us because our high temperature flow system in limited to a maximum reservoir temperature of 150°C, for safety reasons.

14.6.6 Polyacrylic Acid, PAA

The TREPR spectrum acquired during 248 nm photolysis of PAA in aqueous solution at high temperature is shown in Fig. 14.4G. The PAA main-chain radical **7a** is easily identified as arising from direct Norrish I α-cleavage of the carboxyl group, and the CIDEP pattern is again that from the triplet mechanism (net E), the same pattern observed for the ester side-chain analogs. The only major difference between the PAA radical spectrum in Fig. 14.4G and the PEA radical spectrum in Fig. 14.4E is that we have included a small γ-hyperfine coupling to better fit the PAA spectrum. The oxo-acyl counter radical **7b** is not observed in these spectra, even at low temperatures. This is unusual: for the acrylate polymers with ester group side chains, oxo-acyl radicals were observed at room temperature just to the high-field side of the center of the polymer radical signal. Acyl radicals are known to have fast relaxation times due to spin–rotation interaction,[72] and so a lack of observation of radical **7b** in our system at high temperatures is understandable. However, the absence of any signal from radical **7b** at room temperature suggests that this relaxation mechanism may be much more efficient in the smaller carboxyl radical **7b** than in its ester analogs. We are currently exploring methods to test this hypothesis using isotopic labeling experiments.

Another interesting feature of Fig. 14.4G is that there is second signal carrier present, with less intensity but transitions that are clearly visible in between those of the main-chain radical **7a**. This is also a departure from the behavior of the acrylate polymers such as PEA and PMMA. This unknown radical probably has an even number of hyperfine interactions and a similar g-factor to the main-chain radical, because the observed lines are evenly spaced between the lines of main-chain radical **7a** and are symmetrically distributed about them. Unambiguous identification of this radical has not been achieved to date, but some candidates include (1) the propagating radical **7c** (Scheme 14.1), (2) the product of a rearrangement, similar to that described for the radical anions of carboxylic acids and esters,[73] (3) a radical from photochemistry at branched sites caused by backbiting of the propagating polymer chain during the synthesis, or (4) a radical produced by photochemistry at a tail-to-tail or head-to-head defect site in the polymer chain. The radical arising from Norrish I cleavage at a head-to-head defect site in PAA is a strong candidate for the origin of this signal because it has four hyperfine coupling constants that are all expected to be different due to stereochemical issues *and* the disruption of symmetry due to the defect.

14.6.7 Polymethacrylic Acid, PMAA

Quantum yields, molecular weight distributions, and product analyses from the photolysis of aqueous solutions of PMAA have been reported by several research groups.[74,75] Side-chain cleavage and β-scission of the main-chain radical **8a** for PMAA were proposed as the major mechanistic events of degradation, with a particular emphasis on reactivity with H_2O_2.[76] Reactions of the acrylic acids with reactive oxygen species such as hydroxyl radical are well understood and are the focus of biodegradability studies of these polymers.[77] Low-temperature (frozen) SSEPR spectra have been reported from X-ray[78] and γ-radiation[79–81] of several homopolymers and copolymers.[82] In all cases, these reports contain spectra that carry the typical burdens of low-temperature EPR spectroscopy, namely, broad lines and overlapping signal carriers that make assignment and interpretation difficult and/or ambiguous. However, the main conclusion from much of this previous work was that the β-scission process was operative even in neat polymer samples at low temperatures. There have also been studies of PMAA degradation in the presence of additives such as iron salts[83] and acridine,[84] and one report of TREPR experiments during photolysis of an initiator-labeled methacrylic acid sample.[85] This latter work showed an interesting pH dependence of the signal carriers that is relevant to the results reported here and this will be elaborated below.

Figure 14.4H shows the TREPR spectrum acquired during 248 nm photolysis of PMAA in aqueous solution at high temperature. A small amount of ethylene glycol was added to the solutions to go over 100°C. Compared to PAA, it is notable that the observed photochemistry of PMAA is very clean, with no additional radicals present, although such radicals may be masked by the large number of transitions and by the high intensity of this signal, which is assigned to main-chain radical **8a**. Similar to PAA, the oxo-acyl counter radical **8b** is not observed in TREPR spectra of PMAA at any temperature. The spectral simulation of radical **8a** is shown to the right of the experimental spectrum. While the positions of the transitions line up quite well, it is clear that the linewidths for each transition are not all equal in the experimental spectrum. We conclude that while we have confidence in the structure of the radical being created, the polymeric radical has not reached a true "fast motion" state even at the highest temperature accessible with our current apparatus. Nonetheless, each transition in the experimental spectrum does appear to be accounted for in the simulation. One striking difference between the PAA and PMAA radicals is the magnitude of the β-hyperfine interactions (Table 14.1). For PAA these coupling constants are almost identical, whereas for PMAA they differ by more than a factor of 2. This is a much larger difference in coupling constant than observed in PMMA and clearly reflects the asymmetry imposed on the radical conformations by the neighboring stereogenic centers. This effect may also be amplified by solvent effects and by the repulsion of charges on the carboxylate anions along the backbone.

14.7 OXO-ACYL RADICALS

Transitions representative of oxo-acyl radical **b** are not generally observed in high-temperature TREPR experiments, although they often appear at room temperature and below for acrylate polymers. When observed, the oxo-acyl radicals typically exhibit a single emissive peak slightly to the right of the center of the spectrum (the g-factor for σ-type acyl radicals is slightly lower (approximately 2.0009[86–89]) than that of the 3° alkyl polymeric radical (about 2.0026[90–93]). Radical **b** has two unimolecular decomposition pathways available to it: loss of CO and loss of CO_2. In the previous studies on the photoablation of PMMA using high-resolution mass spectrometry, both gaseous products were detected.[94] Griller and Roberts determined the decarboxylation rate of an oxo-acyl radical of similar structure to **b** to be $2.1 \times 10^6 \, s^{-1}$ at 373K.[95] This corresponds to a decomposition time of 500 ns, which is well within the resolution time of our instrument. However, with only one exception (see below), radical **b** is not observed at high temperatures by TREPR at any delay time from 0.1 to 10 μs.

The most likely reason for the absence of signals from radical **b** in the high-temperature TREPR spectra is fast spin relaxation. Paul has shown by line shape analysis that relaxation of an acyl radical is on the order of 10 ns or faster in solution,[96] and this has been confirmed by the kinetic analysis of acyl–alkyl biradicals by Tsentalovich et al.[70,72] The mechanism of this relaxation is through spin–rotational interaction in the radical, which in this case is equal to the longitudinal relaxation time T_1. Electron spin relaxation is much slower in polymeric radical **a** due to hindered rotation, and therefore it remains spin-polarized long enough to be observed by TREPR. It should be noted here that radical **b** is observed at room temperature in acrylate polymer photochemistry, that is, when the oxo-acyl radical contains the ester functionality. Their signals can be seen in the lowest temperature spectra at the top of Fig. 14.2. They are not observed when the functional group is an acid, at any temperature we have accessed to date. It is possible that because of their smaller size, the spin–rotation interaction responsible for the fast spin relaxation is even more dominant in these structures.

There is of course a major exception to this trend in oxo-acyl radical behavior as observed by TREPR and that is for PFOMA (**6**). Figure 14.5 shows the complete temperature dependence of the TREPR signals observed when PFOMA is irradiated at 248 nm. The defining feature of these spectra is the large emissive triplet superimposed on a broad emissive background. The intense triplet arises from hyperfine couplings of the unpaired electron to the protons and/or fluorines in the alkyl tail of the oxo-acyl radical **6b**. As discussed above, the broad emissive signal in the background is assigned the main-chain polymeric radical of PFOMA, **6a**. As the temperature is increased, the general spectroscopic features of the polymeric radicals of methacrylates become apparent on the perimeter of the spectrum (compare to the low-temperature spectra at the top of Fig. 14.2), while the central signal remains unchanged. The observation of radical **6b** at room temperature is remarkable by itself. Even more unusual is that the signal remains strong at high temperatures as well. Both observations are anomalous but perhaps understandable in terms of the steric bulk and conformational stiffness of the perfluoroalkyl ester side chain. Spin–rotation-induced relaxation may be very inefficient for such a structure.

Upon expansion of the highest temperature spectrum to a sweep width of 50 G, it can be seen that the signal in Fig. 14.5 is actually a triplet of triplets (Fig. 14.6a). A simulation of this spectrum is shown in Fig. 14.6b. For this simulation, a ^{19}F hyperfine

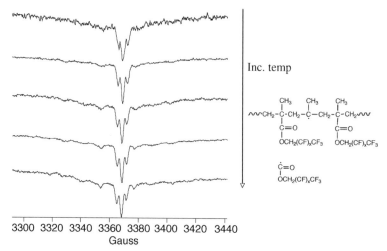

FIGURE 14.5 TREPR spectra of the polymeric (**6a**) and oxo-acyl (**6b**) radicals from 248 nm laser flash photolysis of PFOMA in the FC-70 solvent system (1.72 g in 40 mL). Delay time is 0.3 μG. (A) 25°C, (B) 51°C, (C) 65°C, (D) 80°C, (E) 110°C.

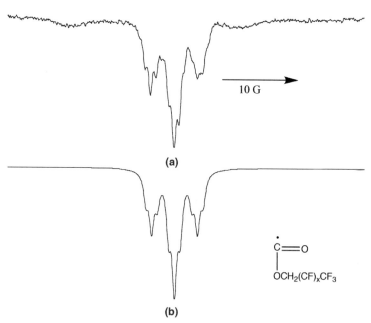

FIGURE 14.6 Experimental (top) and simulated (bottom) TREPR spectra of oxo-acyl radical **6b** from laser flash photolysis of PFOMA in the solvent system FC-70 (3M Corporation). Sweep width is 50 G. Simulation parameters are listed in Table 14.1.

coupling of 3.2 Gauss and a proton coupling of 0.8 G were used. Krusic and coworkers have shown that γ- and even δ-hyperfine couplings from fluorinated radicals can be observed in SSEPR spectra due to spin polarization through C–F bonds.[97] The coupling constants for long-range ^{19}F hyperfine interactions are typically small, from 0.1 to 2.5 G. Our values agree reasonably well with the values of 2.6 and 0.1 G reported by Krusic et al. for γ- and δ-hyperfine ^{19}F couplings, respectively, in perfluoroalkyl radicals.[98,99] The hyperfine constants reported here are slightly larger, probably a consequence of having a σ-type radical rather than a π-type radical as was the case in Krusic's work.

14.8   SPIN POLARIZATION MECHANISMS

The spectra in Fig. 14.2 and Figs. 14.4–14.6 show strong net emission from the TM of CIDEP.[100,101] It is revealing that the dominant polarization mechanism is the TM and not the RPM[102,103] or the spin-correlated radical pair mechanism (SCRP).[104,105] The magnitudes of both SCRP and RPM polarization patterns depend on the rate of encounters taking between the two radical centers. Polymeric radical **a** and oxo-acyl radical **b** have drastically different diffusional properties in solution. Radical **b** is small and will undergo much more rapid diffusion than **a**. An immediate consequence of this is that RPM and SCRP polarization mechanisms may be quenched by relatively slow reencounter rates and will therefore be obscured in this case by the very strong TM. In all TREPR spectra detected here, emissive TM is always observed, which is unusual for aliphatic carbonyl compounds, although not unprecedented for esters.[106]

While we cannot directly correlate signal intensities to radical populations because of the polarization, we know that the photodegradation of our samples over the time taken to collect our data is extremely slow. This suggests that either the polymers do not absorb very much light, or the quantum yield for the production of radicals is quite low. In fact, both situations exist for our samples. The polymer solutions begin to absorb in the UV at about 250 nm, and we are exciting them at 248 nm. We are therefore just on the edge of the n–π^* excitation of the ester carbonyl group. Guillet and coworkers[107] and others[108–110] have measured overall quantum yields of about 0.1 for degradation of acrylic polymers in solution. We conclude that we are not creating many radicals with each laser flash, but the TM polarization they carry is extremely intense.

For TM polarization to be strong, several physical and magnetic parameters must fall in certain ranges. For example, the cleavage rate should be faster or comparable to the electron spin relaxation time in the excited triplet state, typically 1–10 ns for carbonyl triplet states. However, a major factor is the orbital symmetry of the triplet state and its tumbling rate with respect to the external magnetic field. If an excited state tumbles quickly, the intersystem crossing process loses selectivity and all three triplet levels are populated more or less equally from the singlet excited state. The symmetry of the orbitals is not really an issue for carbonyl n–π^* states, but the tumbling rate can be very important. We suggest that (1) the polymer rotational motion along the main axis occurs on a slower timescale than rotation of the individual ester side-chain groups, (2) this asymmetric motion leads to higher selectivity in the intersystem crossing process, (3) a somewhat slower relaxation time of both the excited triplet state and the ensuing polymeric radical exists for the same reasons, and (4) that the overall effect leads to an optimal situation for the creation of TM polarization.

A final noteworthy feature of these spectra is the lack of RPM polarization, which for these radicals would appear as low-field emissive, high-field absorptive transitions. It is curious that such polarization never develops at any delay time, even out to 20 μs where we have observed only TM polarization. The creation of RPM polarization requires reencounters of radicals on a suitable timescale and modulation of the exchange interaction between the unpaired electrons. This is normally accomplished by diffusion of the radicals between weak and strong exchange regions. That it never develops indicates that either these radicals do not make a significant number of reencounters, or perhaps it is due to the fast spin relaxation in the oxo-acyl radical. It may also be that the TM is simply so dominant that the RPM intensity is always much weaker and is never observed. At lower temperatures (cf. Fig. 14.1, top spectrum at 25°C), there does appear to be a slight superposition of an E/A pattern on top of the emissive TM polarization, but it has a very small effect.

Regarding the relative intensities of the observed spin polarization mechanisms, it is also important to note that the β-hyperfine interactions in these radicals are conformationally modulated, and this process can also quench RPM polarization. In a qualitative way, we can consider the modulation process to be a relaxation mechanism that exchanges magnetization between different nuclear spin orientations. Since these different orientations can have opposite phases of RPM polarization, the exchange of emissive and absorptive lines can cancel the intensity of the transitions.

14.9 SOLVENT EFFECTS

14.9.1 pH Effects on Poly(acid) Radicals

Figure 14.7 shows the TREPR signals from the PAA main-chain radical as a function of the pH of the solution. The spectrum of this radical completely disappears in basic

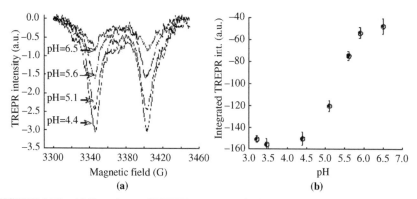

FIGURE 14.7 (a) Experimental TREPR spectra (only the two intense central transitions are shown), obtained upon 248 nm photolysis at 122°C of PAA (leading to radical **7a**) in water, at the pH values indicated. (b) Integrated intensities of these transitions plotted as a function of solution pH.

solutions. This is may be due to a change in the side-chain chromophore structure from COOH to COO$^-$. The absorbance of the solution at 248 nm changes by only 15% across the whole range of pH values studied (2–10), but it appears that the resonance-stabilized carboxylate anions are much less photoreactive than their conjugate acids. Figure 14.7b shows a plot of the integrated intensities from Fig. 14.7a as a function of pH, showing that the most drastic changes occur at about pH = 5.5, which is close to the reported pK_a value for PAA of 4.8 reported by Mandel.[111] It is interesting to note that titration of PAA using NMR chemical shift information as a function of pH, performed by Chang et al.,[112] produces an inflection point at exactly the same pH value as we report in Fig. 14.7b.

It is also possible that stretched (high pH) versus coiled (low pH) polymer chains may exhibit different photophysics. This was commented by Chou and Jellinek in their early report on PMAA photochemistry,[75] and a detailed TREPR study of pH-dependent cage effect manipulation using initiator-labeled PMAA was reported by Maliakal et al.[114] It has been reported by Mittal and coworkers[113] that for small molecule analogs of PAA and PMAA, the quantum yield for Norrish I cleavage drops to zero for the completely deprotonated structure. For PMAA, we also observe a pH dependence in the TREPR signal intensities (data not shown). The data in Fig. 14.7 are the first experimental evidence for such an effect in PAA radicals.

14.9.2 General Features for Polyacrylates

The dynamics of polymer chains in solution depend greatly on solvent properties. Unfavorable polymer–solvent interactions can cause the polymer chain to collapse, slowing down its internal motion. Favorable interactions have the opposite effect, meaning the polymer will adopt a more extended conformation allowing for greater mobility.[115,116] In the magnetic resonance literature, NMR studies of stereoregular polymers in various solvents have demonstrated that the conformation adopted by the polymer is highly solvent dependent.[62,117] It is of interest to see if similar effects are observed in TREPR spectra of acrylic main-chain radicals. In this regard, the radical center in these macromolecular-free radicals are functioning as a tiny "spin probe."

Figure 14.8 shows TREPR field-swept spectra of the PMMA polymeric radical **1a** in five different solvents at room temperature and similar concentrations. At first glance these spectra look very similar, but there are subtle differences, especially in the broader packets of lines from the polymeric radical. Even the lines not broadened by hyperfine modulation show slightly different linewidths and positions. In Fig. 14.9, kinetic profiles of the TREPR signal of the main-chain polymeric radical for each solvent are presented, obtained by measuring the intensity of one of the strong sharp central transitions in the spectra in Fig. 14.8. There are large differences in the decay rates of these signals at room temperature. The decay is fastest in dioxane (Fig. 14.9A), while in methylene chloride (Fig. 14.9B) and chloroform (Fig. 14.9C) it is slowest.

Analysis of the decay of TREPR signals can be complicated and has been the subject of many previous investigations.[38,40] Qualitatively, we can say the following

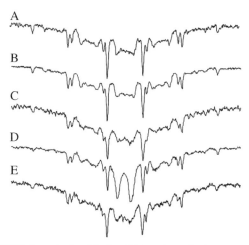

FIGURE 14.8   TREPR data obtained during 248 nm laser flash photolysis of atactic PMMA in five different solvents at ambient temperature. Delay time is 0.6 μs for all spectra, and the sweep width is 150 G. (A) 1.61 g in 20 mL dichloromethane. (B) 1.53 g in 20 mL chloroform. (C) 1.04 g in 20 mL acetonitrile. (D) 1.20 g in 20 mL THF. (E) 1.49 g in 25 mL dioxane. Note the changes in width and shape for the alternating broad and sharp packets of lines.

about the kinetic profiles in Fig. 14.9: (1) the transitions, for symmetry reasons, are not greatly affected by dynamic effects from hyperfine modulation; that is, there are no exchange broadening effects for this line, (2) chemical decay is unimportant on this timescale, and (3) if the decay rates are governed mostly by T_1, we can conclude that we are not in the fast motion regime. The linewidths are almost the same, yet the decay rates are drastically different, that is, $T_1 \neq T_2$. The fact that we do not see the spectrum of the propagating radical from β-scission of the main-chain radical or alkyl radicals from decarbonylation or decarboxylation of the acyl fragment is good evidence that we can ignore chemical decay processes. This, and the fact that the shape of the traces is independent of the microwave power, means that T_1 relaxation is indeed the most dominant mechanism for the TREPR signal decay.

The decay time constants for the polymeric radicals clearly vary from solvent to solvent, ranging over an order of magnitude from about 2 μs in tetrahydrofuran (THF) to 200 ns in methylene chloride. A similar solvent dependence of ^{13}C NMR T_1 values has been reported by Spyros et al.[71] They studied poly(naphthyl methacrylate) using the ^{13}C inversion recovery technique and found that the spin-lattice relaxation time of the polymer varied from 1 μs in chloroform to 5 μs in pentachloroethane. The solvents used to collect the data in Fig. 14.9 represent a much wider range of viscosities and dielectric constants, so it is not surprising that we observe a much larger variation in T_1 in our experiments. If the mechanism of relaxation is hyperfine anisotropy modulation, as is typical for heavily substituted alkyl radicals, then our results suggest that these polymers experience faster motion in methylene chloride versus THF. It is a bit perplexing that there appears to be no direct correlation between the decay constants and either viscosity or dielectric constant in this data set.

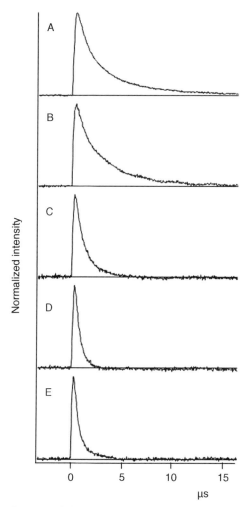

FIGURE 14.9 Kinetic traces of the TREPR signal from main-chain polymeric radical **1a** in various solvents. (A) Methylene chloride, (B) chloroform, (C) acetonitrile, (D) tetrahydrofuran, and (E) dioxane. The dashed line represents the baseline for each kinetic trace.

14.10 DYNAMIC EFFECTS

Dynamic effects have been observed previously in SSEPR studies of acrylic polymers. However, as mentioned above, it is the propagating radical (**1c** in Scheme 14.1) rather than the main-chain radical that is observed in the SSEPR experiment because of the β-scission rearrangement process. In early studies of the photodegradation of acrylic polymers, SSEPR spectra showed an unusual nine-line spectrum with alternating linewidths of five sharp and four broad lines.[118–121] The spectrum was assigned to the propagating radical **c** (Scheme 14.1). As a function of temperature, this spectrum changed reversibly to a 13- or 16-line spectrum depending on the available resolution.

Several theories have been proposed to account for these findings: one is that there exists a superposition of two or more static conformations of the radical and that changing the temperature alters the relative populations of the two conformers.[122–124] A second theory proposes that the propagating radical exists with a Gaussian distribution of dihedral angles, centered about a single preferred conformation.[125]

The third and most recent theory comes from Matsumoto and Giese,[126,127] who suggested that the observed steady-state (SSEPR) spectrum of radical **c** is due to a superposition of two conformations of the same radical, and that one of these structures has a pyramidalized center. Iwasaki et al.[128] invoked a fourth model, hyperfine modulation, to explain the spectra of the propagating radical. They were able to simulate the observed 9- and 13-line SSEPR spectra using a set of modified Bloch equations for a two-site exchange model between two conformations.

For main-chain acrylic radicals, created in solution at room temperature and above, the presence of a superposition of conformations or Gaussian distributions is unlikely. Polymers undergo conformational jumps on the submicrosecond timescale, even in bulk at room temperature.[129–130] The first two theories above require that the radicals be fairly rigid with little (Gaussian distribution) or no (superposition of static conformations) movement around the C_α–C_β bond. The main-chain radical is sterically hindered but still quite flexible, and a dramatic change in the hybridization at C_α is unlikely. We have approached our simulations with the hyperfine modulation model.

14.10.1   The Two-Site Jump Model

The alternating linewidth effect arises because hyperfine couplings fluctuate due to some inter- or intramolecular process. Selective broadening of the lines in the spectrum occurs according to the following equation:[131]

$$T_2^{-1} = \gamma_e^2 \tau \langle \delta a^2 \rangle (M_a - M_b)^2 + T_{2,0}^{-1} \qquad (14.1)$$

where T_2^{-1} is the linewidth after broadening, γ_e is the gyromagnetic ratio of the electron, M_a and M_b are the nuclear spin quantum numbers of the transitions, and $T_{2,0}^{-1}$ is the contribution to the linewidth from other mechanisms. The term τ is the correlation time for the exchange process. The term δa represents the mean square deviation in the hyperfine splitting; that is, it is the difference between the fast motion hyperfine coupling constants.

If the motion of an acrylic polymer radical about the C_α–C_β bond is hindered, changing the temperature should lead to changes in the TREPR spectrum. This is indeed observed for all acrylic polymers we have examined to date. Simulation of the complete temperature dependence of TREPR spectra of acrylic polymer main-chain radicals should allow information regarding the conformational motion of the polymer in solution to be extracted, such as rotational correlation times, spin-lattice relaxation times (T_1), and activation energies for conformational transitions.

Our standard simulation routine for CIDEP can easily be modified to incorporate the hyperfine modulation process according to Equation 14.1. The spectroscopic and

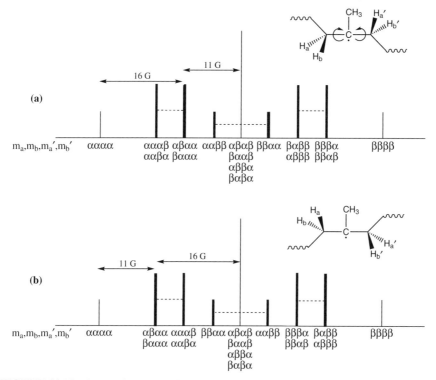

FIGURE 14.10 Streamlined view of the hyperfine modulation model used in the simulation program. The methyl group hyperfine couplings have been removed for clarity. The conformational change shown converts one triplet of triplets from the equivalent β-methylene protons on one side of the radical center (H_a, H_b) into their symmetric counterparts on the other side (H_a', H_b'), and vice versa. The nuclear spin quantum numbers representing each transition in each state (A and B) are shown below the transitions in α and β notation. Dashed lines drawn between thick lines show transitions that exchange due to the motion—these lines broaden in the TREPR spectrum. The transitions represented with thin lines remain sharp through the modulation process, as their total nuclear spin quantum number does not change during a jump between sites.

physical features of the two-site jump model are outlined in Fig. 14.10. The methyl group hyperfine coupling constant in PMMA radical **1a** is assumed to be constant for all conformations and therefore is omitted from Fig. 14.10. It is assumed that one of the two triplets from the β-hyperfines (two diastereotopic H's on either side of the radical center) is being modulated into the second triplet on some timescale, τ. That is, the hyperfine constants for H_a and H_a' are exchanging into those for H_b and H_b', and vice versa. This process is shown graphically on the right-hand side of Fig. 14.10 and clearly indicates the anticorrelation of the two triplets. In the first conformation shown in Fig. 14.10a, the stick plot expected from this triplet of triplets is shown with a given set of hyperfine couplings. Below this stick plot, the total nuclear spin quantum numbers for each transition are listed (our notation reads that α represents $m_S = +1/2$, and β represents $m_S = -1/2$). The lines that do not change total spin quantum numbers upon interconversion to the new triplet of triplets are the transitions that remain sharp in the TREPR spectrum. These are indicated with thin lines in the stick plots. Transitions exchanging different total spin quantum numbers are broadened as predicted by Equation 14.1. Specifically exchanged lines are connected with dashed horizontal lines in Fig. 14.10.

Three sets of hyperfine coupling constants were input to the simulation program. The methyl (or single H α-coupling for PEA) hyperfine coupling constant was assumed to be the same in all conformations and was not considered further. The program is able to calculate the amount of linewidth added to each transition from β-hyperfine modulation separately for each pair of protons according to Equation 14.1. Values for the correlation time (τ) and the hyperfine difference ($a_1 - a_2$) were input as fitting parameters. In all the simulations shown here, the ($a_1 - a_2$) value used was 15.6 G. The program calculates the amount of additional linewidth based on the change in nuclear spin quantum numbers for each transition. Some of the lines in the spectrum received no additional broadening ($M_1 - M_2 = 0$) as per Equation 14.1. As was noted in the discussion on hyperfine modulation, one of the requirements for the hyperfine modulation mechanism to be active is that the sum of the hyperfine couplings must remain constant. A value of 28 G (approximately the sum of the two fast motion (average) coupling constants) was used for this sum.

14.10.2   Simulations and Activation Parameters

The temperature dependence of the TREPR spectra from the d_3-PMMA polymer radical **2a** is shown in Fig. 14.11. The differences between these spectra are less

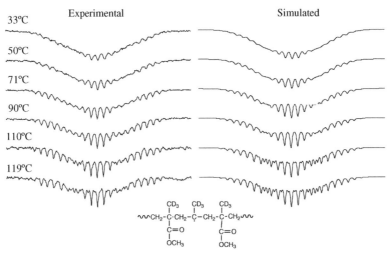

FIGURE 14.11   Experimental (left) and simulated (right) TREPR spectra of the main-chain polymeric radical **2a** from 248 nm laser flash photolysis of d_3-PMMA in propylene carbonate solution at the indicated temperatures. Sweep width for all spectra is 100 G. Simulation parameters are listed in Table 14.2.

TABLE 14.2 Parameters Used in the Simulation of the Temperature Dependence of TREPR Spectra of Radical 2a (from d_3-PMMA) Shown in Fig. 14.11

Temperature °C	a_D (G)	a_{CH2} (G)	a_{CH2} (G)	Line width (G)	$\tau_c \pm 0.2$ (s)
33	3.5	15.7	10.3	2.7	2.5×10^{-10}
50	3.5	15.9	10.5	2.2	2.0×10^{-10}
71	3.5	16.1	10.7	1.7	1.0×10^{-10}
90	3.5	16.3	10.9	1.5	7.0×10^{-11}
110	3.5	16.3	10.9	1.4	–
119	3.5	16.3	10.9	1.3	–

dramatic than that of the nondeuterated polymer (Fig. 14.2) due to the smaller spectral width, however, the alternating linewidth effect is clearly observed at the lower temperatures. The simulations are shown to the right of the experimental data in Fig. 14.11 and the simulation parameters are presented in Table 14.2. In all simulations, a value of 3.5 G was used for the hyperfine coupling constant of the CD_3 group because this group is always in fast rotation about the $C_\alpha - C_\beta$ bond. For the spectra obtained at temperatures of 90, 110, and 119°C, the hyperfine coupling constants are the same. The only difference is the amount of natural linewidth added to each of the spectra and a small amount of hyperfine modulation at 90°C. Increases in the natural linewidth of the spectrum will broaden all of the transitions uniformly, while including hyperfine modulation in the simulation program only broadens selected transitions as described above.

For spectra obtained at 110 and 119°C, additional linewidth to each transition equally is necessary to simulate the experimental data. As the temperature drops below 100°C, incorporation of hyperfine modulation becomes necessary to simulate the data. Simply increasing the natural linewidth for all transitions without addition of hyperfine modulation does not lead to satisfactory fitting. As the temperature drops, the overall linewidth of all of the transitions increases, and also slight changes in the average hyperfine coupling constants from 90 to 30°C were observed. The simulated data in Fig. 14.11 gives values of τ_c that closely fit a modified Arrhenius plot shown in Fig. 14.12.

The activation energy determined from this plot is 22 ± 2 kJ/mol. This value is somewhat higher than that found for the PEA radical (data not shown). This is not surprising because the addition of a methyl group on the polymer backbone will change the conformational energies and in particular it should increase the barrier to rotation of the polymer chain in solution.[132] The activation energy measured here is comparable in magnitude to that measured for protonated PMMA by NMR methods.

The two-site jump model appears to work well for this polymer. However, it should be noted that the density of transitions is large here due to the larger spin quantum number of deuterium ($I = 1$), the fact that there are three of them in the isotopically substituted polymeric radical, and that the coupling constant for each deuterium is smaller by a factor of 6.4 compared to the protonated radical. Coupling these facts to the visual fitting process, these fits may not be unique. In fact, when the same model is applied to the temperature dependence of the protonated PMMA spectra (Fig. 14.2), reasonable visual fits could not be obtained with this model. Deuteration of the

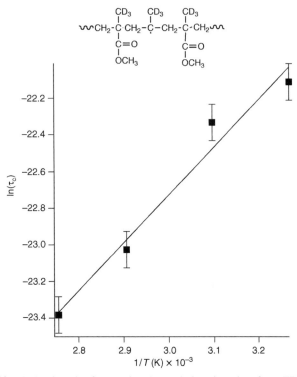

FIGURE 14.12 Arrhenius plot for rotational correlation time data from TREPR spectra of polymeric radical **2a**. Squares are the experimental data, solid line is the linear fit, with $E_a = 22 \pm 2$ kJ/mol.

backbone may have drastically altered the conformational dynamics of the polymer near the radical center. It is also possible that multiple conformations are available to the polymer near the radical center and a four- or even six-site jump model may be more appropriate here. Such a simulation of the dynamics requires a substantial computational effort as well as data from small-molecule model systems; this is the focus of present research in our laboratory.

14.11   CONCLUSIONS

The TREPR experiments and simulations described here have provided an enormous amount of structural and dynamic information about a class of free radicals that were not reported in the literature prior to our first paper on this topic in 2000. Magnetic parameters for many main-chain acrylic radicals have been established, and interesting solvent effects have been observed such as spin relaxation rates and the novel pH dependence of the polyacid radical spectra. It is fair to conclude from these studies that the photodegradation mechanism of acrylic polymers is general, proceeding through Norrish I α-cleavage of the ester (or acid) side chain. Recently, model systems have been reported for both the esters and the acids based on the Kemp's triacid frame-work.[133,134] The molecular weight dependence (or rather lack of one) of this chemistry is currently being studied with experiments on short oligomers of PMMA. Future work will include an examination of block and alternating copolymers, with an emphasis on site-selective photochemistry detected by TREPR.

ACKNOWLEDGMENTS

This work has been carried out in our laboratory since 1995. The initial investigations were performed by Dr. Ruixin Xu, who first discovered that main-chain radicals were observable using TREPR at room temperature. Graduate student Elizabeth Harbron carried out the solvent and tacticity dependence studies, and fellow graduate student Vanessa McCaffrey performed the high-temperature experiments to establish the fast motion limit for these radicals and the nuclear spin symmetry relationships in the main-chain radicals. I thank them both for their extensive effort to characterize these radicals and to gain insight into their dynamics as a function of tacticity and solvent. Coauthor Dr. Natalia Lebedeva performed all of the experiments on the polyacids, including the pH dependence and the necessary simulations. In the early stages the Rohm and Haas Company supported this work, and we thank Dr. David Westmoreland and Dr. Casmir Ilenda for their advice and encouragement as the project developed. We have enjoyed consistent strong support from the National Science Foundation and thank them for their present support of our research program through grant # CHE–0809530.

REFERENCES

1. Schnabel, W. *Polymer Degradation: Principles and Applications*; Hansen & Gardner: 1982.
2. Rabek, J. F. *Mechanisms of Photophysical Processes and Photochemical Reactions in Polymers*; Wiley: New York, 1987.
3. Allen, N. S.; Edge, M. *Fundamentals of Polymer Degradation and Stabilization*; Elsevier Applied Science: Amsterdam, 1992.
4. Swift, G. *Acc. Chem. Res.* **1993**, *26*, 105–110.
5. *Polymers in Defence and Aerospace Applications*, Rapra Conference Proceedings, iSmithers–Rapra Publishing, UK, 2007.
6. Reis, R. L.; San Román, J. *Biodegradable Systems in Tissue Engineering and Regenerative Medicine*; CRC Press: 2004.
7. Addington, D. M.; Schodek, D. L. *Smart Materials and New Technologies: For the Architecture and Design Professions*; Elsevier: Amsterdam, 2005, pp 147.
8. Ryntz, R. A.; *Plastics and Coatings: Durability, Stabilization Testing*; Hansen & Gardner: 2001.
9. Wallraff, G. M.; Hinsberg, W. D. *Chem. Rev.* **1999**, *99*, 1801–1822.
10. Kamat, P. V. *Chem. Rev.* **1993**, *93*, 267–300.

11. Grassie, N.; Scott, G. *Polymer Degradation and Stabilization*; Cambridge University Press: New York, 1985.

12. Grossetête, T.; Rivaton, A.; Gardette, J. L.; Hoyle, C. E.; Ziemer, M.; Fagerburg, D. R.; Clauberg, A. H. *Polymer*, **2000**, *41*, 3541–3554.

13. Nagai, Y.; Nakamura, D.; Miyake, T.; Ueno, H.; Matsumoto, N.; Kaji, A.; Ohishi, F. *Polym. Degrad. Stab.* **2005**, *88*, 251–255.

14. Carswell, T. G.; Garrett, R. W.; Hill, D. J. T.; O'Donnell, J. H.; Pomery, P. J.; Winzor, C. L. *Polymer Spectroscopy*; Fawcett, A. H., Ed.; Wiley: Chicago, 1996; pp 253.

15. Rånby, B.; Rabek, J. F. *ESR Spectroscopy in Polymer Research*; Springer-Verlag: Berlin, 1977.

16. Harbron, E. J.; McCaffrey, V. P.; Xu, R.; Forbes, M. D. E. *J. Am. Chem. Soc.* **2000**, *122*, 9182.

17. McCaffrey, V. P.; Forbes, M. D. E. *Macromolecules* **2005**, *38*, 3334.

18. McCaffrey, V. P.; Harbron, E. J.; Forbes, M. D. E. *Macromolecules* **2005**, *38*, 3342.

19. McCaffrey, V. P.; Harbron, E. J.; Forbes, M. D. E. *J. Phys. Chem. B* **2005**, *109*, 10686.

20. Lebedeva, N. V.; Forbes, M. D. E. *Macromolecules* **2008**, *41*, 1334.

21. Bullock, A. T.; Sutcliffe, L. H. *Trans. Faraday Soc.* **1964**, *60*, 625.

22. Moreau, W.; Viswanathan, N. In *UV Light Induced Reactions in Polymers, ACS Symposium Series;* Gould, R. F., Ed.; Washington, 1976, pp 125.

23. Fischer, H. *Polym. Lett.* **1964**, *2*, 529.

24. Sakai, Y.; Iwasaki, M. J. *Polym. Sci. A-1* **1969**, *7*, 1749.

25. Odian, G. *Principles of Polymerization*; John Wiley & Sons: New York, 1991, pp 198.

26. Liang, R. H.; Tsay, F.-D.; Gupta, A. *Macromolecules* **1982**, *15*, 974.

27. Harbron, E. J.; Forbes, M. D. E. *Encyclopedia of Chemical Physics and Physical Chemistry*; Institute of Physics Publishing: Philadelphia, 2001; Vol. 2, pp 1389.

28. Vert, M. *Biomacromolecules* **2005**, *6*, 538.

29. Hu, S.; Ren, X.; Bachman, M.; Sims, C. E.; Li, G. P.; Allbritton, N. L. *Anal. Chem.* **2004**, *76*, 1865.

30. Yang, S. Y.; Schultz, G.; Green, M. M.; Morawetz, H. *Macromolecules* **1999**, *32*, 2577.

31. Porasso, R. D.; Benegas, J. C.; van den Hoop, M. A. G. T. *J. Phys. Chem. B* **1999**, *103*, 2361.

32. Mittal, L. J.; Mittal, J. P.; Hayon, E. *J. Phys. Chem.* **1973**, *77*, 1482.

33. Tonelli, A. E. *NMR Spectroscopy and Polymer Microstructure: The Conformational Connection*; Wiley: New York, 1989; Chapter 6.

34. Weil, J. A.; Bolton, J. R.; Wertz, J. E. *Electron Paramagnetic Resonance: Elementary Theory and Applications*; Wiley–Interscience: New York, 1994.

35. Piette, L. H.; A new technique for the study of rapid free radical reactions. In *Sixth International Symposium on Free Radicals*; Cambridge University Press: 1963.

36. Avery, E. C.; Remko, J. R.; Smaller, B. *J. Chem. Phys.* **1968**, *59*, 951.

37. Fessenden, R. W.; Verma, N. C. *J. Am. Chem. Soc.* **1976**, *98*, 243.

38. Verma, N. C.; Fessenden, R. W. *J. Chem. Phys.* **1976**, *65*, 2139.

39. Trifunac, A. D. EPR and NMR detection of transient radicals and reaction products. In *The Study of Fast Processes and Transient Species by Electron Pulse Radiolysis*; Baxendale, J. H. Busi, F.; Eds.; Proceedings of the ASI (NATO), Capri: Italy, 1982.

40. McLauchlan, K. A.; Stevens, D. G. *Acc. Chem. Res.* **1988**, *21*, 54.

41. Gonen, O.; Levanon, H. *J. Phys. Chem.* **1984**, *88*, 4223.

42. Levstein, P. R.; van Willigen, H. *Colloids Surf A* **1993**, *72*, 43.

43. Nakamura, H.; Terazima, M.; Hirota, N. *J. Phys. Chem.* **1993**, *97*, 8952.

44. Hoff, A. J.; Deisenhofer, *J. Phys. Rep.* **1997**, *1*, 87.

45. Stehlik, D.; Mobius, K. *Annu. Rev. Phys. Chem.* **1997**, *48*, 745.

46. Lubitz, W.; : Lendzian, F.; Bittl, R. *Acc. Chem. Res.* **2002**, *35*, 313.

47. Wasielewski, M. R.; Bock, C. H.; Bowman, M. K.; Norris, J. R. *J. Am. Chem. Soc.* **1983**, *105*, 2903.

48. van Willigen, H.; Levstein, P. R.; Ebersole, M. H. *Chem. Rev.* **1993**, *93*, 173.

49. Jeevarajan, A. S.; Fessenden, R. W. *J. Phys. Chem.* **1992**, *96*, 1520–1523.

50. Jent, F.; Paul, H.; Fischer, H. *Chem. Phys. Lett.* **1988**, *146*, 315–319.

51. *Chemically Induced Magnetic Polarization*; Muus, L. T.; Atkins, P. W., et al. Eds.; Proceedings of the ASI (NATO), Sogesta/Urbino, Italy, 1977.

52. Salikhov, K. M.; Molin, Y. N.; Sagdeev, R. Z.; Buchachenko, A. L. *Magnetic Spin Effects in Chemical Reactions*; Elsevier: Amsterdam, 1984.

53. Forbes, M. D. E. *Photochem. Photobiol.* **1997**, *65*, 73.

54. Trifunac, A. D.; Lawler, R. G.; Bartels, D. M.; Thurnauer, M. C. *Prog. React. Kinet.* **1986**, *14*, 43.

55. Goudsmit, G. H.; Paul, H.; Shushin, A. I. *J. Phys. Chem.* **1993**, *97*, 13243.

56. Tarasov, V. F.; Forbes, M. D. E. *Spectrochim. Acta A* **2000**, *56*, 245.

57. Tanaka, H.; Niwa, M. *Polymer* **2008**, *49*, 3693.

58. Apel, U. M.; Hentschke, R.; Helfrich, J. *Macromolecules* **1995**, *28*, 1778.

59. Johnson, J. F.; Porter, R. S. In *The Stereochemistry of Macromolecules*; Ketley, A. D., Ed. Marcel Dekker: New York, 1968, Vol. *3*.

60. O'Reilly, J. M.; Teegarden, D. M.; Wignall, G. D. *Macromolecules* **1985**, *18*, 2747.

61. Lambert, J. B.; Shurvell, H. F.; Lightner, D. A.; Cooks, R. G. *Organic Structural Spectroscopy*; Prentice Hall: New Jersey, 1998.

62. Spevacek, J.; Schneider, B. *Adv. Colloid Interface Sci.* **1987**, *27*, 81.

63. Paul, H.; Fischer, H. *Helv. Chim. Acta* **1973**, *56*, 1575.

64. Gilbert, B. C.; Holmes, R. G. G.; Laue, H. A. H.; Norman, R. O. C. *J. Chem. Soc., Perkin Trans. 2* **1976**, 1047.

65. Bascetta, E.; Gunstone, F. D.; Walton, J. C. *J. Chem. Soc., Perkin Trans. 2* **1984**, 401.

66. Fiebig, M.; Kauf, M.; Fair, J.; Endert, H.; Rahe, M.; Basting, D. *Appl. Phys. A: Mater. Sci. Process.* **1999**, *69*(Suppl.).

67. DeSimone, J. M.; Romack, T.; Betts, D. E.; McClain, J. B. U.S. Patent 5,783,082, 1998.

68. Hoffman, H.; Kalus, J.; Thurn, H. *Colloid Polym. Sci.* **1983**, *261*, 1043.

69. Yee, G. G.; Fulton, J. L.; Smith, R. D. *J. Phys. Chem.* **1992**, *96*, 6172.

70. Tsentalovich, Y. P.; Forbes, M. D. E. *Mol. Phys.* **2002**, *100*, 1209.

71. Spyros, A.; Dais, P.; Heatley, F. *Macromolecules* **1994**, *27*, 6207.

72. Tsentalovich, Y. P.; Forbes, M. D. E.; Morozova, O. B.; Plotnikov, I. A.; McCaffrey, V. P.; Yurkovskaya, A. V. *J. Phys. Chem. A* **2002**, *106*, 7121.

73. Koppe, R.; Kasai, P. H. *Chem. Phys.* **1994**, *189*, 401.

74. Baxendale, J. H.; Thomas, J. K. *Trans. Faraday Soc.* **1958**, *54*, 1515.

75. Chou, C. H.; Jellinek, H. H. G. *Can. J. Chem.* **1964**, *42*, 522.

76. Kaczmarek, H.; Kaminska, A.; Swiatek, M.; Rabek, J. F. *Angew. Makromol. Chem.* **1998**, *261/262*, 109.

77. Ulanski, P.; Bothe, E.; Hildenbrand, K.; von Sonntag, C. *Chem. Eur. J.* **2000**, *6*, 3922.

78. Saraev, V. V.; Alsarsur, I. A.; Annenkov, V. V.; Danilovtseva, E. N. *Russ. J. Appl. Chem.* **2001**, *74*, 1585.

79. Ormerod, M. G.; Charlesby, A. *Polymer* **1964**, *5*, 67.

80. Abraham, R. J.; Melville, H. W.; Ovenall, D. W.; Whiffen, D. H. *Trans. Faraday Soc.* **1958**, *54*, 1133.

81. Seropegina, E. N.; Kochetkova, G. G.; Fock, N. V.; Mel'nikov, M. Ya. *Polym. Photochem.* **1985**, *6*, 195.

82. Fox, P. A.; Hill, D. J. T.; Lang, A. P.; Pomery, P. *J. Polym. Int.* **2003**, *52*, 1719.

83. Toptygin, D. Ya.; Pariiski, G. V.; Davydov, E. Ya.; Zaitseva, N. I.; Pokholok, T. V. In *Proceedings of 9th Donaulaendergespraech Conference*, **1976**, *1*, pp 26.

84. Tsuchiya, M.; Oouchi, Y.; Tanaka, S.; Kojima, T. *J. Appl. Polym. Sci.* **2005**, *97*, 1209.

85. Maliakal, A.; Weber, M.; Turro, N. J.; Green, M. M.; Yang, S. Y.; Pearsall, S.; Lee, M.-J. *Macromolecules* **2002**, *35*, 9151.

86. Griller, D.; Roberts, B. P. *Chem. Commun.* **1971**, 1035.

87. Griller, D. J. *Magn. Reson.* **1972**, *6*, 402.

88. Hefter, H.; Fischer, H. *Ber. Bunsenges. Phys. Chem.* **1970**, *74*, 493.

89. Metcalfe, A. R.; Waters, W. A. *J. Chem. Soc. B* **1967**, 340.

90. Paul, H.; Fischer, H. *Helv. Chim. Acta* **1973**, *56*, 1575.

91. Kochi, J. K.; Krusic, P. J.; Eaton, D. R. *J. Am. Chem. Soc.* **1969**, *91*, 1879.

92. Ohno, A.; Kito, N.; Ohnishi, Y. *Bull. Chem. Soc. Jpn.* **1971**, *44*, 470.

93. Chen, K. S.; Kochi, J. K. *Can. J. Chem.* **1974**, *52*, 3529.

94. Krajnovich, D. J. *J. Phys. Chem.* **1997**, *101*, 2033.

95. Griller, K.; Roberts, B. P. *J. Chem. Soc., Perkin Trans. 2* **1972**, *6*, 747.

96. Paul, H. *Chem. Phys. Lett.* **1975**, *32*, 472.

97. Fagan, P. J.; Krusic, P. J.; McEwen, C. N.; Lazar, J.; Parker, D. H.; Herron, N.; Wasserman, E. *Science* **1993**, *262*, 404.

98. Krusic, P. J.; Kochi, J. K. *J. Am. Chem. Soc.* **1968**, *90*, 7155.

99. Krusic, P. J.; Chen, K. S.; Meakin, P.; Kochi, J. *J. Phys. Chem.* **1974**, *78*, 2036.

100. Atkins, P. W.; Evans, G. T. *Chem. Phys. Lett.* **1974**, *25*, 108.

101. Wong, S. K.; Hutchinson, D. A.; Wan, J. K. S. *J. Chem. Phys.* **1973**, *58*, 985.

102. Pedersen, J. B.; Freed, J. H. *J. Chem. Phys.* **1973**, *59*, 2869.

103. Adrian, F. J. *J. Chem. Phys.* **1971**, *54*, 3918.

104. Closs, G. L.; Forbes, M. D. E.; Norris, J. R. *J. Am. Chem. Soc.* **1987**, *91*, 3592.

105. Buckley, C. D.; Hunter, D. A.; Hore, P. J.; McLauchlan, K. A. *Chem. Phys. Lett.* **1987**, *135*, 307.

106. Wymann, L.; Kaiser, T.; Paul, H.; Fischer, H. *Helv. Chim. Acta* **1981**, *64*, 1739.

107. Guillet, J. *Polymer Photophysics and Photochemistry: An Introduction to the Study of Photoprocesses in Macromolecules*; Cambridge University Press: 1987.

108. Fox, R. B.; Isaacs, L. G.; Stokes, S. *J. Polym. Sci.* **1963**, *1*, 1079.

109. Gupta, A.; Liang, R.; Tsay, F.-D.; Moacanin, J. *Macromolecules* **1980**, *13*, 1696.

110. Shultz, A. R.; Frank, P.; Griffing, B. F.; Young, A. L. *J. Polym. Sci., Polym. Phys. Ed.* **1985**, *23*, 1749.

111. Mandel, M. *Eur. Polym. J.* **1970**, *6*, 807.

112. Chang, C.; Muccio, D. D.; Pierre, St. T. *Macromolecules* **1985**, *18*, 2154.

113. Mittal, L. J.; Mittal, J. P.; Hayon, E. *J. Phys. Chem.* **1973**, *77*, 1482.

114. Maliakal, A.; Weber, M.; Turro, N. J.; Green, M. M.; Yang, S. Y.; Pearsall, S.; Lee, M.-J. *Macromolecules* **2002**, *35*, 9151.

115. de Gennes, P. G. *Scaling Concepts in Polymer Physics*; Cornell University Press: Ithaca, NY, 1979.

116. Flory, P. J. *Principles of Polymer Chemistry*; Cornell University Press: Ithaca, NY, 1953.

117. Ono, K.; Sasaki, T.; Yamamoto, M.; Yamasaki, Y.; Ute, K.; Hatada, K. *Macromolecules* **1995**, *28*, 5012.

118. Abraham, R. J.; Melville, H. W.; Ovenall, D. W.; Whiffen, D. H. *Trans. Faraday Soc.* **1958**, *54*, 1133.

119. Doetschman, D. C.; Mehlenbacher, R. C.; Cywar, D. *Macromolecules* **1996**, *29*, 1807.

120. Kamachi, M.; Kohno, M.; Liaw, D. J.; Katsuki, S. *Polym. J.* **1978**, *10*, 69.

121. Tian, Y.; Zhu, S.; Hamielec, A. E.; Fulton, D. B.; Eaton, D. R. *Polymer* **1992**, *33*, 384.

122. Sugiyama, Y. *Bull. Chem. Soc. Jpn.* **1998**, *71*, 1019.

123. Harris, J. A.; Hinojosa, O.; Arthur, J. C. *J. Poly. Sci., Polym. Sci. Ed.* **1973**, *11*, 3215.

124. Ingram, D. J. E.; Symons, M. C. R.; Townsend, M. G. *Trans. Faraday Soc.* **1958**, 54.

125. Iwasaki, M.; Sakai, Y. *J. Poly. Sci. A-1* **1969**, *7*, 1749.

126. Matsumoto, A.; Giese, B. *Macromolecules* **1996**, *29*, 3758.

127. Spichty, M.; Giese, B.; Matsumoto, A.; Fischer, H.; Gescheidt, G. *Macromolecules* **2001**, *34*, 723.

128. Iwasaki, M. S. *J. Polym. Sci. A-1* **1969**, *7*, 1537.

129. Bendler, J. T.; Yaris, R. *Macromolecules* **1978**, *11*, 650.

130. Inoue, Y.; Konno, T. *Makromol. Chem.* **1978**, *179*, 1311.

131. Fraenkel, G. K. *J. Phys. Chem.* **1967**, *71*, 139.

132. Fytas, G.; Meier, G.; Patkowski, A.; Dorfmüller, T. *Colloid Polym. Sci.* **1982**, *260*, 949.

133. Lebedeva, N. V.; Gorelik, E. V.; Prowatzke, A. M.; Forbes, M. D. E. *J. Phys. Chem. B* **2008**, *112*, 7574.

134. Lebedeva, N. V.; Gorelik, E. V.; Magnus-Aryitey, D.; Hill, T. E.; Forbes, M. D. E. *J. Phys. Chem. B* **2009**, *113*, 6623.

INDEX

Carbon-Centered Free Radicals and Radical Cations, Edited by Malcolm D. E. Forbes
Copyright © 2010 John Wiley & Sons, Inc.